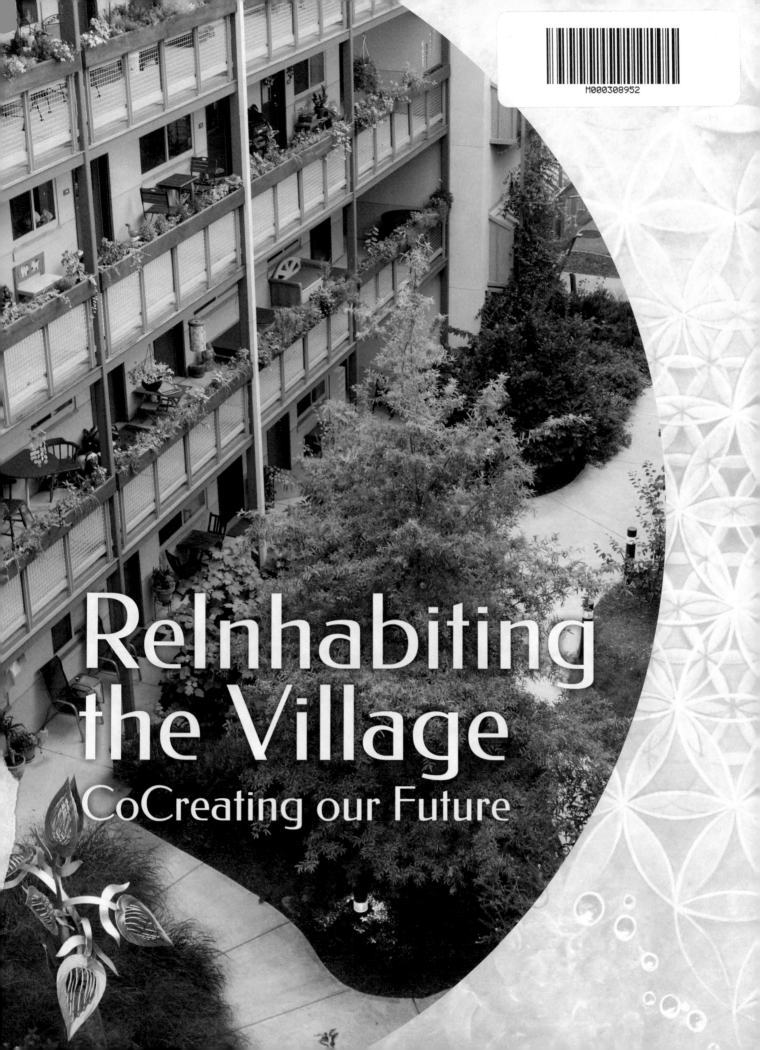

ReInhabiting the Village

CoCreating our Future

ReInhabiting
the Village
CoCreating our Future

ISBN: 978-1-944297-01-5
Library of Congress Control Number: 2016939490
For more information visit:
http://ReInhabitingTheVillage.com
http://keyframe-entertainment.com

Executive Producer

Associate Producer

Robert D. Reed Publishers
PO Box 1992
Bandon, OR 97411
Phone: 541-347-9882; Fax: -9883
Email: 4bobreed@msn.com
Website: www.rdrpublishers.com

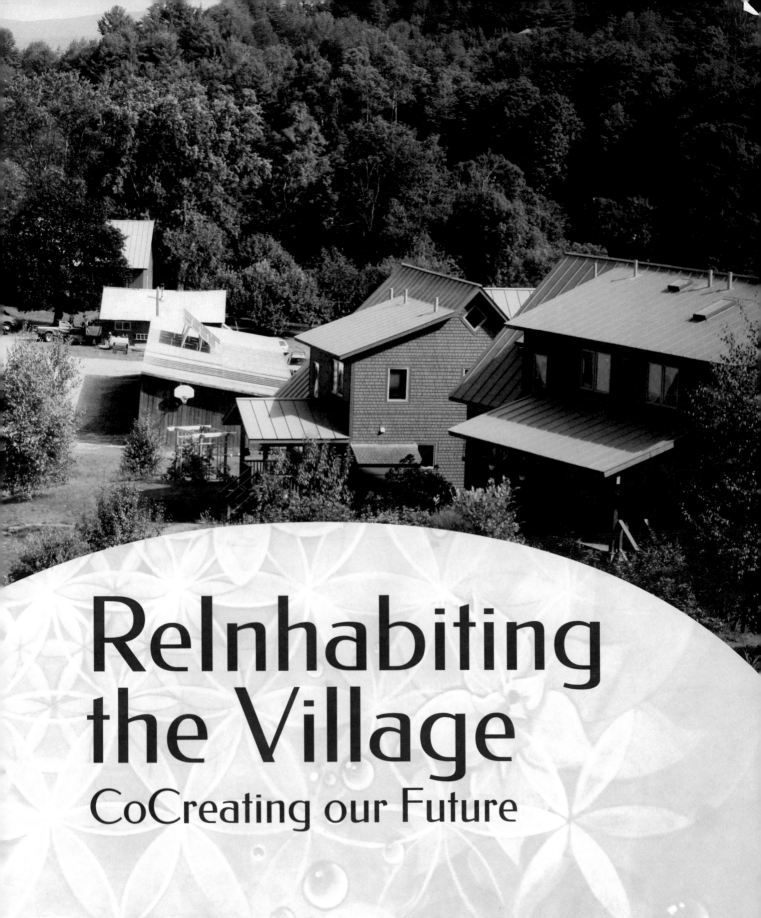

ReInhabiting the Village
CoCreating our Future

JAMAICA STEVENS

Collaboratively authored with many "Voices of the Village"

Contents

Introduction

A DREAM YOU DREAM ALONE IS ONLY A DREAM.

A DREAM YOU DREAM TOGETHER IS REALITY. John Lennon

What is a Village and how do we ReInhabit it? At the simplest level, a "Village" is a Place we belong to. Whether urban or rural, it is the place where we live, the land we share, it is the food we eat, it's our watersheds, it is the ecosystem that we are a part of. A Village is also the People we are connected to, it is our children and elders, our families, our friends, our neighbors; it is the "WE" who share a "common unity," a community of individuals who create a collective culture. It is the system by which we exchange and steward our resources and agree upon value as a means to ensuring our mutual survival.

For most of human history we have lived as Villagers, learning about life from lessons passed down through a lineage. Through stories told to preserve the wisdom of those who had come before, we have shared the successes of time-honored approaches, and offered the hard-earned reflection that challenges provided to our predecessors in order to nurture the continuation of our connection to the plants, to the waters, to the animals, to each other, and to the Spirit of a place.

Yet, in a matter of a few generations, much of the world's populace has become imbalanced and disconnected from our environment and from each other and we are seeing the ramifications of this disconnection across all spectrums of life. This time in history affects all peoples of every culture, gender, socioeconomic class, religion, age, and nationality. We as a species live at a critical crossroads, a time in which the impacts of our past and the potential of our future intersect. A time in which the choices each one of us make every day, and the decisions we collectively agree to influence and create the legacy for future generations.

Will we choose to continue down a precarious road not built to sustain us beyond one or two more generations, or choose to empower ourselves to come together and find a way to live on this planet, to unite our creative minds and open hearts to forge the path ahead, to bring our gifts and our solutions forward to uplift each person, to value not only each human life, yet all of life?

Through our shared fate, we are becoming a "Global Village." Yet how do we maintain a Global Village when we are comprised of disparate nation states struggling with deep divisions, mistrust, and perceived cultural barriers?

A thriving Global Village can only exist if we return to simplicity, beginning with honoring the individual, creating empowerment and vitality in our lives. Just as individual cells comprise more complex organisms, each one of us is a necessary and impactful piece of the whole.

As each human is honored, we can then cultivate interpersonal connections based on mutuality, respect and support, bringing together a greater circle of community based on core values and common purpose. We can create cohesive units of family and friends as chosen family to pool our resources and talents to support greater

abundance for each individual. By also finding connection points to those who may not be our immediately chosen circle, yet share the same streets, the same schools, the same air, water, and lands we create a pathway to vibrant and interconnected local communities.

Thriving local areas create resiliency, a "currency" of exchanging skills, knowledge, resources, service and support, building a strong, yet adaptable connective tissue that uplifts people, uplifts families and focuses on opportunity for all people in that community to have not only their base needs of survival met, but also to create the potential to thrive.

Only through sustaining life at local levels and sharing the sense of mutuality, interconnection, and understanding of our place in a Global context, can we transform the complexity of our current system into diverse, yet cohesively woven micro systems designed with the relevance and functionality that comes from an integrated Whole System perspective.

ReInhabiting the Village: CoCreating Our Future is an invitation to participate in a discussion, an inquiry, and to share reflections of different voices

from different perspectives, discovering how to write a new human story, together. This is not a project focusing on the challenges we face. Those resources already exist in droves. We instead are showcasing solutions, inspirations, touching the edges of the shift in consciousness required to create a new way of living. The collaborative nature of this project features voices that are each a unique individual reflection of varying approaches, wisdoms and opinions about how to live in holistic balance. This project is a composite, with contributions from various co-authors, artists, leaders, Visionaries, and changemakers uniting to explore and celebrate our human potential.

In my work as a community leader and consultant, I have been inspired by the projects, organizations, and allies who every day are finding creative, relevant, and positive ways to engage solutions for balance and justice in their own lives and through their work. I feel surrounded by a culture of service, and honor the lineage of changemakers, activists, educators and everyday heroes who have blazed a trail to pick up and carry forward. I do not pretend that this information is new, or revolutionary.

FAITH IS TAKING THE FIRST STEP,
EVEN WHEN YOU CAN'T SEE THE WHOLE STAIRCASE.

Martin Luther King Jr.

It is simply another drop in the bucket adding to the conditions for a tipping point that will ripple an inevitable shift. I honor the dedication of the many hearts and hands who are endeavoring to create changes that benefit all. As a mother, my own devotion to this work is inspired by my amazing daughter, who became the compelling motivation to give all that I have and all that I am to create a world for her grandchildren to thrive in. As an author, I acknowledge that this project is birthed from the perspective lens of one fraction of the world's populace. As we can only see from the vantage point that we have experienced, I recognize that I live from the vantage point of privilege.

It is with that recognition and humble understanding that I felt the call to bring forward a platform that creates an inclusive playspace for a co-creative discovery of the common root that we all share.

THERE IS NO SINGLE RIGHT ANSWER that encompasses the complexity of the matter before us. This project's core aim is to discover a way forward together that honors the wisdom of the past and the intelligence available to us now in order to create a thriving future. It will take every answer we have, every innovative solution possible, every human's participation to address what requires a deep personal shift so that a collective shift can emerge.

YOU are invited. YOUR voice matters, YOUR needs matter, YOUR gifts are invaluable to the whole. Take this journey with us, leverage the wisdom, tools, and resources shared throughout this multi-media project to bring your service, your creativity, your skill, and your ingredient to the collective feast of potential laid before us. Share your wisdom, your experience, your knowledge with others to support their own discovery. Light a fire of inspiration in your heart SO bright that it catches on the winds and carries your dreams, your passion, your commitment to all of the corners of this precious planet we call HOME.

This is for the ancestors whose bones we dance upon. This is for the future ones, whose songs call us forward. May this be the moment where we choose to answer that call and create a harmonious chorus, a symphony of love.

May my daughter and all of the children of this world continue this journey long after us, being fed by and feeding the miracle of life.

Jamaica Stevens

Editor's Note

Along with our acknowledgments it feels essential to take a moment to breathe in the fullness of the times that we are living in. Indeed, the time is Now, always full and yet ever flowing like a river. Breathing a pause we consider this a potent time for our world and for Life.

This project has been a journey, and in many ways it is just beginning. The content curation process has been organic and emergent and is a story that is telling itself; one that has demanded service, surrender and total commitment. By nature of our mode of the written word and visual representations we have the natural limitations of language as well as the creative expansive potential applied by the style of layout and the variety of voices.

In our work we are listening to the many voices of Gaia, the voices from the Global Village, and we are opening to hear and see our world and ourselves in a new way. What we have assembled here is indeed a collection and represents some of these voices. Our intention is that this dynamic weave of voices, visuals and perspectives invites a harmonious and inspiring sense of what is possible, along with grounded practical tools, examples and resources.

We invite you to take these voices in deeply along with the imagery and to open further to your own creative potential. Where do you resonate with the voices shared here? Are there new and different perspectives expressed? Do you feel where your own voice fits into this shared story? Are you able to see something familiar in a new way? What inspires you?

We want to know your answers to all of these questions and to see your inspiration in action. We invite you to take many journeys through this book and to review the How To Use This Project section in the following pages as a way to continue to access the breadth of the offering.

Thank you for joining us ~ we are honored to share in this unfolding and the emergence of many directions and connections for where our paths lead us home to the Global Village.

Sacred Journeys!

In Loving Service to Life,

A. Keala Young
Atlan, Summer Solstice 2015

Acknowledgements

First of all, I would like to thank the Spirit of Life that permeates all of Creation. Thank you Mama Earth and those you have birthed to whom this whole project is a Love Song of Devotion. Thank you to the Ancient Ones, Thank you to the Future Ones. Thank you to Amelia Hope for sharing your mama and inspiring me to work towards a future worthy of you. Special thanks to Julian Reyes and Keyframe-Entertainment who believed in this project and fully got behind me- wings and all. Thank you for continuing to uplift this culture. To the most amazing project team, A. Keala Young, Kelli Rua Klein, Davin Skonberg, Natacha Pavlov, Jonah Haas, Antje Martin Schaffer, Dana Wilson for your commitment, your creativity, your trust, and your incredible work. NONE of this would be possible without you. To the Contributors and Artists who shared your gifts and wisdom; Aaron Dorr, Adam Apollo, Akira Chan, Alberto Ruiz, Alokananda, Aloria Weaver, Amanda Creighton, Andre Soaras, Andrew Eckert, Autumn Skye Morrison, Barbara Marx Hubbard, Betsy Catchionne, Blake Drezet, Brad Nye, Brandi Veil, Carl Grether, Charles Eisenstein, Chelsea Estep-Armstrong, Clayton Gaar, Cynthia Robinson, Danielle Gennety, Darren Minke, Dave Pollard, Dave Zoboski, David Casey, David Heskin, David Sugalski, Debra Giusti, Delvin Solkinson, Dustin Engleskind, Eric Nez, Ferananda Ibarra, Francesco Tripoli, Geoffrey Collins, George Atherton (Geoglyphiks), Heather Beckett, Ian McKenzie, Ishka Lha, Ivan Garcia Sawyer, Jacky Yenga, Jeet-Kei Leung, Jeff Clearwater, Jessica Plancich, Jessica Perlstein, Ka Amorastreya, Kristen M. Rivers, Krystaleyez, Kym Chi, La Laurien, Lila Cari Bivona, Lucy Legan, Luke Holden, Lunaya Shekinah, Magenta Imagination Healer, Marissa Weitzman, Mark Goerner, Mark Henson, Mark Lakeman, Matheo James, Matthew Finkelstein, Maya Zuckerman, Melanie St. James, Melissa Hall (ALIA), Melora Golden, Mikki Willis, Nature Hogan, Nick Algee, Niema Lightseed, Patricia Ellsberg, Robert Gilman, Robin Liepman, Robroy Rowley, Roman Hanis, Ryan Rising, Samantha Sweetwater, Saphir Lewis, Sebastian Collett, Sheri Herndon, Simon Yugler, Steven Michael Ehlinger II, Wesley Pinkham, Wren LaFeet- Mahalo! To the Muse and Onedoorland Family- Thank you for being so supportive and committed to the journey of intentional chosen family. Thank you to my blood family and to Alex Stevens for the roots to these wings. To the Tribal Convergence family- ALL of you have inspired the Village to be embodied, bringing Visions to Reality through all of our journeys as Souls reuniting. To the Lucidity Festival Family- All y'alls- it is such an honor to Play, Dream, and GROW with you! Thank you for your unwavering support. To all of our Collaborative Partners, Allies, and Sponsors, and Kickstarter Contributors, thank you for being the backbone of Grassroots changemaking. To Carl Weiseth, Jason Ross, Elliot Resnick, Manoj Matthews, Nancy Zamierowski, Jake Musselman, Andres Amador, Kinnection Campout, Republic of Light, Amma Lightweaver, Theo Brama, Brett Lee Gillian, Elijah Parker, Patrick Riley, VillageLab for your special Kickstarter magic making! It truly takes a Village to build a Village. Bless you Raquel Hugo, Erick Gonzalez, Carmen Sandoval, Rex, Alaya Love, Starhawk, Jeanette Acosta, Racquel Palmese and all of the cultural teachers who have humbled me and awakened up my heart, helping me remember the original heart when the "maya" would compel me to forget. Thank you to ALL of the community, friends, production team team members and kindred playmates who are building, birthing, dancing the future into the NOW through our love, our prayers, our hearts, and our courage. Thank you to Nolus and Binah for teaching me the way through the mystery of death into brilliance of life. Thank you to all the original peoples, the unnamed ones, the forgotten ones, the un-sung heros, and the re-evolutionaries who's dreams, hopes and shoulders we are standing on as we CoCreate our Future Now.

We remember you.

Jamaica

How to use this Project

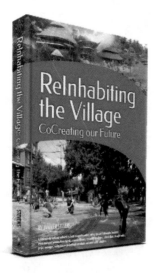

The Book

Welcome to ReInhabiting the Village: CoCreating our Future! This multimedia project includes the printed Book, Workbook, E-book and a Website featuring a Resource Sharing Hub.

This book is the first step on a long journey. With our 7 person production team we have curated 10 Artists and over 60 contributing authors featuring "Voices from the Village" sharing their experience, best practices, strategies and resources to empower through practical wisdom and inspiring perspectives.

HEART OF COMMUNITY

WHOLE SYSTEMS DESIGN

HEALTH & HEALING

APPROPRIATE TECHNOLOGY

ART & CULTURE

MEDIA & STORYTELLING

LEARNING & EDUCATION

LIVING ECONOMY

REGIONAL RESILIENCE

HOLISTIC EVENT PRODUCTION

INHABITING THE URBAN VILLAGE

COMMUNITY LAND PROJECTS

We explore ReInhabiting the Village through the lens of 12 themes, each with an associated color and sigil. You can read this book cover to cover, or jump in at any theme of your interest. This book is meant to be an artifact, a snapshot in time, inviting the continuation of a long movement of changemakers through the ongoing development of the website resources.

Each of the book Themes contain an Introduction from Jamaica Stevens, a breadth of articles, author biographies, visionary art, community photography, informational graphics, inspirational quotes and project features. In closing, the book offers References, Credits, Contributors and a Glossary. The Glossary terms are identified in bold-italics throughout the book for your reference. There may be new words introduced in the following articles, we invite you to be curious.

The Workbook

Bringing idea into form, the Workbook features exercises, how to's, best practices, step-by-step instruction, and practical approaches to integrating the inspiration from the book in order to engage community development, interpersonal communication, project management, regenerative living, holistic design, and more!

The eBook

Taking the pages of the printed book into the digital realm, the interactive eBook will be available to a broader audience. This eBook will provide active hyperlinks to enable direct connections to the Voices of the Village contained within this project.

The Website

Continue the journey with our interactive Online Resource Sharing Hub and participate in the movement. This living library features wisdom of the village through a Directory of Projects, Organizations, and Services, Community Blog and Calendar, Contributor Pages, Holistic Consultants, and more! Find projects and communities to get involved with, share your project and network with like-minded people, learn about things that inspire you. YOU are invited to grow with us and to continue to build this Resource, from community, for community.

Aligning the insight from generations of cultural models and the innovation of up-and-coming thought leaders, this collaborative platform invites open-source sharing of holistic living systems, showcasing simple ways to live in balanced connection with our communities and the places we steward.

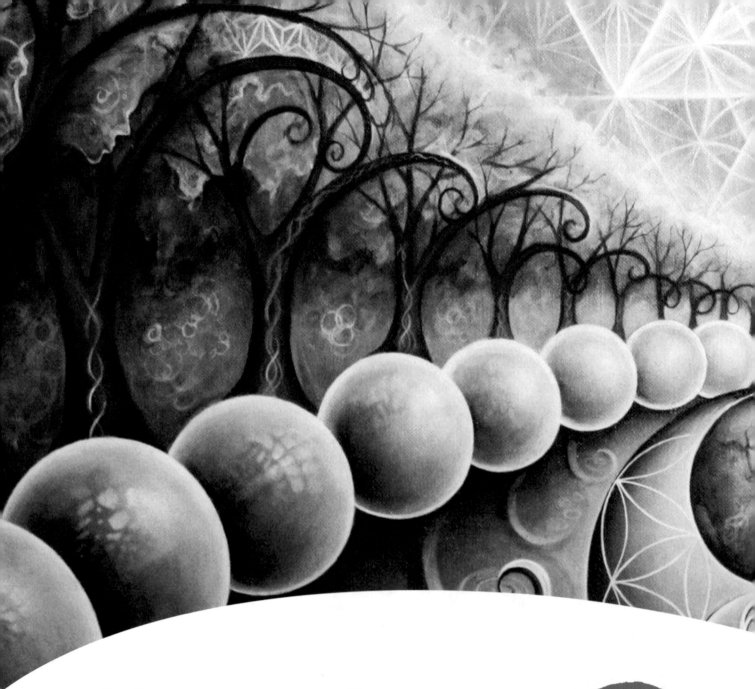

Heart of Community

HUMANKIND HAS NOT WOVEN THE WEB OF LIFE. WE ARE BUT ONE THREAD WITHIN IT. WHATEVER WE DO TO THE WEB, WE DO TO OURSELVES. ALL THINGS ARE BOUND TOGETHER. ALL THINGS CONNECT. Chief Seattle

What is our common thread? What are the unifying factors, the basic foundations of life that connect us all, regardless of the seeming insurmountable differences?

There are simple things like our relationship with food, with our children, some form of language/communication, expression of culture through art/music, and the non-negotiable necessity of Water that are all factors that every culture, from primitive to more advanced, share. And the one unifying factor that unites us all is that we share the same HOME, one planet, a precious blue jewel called Earth set in a vast expanse of endless night.

Yet somehow, so far, that hasn't been enough to lay down our arms and open our hearts to connecting as one species, in stewardship of one home. There is a wound within the fabric of humanity buried deep underneath the layers that separate us.

There is a longing to belong, to be safe, to be valued. Along with that longing is a fear of abandonment and scarcity that can compel actions that take and destroy instead of tend and create. And from that destruction is a great shame, guilt and sadness. With so much grief, how can we begin to forgive, not only each other, but ourselves? How do we see ourselves in the "other" and move forward with commitment to compassion, generosity, and equality? Acknowledging the complexities of social structures, socioeconomic divisions, diversity of culture, violence, environmental degradation, and power struggles within global systems, we must find our way to the heart of the matter and to the essence of our humanity. Now is the time for LOVE.

In this theme we explore what is at the HEART of community and how we can begin to engage the journey of healing by aligning our everyday actions with the spirit of Love in service to Life.

At the Heart of Community is the ethos of TRUST. Trust is built on consistency, a willingness to take ownership for mistakes and the courage to be vulnerable. Trust is also built on accountability which requires transparency. There can be no more hiding from certain core truths. As we face the future before us, together we must build the bridges between our distant shores and recognize just how interconnected we are and how each one of us is responsible for our shared fate. The future of this planet requires justice for all. The vitality of all people, all species, all ecosystems must be included and considered in policies, practices, decisions, and actions.

There is a journey still before us, and when "You" and "I" join together, "WE" will find the way. We will have arrived in the heart when our perspective is holistic, when each member of the community is valued, when we help each other instead of fight, when our leadership works in service to the voice of the people, when resources are shared, when economics include the preservation of seeds, soil, water, forests and human dignity, when basic health needs are met, when city streets are filled with gardens, when learning is a lifelong love affair accessible to all. When we share in celebration, in purposeful work, in play, when we grow from our challenges and triumph over our adversity, when the generations are reconnected and can share their medicine with each other, when skin color is a work of art, when all religions are in honor of the Sacred in all things, when we reweave the frayed edges of the web of life and become the ancestors that the next seven generations will thank. This is not an impossible dream, this is a potential waiting for us to choose. So what do you choose?

Jamaica Stevens

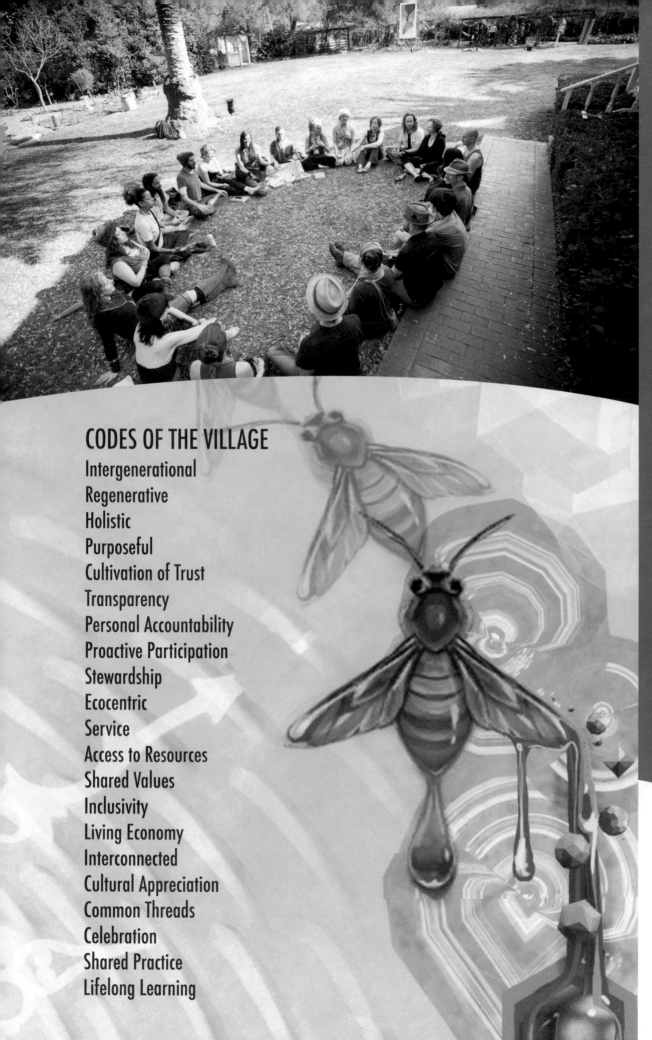

CODES OF THE VILLAGE
Intergenerational
Regenerative
Holistic
Purposeful
Cultivation of Trust
Transparency
Personal Accountability
Proactive Participation
Stewardship
Ecocentric
Service
Access to Resources
Shared Values
Inclusivity
Living Economy
Interconnected
Cultural Appreciation
Common Threads
Celebration
Shared Practice
Lifelong Learning

The Spiritual Heart of Community

"ReInhabiting the Village" is a powerful movement now taking place within progressive cultures and hearts of evolving souls throughout the world. This dynamic shift is an evolutionary cycle that is guiding us away from isolation and separation—which have been the side effects of rapid expansion of our developing societies—and toward a culture that prioritizes community, high-level collaboration and fosters connections and personal support. Whether we are attending a community gathering, participating in a large festival, or creating an intentional community, as we come together in like-minded groups, we experience belonging to something greater than our personal world.

While connecting and collaborating with our tribe, our chosen family that we create and experience life with, we become personally empowered and gain a greater understanding of ourselves. We discover a deeper level of who we are, why we are here and how we can support one another to express our gifts for the benefit of all. As part of a highly aligned group, we have the ability to co-create at powerful levels and together we can create a sustainable lifestyle and share gifts of creativity and healing. And most important, we are surrounded by people who align together with an essential core value: to share deep experiences of the heart.

Throughout humanity's 200,000 years of evolution, our ancestors lived in tribes. The tribe helped keep us safe and allowed us to survive the perils of each period in humanity's history. The challenges of the modern times now demand that we take another evolutionary leap forward, while at the same time taking a step back, to transform both our

local and planetary cultures. In honoring the ways that indigenous people have lived, and today still carry this wisdom, we need to return to their core values of living in sustainable and healthy ways. This recent "back to the tribe" movement, which began in the early sixties, is now empowered with the talents, experience, wisdom and visions of four generations. It is a movement that is now ripe for the next level—to truly anchor a dynamic lifestyle of sustainability, health and harmony in both our personal lives and in our communities. And it's truly the deep experience of personal connection, shared experiences, bonding and ultimately love, that cements relationships, enriches our daily lives and strengthens the spiritual heart of community.

Tribal living is still present in some regions around the world, but modern society has replaced tribal values with individualistic values. We traded the extended family of the tribe for a nuclear family, intent on acquiring wealth and all the comforts the developed world has to offer.

We no longer enjoy an interwoven and interdependent community, but rather live with a persistent disconnection in modern cities and towns. It is not uncommon to be surrounded by neighbors with whom we don't know or relate. Even biological families often fail to collaborate and support one another. Siblings and parents who live on the other end of town or the other side of the country may rarely see each other. Each nuclear family accumulates their own supply of personal conveniences, supplies and tools purchased solely to serve a single-family unit. The cost of sustaining such a lifestyle creates tremendous stress, which breaks down the family even more as now both parents are forced to work.

My personal desire to live in community began during the 1960's when I was a teenager. An entire generation was rising up as an emerging new consciousness created a wave of new lifestyle choices that rippled through all aspects of life. A loud call for change sounded through the culture and gave birth to holistic health, the environmental movement, progressive politics, free love, women's liberation, the arts, and alternative music with the popularity of rock 'n roll. Like an explosion with massive ripples, this new wave of consciousness broke the mold of the previous generation's ways of living. Instead of following the rules, young people claimed the absolute right to experiment with what's possible. Communal living, co-creating and celebrating in tribe, and free love was a central theme, and the "power of love" was the true bottom line.

Inspired by the revolutionary power of the Woodstock Festival, I created the Health and Harmony Festival in 1978, a music, arts, eco-spiritual community festival in Santa Rosa, CA. The Harmony Festival has touched over one million people during its 33 years, and in 2011 alone attracted 30,000 people (www.harmonyfestival. com). Every year, a multi-generational crowd came together for three days and nights to share the power of community, and explored holistic health, ecological and sustainable living, spiritual teachings, and enjoyed a massive celebration of art and music. The power of community co-created villages within the festival grounds including an Eco-Village, Goddess Grove, Healing Sanctuary, Health Expo, Kid's Town, and Techno Tribal Community Dance.

Over the years, I witnessed firsthand the power of community collaboration as thousands of people attended, participated, and co-created the Harmony Festival, where the magic of community inspired deep connection and unlimited creativity. Like-minded groups of people found their soul-mates and soul-friends, established their tribe, co-created miracles at major levels, and truly transformed the consciousness of the northern California community.

And now I witness and participate as the next level of conscious community is birthed. Not only does the new intentional community model support the quality of life, it has become important for humanity's survival. So much has been discovered, learned, and healed in the last 50 years. Now with the wisdom, experience, and vision of four generations, we have the maturity, inspiration and new vision to manifest at a whole new level. This new movement now incorporates the latest technology, new levels of governance and management, developed systems for collaboration, and created proven programs of sustainable living, permaculture gardening, and a healthy protocol for high-functioning, deep relationships.

I believe that "ReInhabiting the Village" is the next chapter that will usher in what Eckhart Tolle has termed "The New Earth." I see a new form of tribal consciousness emerging to establish and anchor new templates for multi-generational living in healthy, sustainable ways that honor our relationship with the earth as well as empowering our next evolutionary cycle. This movement supports, acknowledges and utilizes the talents and the wisdom of everyone in the tribe, supported by the power of the collective. It allows for extreme creativity bolstered by technology. It empowers both the masculine and the feminine, celebrates all the generations, and offers them the opportunity to vision, create, and manifest their dreams together.

LOVE is the glue, love is the power, love is the overriding all-pervasive force. Compassion, deep respect, and spiritual alignment are the essential juices of life in community. But love is the bottom line; the magnetic force that draws us together; the light that allows us to heal each other; and the sweetness that finds us celebrating life.

Love is the Spiritual Heart of Community.

Debra Giusti

Debra Giusti has spent the last 35 years on the leading edge of the ever-emerging progressive culture, creating and supporting "new paradigm" evolution. She has been an entrepreneur since 22 years old, growing an ongoing series of businesses that support personal growth, transformation, healthy and sustainable lifestyles, cutting-edge spirituality, and community connection. She is the founder and co-producer of 33rd Annual Harmony Festival, which took place from 1978 – 2011 and drew 30,000 people to Santa Rosa California. She also produced/created Wishing Well Distributing Company (featuring 3000 nationally distributed holistic health products), Spirit of Christmas Crafts Faire, Well Being Community Center, Well Being Community Calendar, Transforming Through 2012 Multi-Media e-Book, and hosted multiple workshops, transformational events, and celebrations.

She now produces Debra Recommends eNewsletter (with 50,000 subscribers) and Harmony Connections eNewsletter (75,000 subscribers), which inform and network northern California and the San Francisco Bay Area community with progressive events and resources. She is a prominent positive force in business and community, continuing to develop new businesses that support the emerging new culture, and inspire the development of deep community.

http://debrarecommends.com

http://debragiusti.com

http://harmonyconnections.com

So Many Roads
A Retrospective

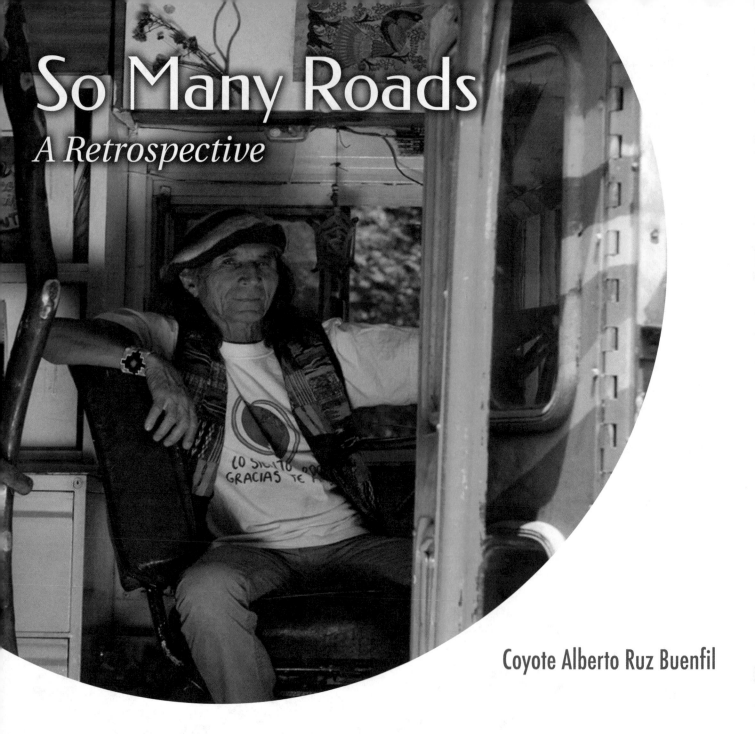

Coyote Alberto Ruz Buenfil

Back in 1968, when I was 23, I decided to be one of the hundreds of thousands of young people in my generation to "turn on, tune in and drop out" from the system. My serious game of life consisted in changing the system, and when I definitively left the University in Mexico City, I hitchhiked with a group of my friends to San Francisco and began my quest for the tools, contacts, and ideas to fulfill that purpose. I took part in the occu-pation of Washington DC during the Poor People's Campaign, and became acquainted, while in the States, with the principal leaders of the Black Panthers, the Brown Chicano Panthers, the American Indian Movement, and especially in New York, with the street band Up Against the Wall Motherfuckers, the SDS occupiers from Columbia University and the publishers from The Rat. In the West Coast, I associated with The Berkeley Commune, Abbie Hoffman, Jerry Rubin, the Berkeley Barb and the SF Oracle. When I left the States, my quest was clearly defined: I was going to create an intentional community, learn more about that idea, gain experience and start materializing my vision.

I continued my journey visiting and living in several *kibbutz* and *moshavs* in Israel, took part in the first urban occupations in London, Amsterdam and Copenhagen, spent a year at the Bauhaus Situationist

Drakabygget in the south of Sweden, stayed in numerous crash pads in Germany, Switzerland and Austria, and ended up participating in some of the early attempts to create urban radical communities in France and Italy. By this time, my friends and I had become a permanent group of roving artists, inspired by ***The Living Theatre***, and using the streets, parks and public places to perform events and spontaneous performances to provoke the passive audiences and turn them into agents of change in their own cities and towns. This was the way we were contributing to raising the social consciousness of thousand of people on issues such as; the refusal of war as a solution to conflicts, injustice, racism and segregation, gender and economic inequality, and the rejection of the dominant model of society and its values. Through our plays and actions we were also supporting the revolutions and Independence Movements of the 60's, and especially advocating for a new society where justice, freedom, solidarity, peace and love would be the main principles to govern our lives.

In the year 1973, our tribe, first called the Chaos Affinity group, consisting of seven or eight members and three children, moved a step further in that same direction, and started traveling and living in another kind of traditional and spiritual community. In that second phase, we lived with Bedouins in Palestine, Berbers in Morocco, Kurds in Turkey, tribes of Tuaregs in the Sahara, and finally with many different ethnic groups and in several Ashrams during our one year sojourn in India. Most of the group continued the odyssey to Indonesia and Australia, but unfortunately, myself, my partner Gerda and our two children--Odin, born in Sweden, and Mayura in the Canary Islands--were deported back to India, and from there to Europe and from Paris to México City, where we returned almost eight years after our first trip hitchhiking to California in '68.

Our Group left Asia with a new name, Hathi Baba´s, the Transit Ashram Commune and a rich experience of Eastern life, Hindi and Buddhist spirituality, and mostly having solidified our collective vision to become an extended family, a true stable community. We continued doing theater as our way of living and sharing all these experiences, now with an emphasis on using legends, mythologies, music, songs, dances and rituals from the many cultures we had visited, as a way to convey our message that beyond the geopolitical borders, we are one humanity, with many colors, languages, ways, and that richness should be defended to avoid the desertification that globalism is introducing and imposing in all the corners of the Earth.

After Asia and Oceania, we all convened again in Mexico, where

We left Mexico with ten adults and five children, and arrived to California just in time to take part in our first Rainbow Gathering. After all those years around the globe, we had lost contact with the countercultural movement in the United States, and as we camped in the middle of the forests, we found that several thousands of other people and extended families like ours were part of this emerging Rainbow Nation, in which we immediately felt we'd arrived home after a long exile.

From 1976 to 1980, our tribe, renamed The Illuminated Elephants, took part in the emerging **Ecotopian**

ed, and opened many doors for us everywhere, as well as a growing number of friends all across North America.

In 1980 we decided to start a new pilgrimage, this time with six buses, 28 adults and a dozen children, with the purpose of visiting all the traditional nations that we found in our path; ranging from the redwood forests of California, through the deserts of the US Southwest and North of Mexico, up to the top of the Sierras and down to the tropical jungles of Chiapas in Mexico. We shared camps with Pomo people; Hopis, Pueblo and Navajos, Yaquis,

"My advice to people today is as follows: if you take the game of life seriously, if you take your nervous system seriously, if you take your sense organs seriously, if you take the energy process seriously, you must turn on, tune in, and drop out." Timothy Leary

we established our community in the south of the city, and connected with other groups and communities that had survived the repression that the Mexican government had unleashed against the youth and the social movements. We managed to survive in the City for another year, doing theater, selling home-baked bread, learning more drama techniques, holding a safe meeting place for the alternative movement, and preparing for the next chapter of our journey. Jan Svante and I, from the Hathi Baba´s tribe, went to California and came back with a huge converted bus, a motorbike and a small rubble boat, with which to pick up the rest of the family and get back on the road together for the North.

movement in the West Coast--notably in northern California, Oregon and Washington State--living in several rural and urban communities, growing in numbers, children and vehicles, and becoming part of the Federation of Intentional Communities, a network listing several hundreds of projects in the States and the rest of the world. We took part in New Earth Festivals, Simple Living Workshops, Rainbow Gatherings, Pow Wows in indigenous lands, peaceful actions against nuclear energy, and dozens of cultural activities everywhere we went.

I personally continued writing about our history, the antecedents of the countercultural and alternative movements we were part of, the communities we had visited or lived in, and the scripts for our multimedia plays, which became more and more professional and sophisticat-

Tarahumaras, Huicholes, Nahuatl, Totonaca and Mayan communities; the journey taking us almost two years. In Mexico City, as a theater group we received support from the Cultural Ministry, allowing us to perform in the cities, plazas, and theaters from the most remote rural communities to the capitals of most states in the country.

At the beginning of 1982, after the oil crisis, the devaluation of the peso in Mexico, and the lack of government funding, we decided that moving in our old Ford bus or "mechanic elephant," was too expensive and that it was a time for a change in our lifestyles. I had a new companion, Sandra, and a new daughter, Ixchel, and as our children were reaching school age, some of the members of the family needed a good rest after all these years of countless adventures and difficulties.

A group of us spent a few months scouting for a place for a more permanent base, and in March 1982, we finally found one that the rest of the tribe agreed was the collective nest we were looking for, and we started payments for a piece of land of five acres in the mountains of the state of Morelos. The place had no water, electricity, or access, but in the center was a hundred-year old Amate tree, whose roots and branches made for a natural magnificent sculpture, and in one of the mountain cliffs we found out that a waterfall would undoubtedly be flowing in the rainy season. These two main reasons gave us the thumbs up signal to start building our settlement, Huehuecoyotl, the place of the Old Coyotes, which today, 33 years later, is considered to be the oldest ecovillage project in Latin America.

Thanking the Elephants "*naguales*" for their protection and accompanying us for all the roving years, and after a public *psycho-magic* performance in Mexico City, we became officially "Los Viejos Coyotes." Living in our old, decaying buses, we began constructing our homes and communal places, using local materials such as adobe bricks, lava stones, ocote wood from the neighboring forests, waterproofing our walls and cisterns with "baba de nopal" (cactus slime) and applying all the knowledge that our journey provided us in terms of simple living, adapting and taking from Mother Nature just what is needed and returning her only those materials that can be recycled. We learned how to collect, contain, use rationally and distribute the rain waters from our yearly crop, and how to use the grey waters and dry toilets to avoid spoiling clean water, sending our excreta to the aquifers and producing humus for our trees and gardens.

Visiting other similar projects in the rest of the country, we learned from their experiences and shared with them what we had found out in other countries and communities. Without electricity, with a manual mimeograph machine we produced "Arcoredes" in Huehuecóyotl, the first artisanal magazine with eco tips, stories from different projects, advertizing for local products and information on future events. In three years, with a couple of dozens of other organizations, we had created the first green network in Mexico, and called in 1985 the first National Encounter of Ecologists, an event that has never been repeated since. We also joined and took leadership of the marches against nuclear energy--when the government was planning to open Laguna Verde, the only nuclear plant in Mexico--and were able through our actions and protests to stop any further project of this kind in the country.

In 1987, year of the Planetary Harmonic Convergence, Huehuecoyotl hosted the 1st Fifth Sun Festival, honoring the beginning of a new cycle of time according to the Maya-Aztec tradition. It was a festival of arts, healing, ceremonies, workshops and organic market, to which several hundred people came, camped and visited our up-and-coming community. It was the first of seven similar events, which attracted more and more people every year, and a space where many children and youth got to experience the manifestations of a new form of culture of peace, created by two generations of *eco-artivists*, a term that designated the fusion of ecology, art and activism, and brought us together regardless of our age.

In the Spring of 1990, Huehuecóyotl hosted the "Encounter for Nature of the Guardians of Sacred Traditions," consisting of spiritual leaders, ecologists, artists and scientists, and a seed was planted to continue doing this kind of events in the future, which were to be called "Vision Councils of Earth Keepers." One year later, together with the network Arcoredes, Rainbow Nation representatives from various indigenous nations of Mexico, and well-known artists that offered their work in benefit of the event, held the 1st Vision Council." Held at the ceremonial center of Temoaya, from the Otomí-Ñañu nation, this started a new tradition that has not only lasted up to our present year 2015, but has been reproduced in a dozen other countries of Latin America and Europe.

The design of the "Consejos" (cf. article of Ivan Sawyer, Laura Kuri and Coyote Alberto in this book), has evolved over the years and has adopted particular bioregional characteristics in the different countries where it has since taken place. This evolution went parallel to the growth of our ecovillage, where not only thirteen new houses and a beautiful communal theater were built, but from which also came Sandra's idea to establish "Cetiliztli," an alternative holistic school where our children, our friends' and children from the nearby villages like Tepoztlán, Amatlán and Santiago could go, where a curriculum could be offered according to our visions, values and means. Cetiliztli lasted fifteen years, and several of our kids grew up to be teenagers, and then moved to higher levels of education in the closest city to our community, Cuernavaca.

The green network of Arcoredes and the founders of Huehuecoyotl also had its own process. Some of the early ecologists who attended our events, meetings and actions became green politicians and created the Mexican Green Party, and the National Movement of Ecologists, which continue to this day to hold posts in different state and national administrations. Others became scientists, and are now experts in alternative energies, reforestation, echo technologies and environmental laws. Some of them are developing new curriculums of *permaculture*, *agro ecology, bio construction, ecovillages* design, facilitation for decision-making for groups, etc. A few more are activists for the defense of the *non-transgenic seeds*, work with local farmer communities, and are present in all the different actions against Monsanto and other agribusiness companies.

In 1996, after finishing and publishing my book "Rainbow Nation Without Borders," translated into Spanish, Italian and Portuguese, and after the birth of my youngest child, Solkin, from Lourdes, my then Basque companion, I decided to start a new nomadic project that I called The Rainbow Peace Caravan, left Huehuecoyotl one year later with a crew of enthusiastic volunteers, and began a journey that ended up being a little longer than planned. Originally, we began as an outreach for the Bioregional Movement, the recently created Global Ecovillage Network (GEN) and the "Consejo de Visiones" to help promote our common values in the form of various types of multicultural activities, workshops and courses, audiovisuals and conferences, ceremonies and rituals with the original nations that we visited on our way. But eventually, and mostly, it set a foundation for what came to be labeled "a mobile Living and Learning Center," that was recognized in the year 2000 in Colombia as an integral part of the Ecovillages Network of the Americas (ENA), part of GEN itself.

The Caravan traveled first through Mexico and all the different countries of Central America, crossed Panama to disembark in Colombia in 1997, spending three years in the most remote and violent zones of the country. Then in Venezuela and the Amazons in Brazil; teaching the first courses of permaculture, consensus decision making, and

ecovillage design, creating the first Networks of Ecovillages in Colombia and Venezuela. Then continuing its journey through Ecuador, Peru, Chile, Argentina and Uruguay, where among many other activities, the Caravan organized and hosted the International Call of the Condor, near Machu Picchu in Peru. This event brought together nearly 1,000 participants from 36 countries, and dozens of representatives of the most recognized environmental and ecologist networks, including those from GEN and ENA, the Bioregional Movement, the International Rainbow Nation, and many others from most of the Central and South American countries. Indigenous leaders were also present, and many artists which came once again to offer their work for the benefit of the event and its ideals.

Similar events were organized in Ecuador, (Aldea de Paz para Mujeres Líderes), where I met in 2002 my life companion, Veronica; and then in Chile (Llamado del Arcoíris in Santiago de Chile and Llamado del Aconcagua in Viña del Mar). It was only in 2005, nine years after we left Mexico, that we reached Tierra del Fuego and Ushuaia, the foremost town in the continent, thus concluding our commitment to cross the Americas as it was planned in our original vision.

But the Great Mystery was not satisfied, and we were immediately invited to organize another international encounter, this time in Brazil, which was called The Call of the Hummingbird. The Caravan was thus reorganized to continue the pilgrimage, and arrived north of Brazil and then to the Chapada dos Veadeiros, in 2005 to host--with the help of local Brazilian activists, in a beautiful place shared by an afro-Brazilian community and an ecovillage--an event that this time brought together more than 1,500 people from all over the world.

Right after this historical event, Brazil's Minister of Culture, Mr. Gilberto Gil, invited us to spend a few years in his country, doing the work we had done for the last nine years in the rest of the continent. We ended up receiving funding to transform the Rainbow Peace Caravan into the Caravana Cultura Viva (Living Culture Caravan), a project that lasted from 2006 until 2009, and took us to more than 150 different communities from the whole subcontinent that is actually Brazil. Each of the chapters of this story could be extended to the length of a book--in fact I have four books written and published on this subject: "Hay Tantos Caminos," (So Many Roads), "La Leyenda del Cuarto Mago" (The Legend of the Fourth Magician), "Historia de los Movimientos Comunitarios del Brasil" (The History of the Communitarian Movements in Brazil), and "De Punto en Punto" (From One Point to the Other).

Today, almost 70 years old, having lived most of my dreams in this lifetime, Veronica and I are back in Mexico. After the Caravan, we worked three years to form a first generation of "Promoters of Ecobarrios" in the City. But right now, we have found a much more ambitious purpose: to help articulate a global campaign for the adoption of a Universal Declaration of the Rights of Mother Nature. If I can see this happen in the years to come, I will know that the purpose of my existence has been achieved.

I hope this story inspires you to dare to live your dreams. Have a great journey!!

Ecovillage Huehuecóyotl, Mexico
15/02/15

Coyote Alberto Ruz Buenfil

Forty years dedicated to studying, creating, promoting and serving as an international networker, make Coyote Alberto Ruz a first line pioneer, veteran and historian from the intentional communities, ecovillage and bioregionalist movements.

• Co-founder from Huehuecóyotl ecovillage in Mexico (1982)

• Originated in 1996 the "Caravana Arcoris por la Paz."

• Itinerant focalizer for ENA in South America since 2000.

• Ashoka fellow (2002), and adviser to GEN (2003)

• Partner to the Brasilian Ministery of Culture´s program "Cultura Viva" (2006-2007)

• National award "Escuela Viva" given to the "Caravana Cutura Viva" by the Government of Brazil in 2007

• Originator of program "Ecobarrios" in Delegación Coyoacan, with support of Government of Mexico City (2009-2012)

• Nominee to the National "Premio al Mérito Ecológico" in México, given by Secretaría de Medio Ambiente y Recursos Naturales (2011)

• Co-creator of C.A.S.A at the 1st Iberoamerican Summit of Ecovillages, Colombia, 2012.

• Director of Environmental Culture in the State of Morelos, for Secretaría de Desarrollo Sustentable, (2013)

• Nominee and finalist for "The Kozeny Communitarian Award 2015", given by Federation of Intentional Communities.

• Adviser to the Asamblea Legislativa Constituyente (Deputies Chamber) from Mexico City for the promotion of the Law of Rights of Nature(2014)

• Representative from C.A.S.A Continental at the 1st.Global Summit of Ecovillages in Senegal, 2014

http://caravanaarcoiris.blogspot.com

The Spirit of the Village

Jacky Yenga
The Village Wisdom Messenger

In the West, we live disconnected lives. People live in isolation, or with unauthentic and unfulfilling relationships. This creates stress, loneliness and a deep sense of separation that often leads to major cities around the world identifying social isolation as one of the biggest issues for their populations, above hunger and homelessness.

We all know that authentic relationships are essential for our growth, and many of us believe that without heart to heart connections, our soul suffers and we cannot be truly happy. Unfortunately, most people in the West live a disconnected lifestyle.

Many research, documentaries and direct experiences have shown us how traditional cultures find happiness in simply being together and feeling connected to one another.

But in the West we are deeply conditioned to live our lives based on financial considerations and define "success" in terms of material achievements.

We love and praise those quotes and stories that deliver inspiring messages for a more balanced lifestyle, but we don't actually live our everyday lives according to those values. Instead, we are too busy chasing after the North American model of success: a big empty house, with a huge lawn and a big fence to separate us from others, and of course lots of

privacy and a fat bank account.

It's OK to want material success and social status, but making that the primary focus of one's life leads to an empty life.

Connection to community is one of the key components to happiness. We know that intellectually, but we do not make it a real priority to live less isolated from one another.

What would happen if we decided to change that and to really become each other's family? What would happen if we decided that those values that promote autonomy and individuality over collaboration and solidarity are no longer how we want to live our lives?

The opposite of happiness is not sadness... it's loneliness.

Typically in Africa, our foundation for life is rooted in togetherness. Our traditional teachings focus on how to live with each other, before anything else. We grow up listening to stories that promote togetherness, solidarity and collaboration, respect for the elders, for nature and the ancestors. As a result, we develop a very strong sense of belonging to our community and our land, and we are taught to see others as potential members of our family from a young age. All the adult women are aunties, the men are uncles, and the children are brothers and sisters. We have no choice but to live a connected life, and that's all we know to be true.

Choosing a connected life in a culture that promotes isolation and autonomy as a way of life is challenging.

In the West we have accumulated lots of knowledge about the importance of living a more community-oriented life to be happier. We know what to do and why to do it. But taking consistent steps towards more togetherness and turning the knowledge into action is so challenging to do with a mindset that is set on individuality and financial profit!

Yes. In our efforts to redefine our lifestyle choices, we must realize that we are going against generations of conditioning and deeply ingrained values and habitual ways of living. Today we have enough evidence of how we are killing the planet and each other, we can no longer deny it, and yet we are so used to our comfort and way of life that defending our "right" to consume whatever we want often takes precedence over fighting for what we know is right, like democracy or basic human rights. That's our current reality. Thankfully, it is not the only one. Everywhere in the Western world, many people are gathering and reclaiming community as our natural state, and are making commendable efforts to live according to more humane values.

However, these efforts remain isolated and do not seem to shift the balance in a significant manner. That's because what is required is a total mindset shift, not just clever strategies to better use our resources and be better neighbours. What would that look like?

First, we would have to stop trying to define what a community is and actually start living like one. In African villages, we do not hold debates about what a community is because we are too busy living as a community! Also, we have a larger definition of community than most people in the West understand. Our community not only includes the people near and far, but also nature and those who have now rejoined the land of our ancestors!

Second, we would have to re-align our values.

In the West, we value academic intelligence to the point that we seem to believe that we can solve everything with smart tactics and strategies, including our relationships …

We need to learn to value wisdom again.

What is wisdom? It is having experience and knowledge and the ability to apply them in an appropriate manner. Wisdom is not logic or intelligence, and it is not intuition.

To put things into perspective, intelligence allows you to be successful at living a practical life, like investing money or planning for the future. Wisdom, not success strategies, is essential in developing meaningful connections.

Without the guidance of wisdom in our relationships, we feel lost and seek out a system to tell us what to do. We even learn techniques that teach us what to say to each other and when to say it, but then we lack the sincerity and authenticity that can only come from a much deeper place.

A common African proverb says that "Wisdom does not come overnight," meaning that it grows with maturity. However, Western culture teaches us to use our intellect and to be logical and rational all the time. We go to school to learn to be smart, but we are not taught to be wise. Then we confuse the two to mean the same thing.

In African culture, we believe that knowledge without wisdom is like water in the sand, and in our tradition, it is when a person is older that

they are the most valuable to their community, precisely because of the level of wisdom they have reached, not because of the number of diplomas that decorate their wall.

By considering these 2 points to begin with, we can have a good start in implementing real change and ReInhabiting the Village.

Africa has a lot to offer to the world. More and more people in the West are opening up to the spiritual depth of African culture, and are recognizing that "Indigenous" peoples carry a way of being that is sustainable, harmonious, healthy and just. This wealth of wisdom is accessible to you if you want it, however it is time to put into action all the academic knowledge that we have proudly accumulated.

Africans have a thing called Ubuntu. It is about the essence of being human; it is part of the gift that Africa will give the world. It embraces hospitality, caring about others, being willing to go the extra mile for the sake of another. We believe that a person is a person through other persons, that my humanity is caught up, bound up, inextricably, with yours. When I dehumanize you, I inexorably dehumanize myself. The solitary human being is a contradiction in terms. Therefore you seek to work for the common good because your humanity comes into its own in community, in belonging. Archbishop Desmond Tutu

In the end, it is our friendships that count the most and connection to community is essential to our happiness. We have to believe that it is possible to live a life where we have it all. And that happens when we combine the material abundance found in the western world with the high levels of community and connection we find in African villages. That's how we authentically take care of each other and our planet. That's how we experience the Spirit of the Village, where nobody feels alone and the earth is not abused for profit.

Jacky Yenga

Author of the upcoming book: The Spirit of the Village: How to break the habit of living disconnected and experience the joy of consciously living a more authentic and connected life where you belong.

Originally from Cameroon, Central Africa, Jacky Yenga grew up in Paris and now lives in Vancouver, BC. Her foundation for life was rooted in togetherness and collaboration, respect for the elders, for nature and the ancestors. Living a connected life was her only reality, until the age of 9. She experienced the trauma of disconnection when she was sent to the West to "live a better life". Now she is an inspiring speaker and an enthusiastic ambassador for the wisdom of Africa and their message of togetherness, which she shares around the world in various forms, including as a singer, music and dance performer.

"I believe in contributing and making a real difference in the lives of the people I meet. I am here to help connect people to themselves and to each other. I picture a world where your life is filled with nourishing connections, a real sense of belonging, more authenticity and more humanity – because that's the choice we make when we come from a place of deep inner wisdom. I believe in adding value and sharing with you that there's a better, healthier and more compassionate way to live that takes care of everybody."

http://jackyyenga.com

NeoTribal Culture

My name is Andrew Ecker. I am a drum circle facilitator, artist, poet and cultural architect. For the past 20 years I have been associated with a cultural movement I call "Neotribal Culture." I self-identify as "Neotribal" in hopes that we as Neotribal people one day have a voice in the contemporary culture and hopefully gain social presence for the issues that pertain to our self identified culture; my own experience being the son of two self identified hippies, one native american and the other white. I grew up in a home that was diverse and oftentimes in midst of a culture war. It is hard to share something that has molded a person's life and given rise to self-identity; it takes vulnerability and exposure. But this cause is so great and the hour so close that it is my hope that as we together define a culture that is synonymous with the concept of "production for the means of restoration" that we may see a deeper unity amongst the many subgroups within the space of Neotribal culture.

The term Neotribalism was possibly first used in a scholarly text by a French sociologist named Michel Maffesoli. In his book the *"The Time of the Tribes,"* Maffesoli eloquently predicted that in the postmodern era a great time of social awareness would occur. This awareness would dramatically change the social climate by fostering a time in the human social evolution where key factors of separation based around self-identity such as class and race would fade to an identity based around experience. This for me helps to bring light to the concept within "Neotribal Culture." Although Maffesoli gives us a direction and reason for why we as people are drawn to a tribal experience, he doesn't actually define the cultural movement as I see it.

For me, Neotribal Culture is based around music, food, art, fashion, ritual, spirituality and the basic communal action of gathering to share in ideas, collective awareness and consciousness. This differs from the concept of Neotribalism which in turn is a social phenomena. Instead Neotribal culture is generated from the utilization of key ancient tribal technologies of social engagement dance music art food and ceremony.

There are also several indicators that differ from the traditional forms of tribal peoples. Some of the significant differences in Neotribal Culture that are distinct from traditional tribal culture are based around the concept of a blood connection to clan or tribal membership. Neotribal communities, for the most part, do not set apart people of different race or even any ethnocentric form of belief systems such as religion or worldview. In contrast, the one primary foundation with Neotribal culture is acceptance of people for who they are. There is the potential that as Neotribal groups become more realized there may be an emergence of blood connection and clan systems as membership to these communities grows and evolves.

Neotribal individuals are people who desire community in its many forms of communal expression. It is a common practice in the Neotribal arena to simply refer to a peer group as a "tribe." This need to identify with "tribe" is based around the concepts within the contemporary terminology associated with tribal membership in association within the indicators of artistic and spiritual expression. As people gather in cultural development through the foundational principles within defining a culture, the foundations of that cultural connection emerge and become more defined. This defining of tribal connection then takes on a deeper purpose and meaning as people engage with one another and continually reinforce the foundation of this emergent experience of community. At some time within this process, people begin to define the "tribe" they are associated with. These new tribes that are formed are the foundation of the Neotribal cultural movement. There are many tribes within the Neotribal Culture.

The "Rainbow Tribe" is a global group of people self-identifying with several Native American prophecies concerning the return of the ancient tribal technologies after a long period in which people would not be able to freely worship in a tribal sense. The "Rainbow Tribe" or commonly referred to as the "13th Tribe" or the "Rainbow Family of Living Light" is a foundational part of Neotribal culture and many within Neotribal culture believe this is the foundation of the entire cultural movement. The cultural practices of people within the Rainbow Tribal experience vary significantly from one person to person but there is one primary foundation, and that is the vision of world peace and the need to reconnect with nature in order to obtain global peace.

"Moon Tribe" is a community of moon-loving nature-loving southern California desert dancers who have been gathering under the full moon since 1993. To dance and take part in the traditional tribal ritual of celebrating the full moon, "The Moon Tribe" has had several transformations and is currently survived by a second generation of tribal members. These new members have taken on the traditions set before them by the many founding members of "Moon Tribe." "Moon Tribe" is a great example of group of people gathering together to engage in the ancient practices and technologies of ancient tribes in a contemporary setting. The use of dancing, nature and celestial cycles and phases of the moon to create a sacred space is one of the oldest traditions and dates back thousands of years.

"One Dance Tribe" is a global conscious movement-based community. It is a portal to a new unified field where teachers and dancers from different countries, modalities and walks of life come together as ONE TRIBE to celebrate the power of MOVING AS ONE. Their mission is to awaken love in all hearts and on the planet, and contribute to the creation of a peaceful and prosperous global community.

Deer Tribe Metis Medicine Society is a group of Rainbow SunDance Warriors brought together to fight against ignorance, slavery, bigotry, racism, war, dogma and superstition using the Wheels and Keys of the Sweet Medicine SunDance Path to seed the future generations with Beauty, Power, Knowledge and Freedom.

All people within Neotribal culture have an affinity to the utilization of ancient tribal technologies in this contemporary world and for us the task at hand is great. Now more than ever our world is being defined by a time of global industries fueled by greed. The medicine of the drum, the dance and ceremony is being replaced with sounds of 60 second commercial jingles. In the place of looking into fire most are now looking into the TV. In the midst of this greed the time is now for all TRIBAL people to define our way to live, to chant our chants, to play our drums, to dance our dances. Two thousand years ago when the people gathered by the fire and the great shamanic seers and tribal prophets knelt down to the ground and looked into the infinite vibration, they saw this time that we are now living in and they wrote it, drew it and told it. These teachings are for us NOW to wake up! The great holders of the ancient tribal wisdoms are sharing the medicines and teachings of the original peoples once again and we don't know if this convergence of new and old will ever happen again. So it is up to us to use what ever tools we can to make a difference, to produce, to restore people, families, communities and the planet. We are the living definition of our culture, ***WE ARE ALL ONE TRIBE.***

Andrew Ecker

Andrew Ecker, owner of Drumming Sounds based in Phoenix, AZ, is a Poet, Artist, Drum Circle facilitator and architect of Neotribal culture. Born in Portland, OR and raised in the Hawthorne district, he is the grandson of Cresencia "Cora" Alvarez, an Apache Curandrea "Herbalist and Healer" from the southwest.

Andrew has continued this family tradition of transformative healing work by honoring the ancestors in the work he does with the drum. It's only through the intentional movement of energy that we as a Neotribal people will see our world transformed to serve our awakened culture. This art of community transformation and personal healing has allowed Andrew the honor of working with many amazing nations, organizations, hospitals, festivals and businesses including The Havasupai Nation, The Hopi Nation The Tohono O'odoham Nation, The Navajo Nation, Cancer Treatment Centers of America, John C Lincoln Hospital, The Lucidity Festival, Firefly Festival, Blue Star Gate spiritual gathering, City of Phoenix, AZ and the Veterans Association, to name only a few. Andrew has been participating or facilitating the Neotribal expression of the drum circle for close to 20 years. He brings wealth of knowledge from personal experience and the amazing training of his mentors The DRUM, Air Water Fire Earth, Christine Stevens, Arthur Hull, Jim Boneau, TINY Hanna, Dr. Barry Bittman, Linda Rettinger, and Teddy Begay.

Finding our way home

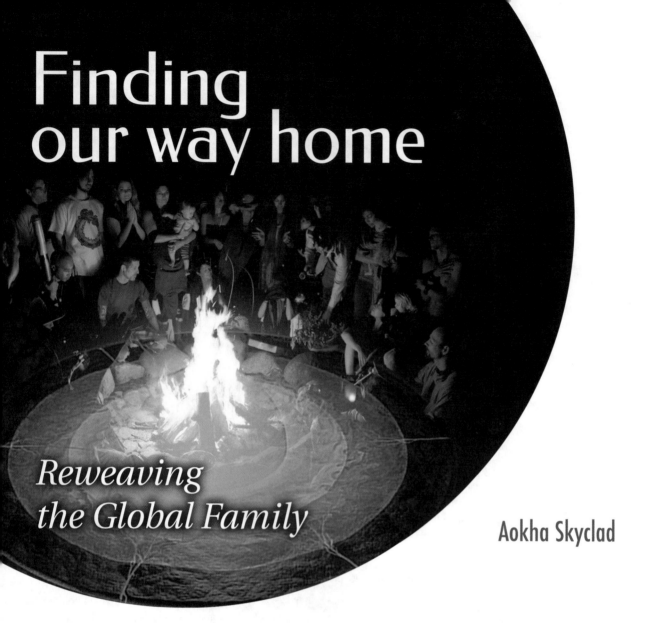

Reweaving the Global Family

Aokha Skyclad

Family

There was a time when we all had a village, so many eons ago. We all had a village council of elders who sat together and shared the needs of the community and made group decisions based on those needs after all voices were heard. Those decisions were made by local leaders who were well aware of and had a stake in making sure that the community's needs were being met, because they were a part of that community and knew and loved the rest of the community members. Because a village is a family. And the people making decisions for the family should know and love the family, or else how can they possibly speak for them? This village family was aware of all the pressing issues facing the community. No one was hungry. No one was without health care, because every person was known and the group could come together if an emergency arose. No one fell through the cracks. We remembered our place in the ecosystem and knew how to live in balance with the earth so that our resources stayed abundant and thriving.

Then somewhere along the way, as cities arose, people migrated away from their villages and became anonymous numbers in the throng. People started falling through the cracks as there were too many in one space, and there was no village to remember them or look out for them. As industrialization increased, a carefully crafted concept of the "nuclear family" further eroded the village concept by making the extended family obsolete and unpopular. This nuclear family was made to be seen as the preferred modern way of living, disconnected from our elders and family and easily manipulated or swayed by the propaganda of corporations, whose only motivation lay in creating a compliant and pliable workforce and economic base from which to draw its labor. This disconnection has only increased, to the point that we now have nations trying to make decisions for such large groups of people that there is

no way they could ever truly say they know the will of the people they claim to speak for, not only because they have never met them, nor do they love them, but because the group itself is so large that it contains opinions that are infinitely varied, and no one decision could possibly be right for everyone.

Progress

People sometimes claim that those of us who promote the village concept are enemies of progress. I have no issue with progress, as long as the definition of progress includes that the lives and livelihood of all people are improved, to the detriment of none, and without destroying the earth, our home, as a byproduct. Moving from the village to cities and the industrialized world is not automatically "progress" just because they happened in sequential order. And the village concept can be adapted and updated to include modern technology and conveniences. Clearly from my article "Connectivity in the Digital Age," it can be seen that I am no enemy to progress. But sometimes in order to find the true path to "progress," we must carefully observe the old ways, combine them with the new ways, and analyze what actually worked well for all of the people, with the least collateral damage. And that means revisiting the village, community, connection, and family.

So how do we find our way back to the Village? Who can we ask to learn how to do this? We naturally turn to the indigenous for guidance, the ones who still remember and sometimes still practice the old ways. But there is so much wounding there. We, of the Western World, are like orphaned adolescents innately drawn to a culture which makes sense to us but into which we were never invited; lacking lineage and disconnected from a sense of family or cohesion. We find ourselves living in physical villages, but with walls and fences and constructed privacy to make sure that we do not get to know our neighbors, and instead think of them as strangers. Inwardly we yearn for our elders and our wisdom keepers, our teachers and support. A family who sees us, and recognizes our value within the group. We yearn for the village and those who can teach

us how to live in community again, but without any idea where to find them. We look around to see such great ancestral grief carried among the indigenous who remain, and the wounding continues still. How can we even approach that? We, ourselves, carry an ancestral shame that has nothing to do with the reality of our present incarnation. How can we heal this rift and invite all of our relations back to the table to find a common way forward?

Knowing Self

I am a chameleon and anthropologist. I am also a mutt. Of my genetic coat of many colors, some small part is Native American. Not a big enough piece to be recognized by my government, but enough that many people have seen it in my face and commented on it. Some practices of the indigenous have always been my ways, since before I knew I had ways. I communed with animal spirits, connected with nature and plants, and understood the intelligence of talking circles before I knew

that those were indigenous practices. My indigenous heritage is the part of myself I have always been most proud of, and also felt the most alienated from. My call, throughout my years, has been a vision of a global family. As a nomad, I see no borders or nations. In my travels I have been welcomed into families of indigenous and non-indigenous in the same manner of warmth, compassion and generosity. I have been treated as one of the tribe and family. All are HOME to me. In every place I have traveled, I hear them tell their creation stories and at the heart they all tell the same story. All of humanity carrying one shared heart, only divided by an illusion of separateness as if our differences must divide us instead of simply honoring the perfection of our unique and varied distillations, and recognizing the immense beauty that is created by the rich textures of this variance in the greater weave… All my life I have been collecting the shards, finding the points of light and connecting the dots. Doing my best to reweave the hoop and find our way back to the remembrance

of the global village; the knowledge that in the end, we are all one human family.

So how can we heal this? How can we come back in a good way, with respect and honor, showing that we are HERE, full of heart, with humility, and ready to learn? Yearning for teachers and elders to show the way, but also, with much to offer. Scientific discoveries and studies providing methods to improve our ways of life combined with full readiness to temper that knowledge with integrity, balance, and forethought. One of the greatest barriers to reweaving the indigenous peoples and their wisdom with the Western cultures is that, besides past wounding, there has also been so much blind forward motion perpetrated by non-indigenous without thought of repercussions. Indigenous groups are understandably wary.

Cultural Appropriation

One of the biggest wounds lately has been the question of cultural appropriation. Indigenous tribes have already suffered so many injustices, so much of their culture stomped on. The earth, which they had stewarded with great care and consciousness sold off in parcels, fenced and caged, dissected and disemboweled. Their children stolen and imprisoned in schools intended to educate the "savage" out of them. The spirit of these earth guardians has been squashed into smaller and smaller boxes by so-called "progress," and yet they stand firm in their outrage over the pollution of their lands and waters… And now, to add insult to injury, it has become fashion and fad among certain non-indigenous folks to wear the garb and mimic the ceremonies and practices that they believe the indigenous perform. To many indigenous this is the highest form of insult, whether it be new age spiritualists or fashion-frenzied festival goers. Non-indigenous self-taught and self-titled "shaman" performing ceremonies with sacred medicines, often never having trained with any indigenous elders nor having the experience or training or the humility to take the time needed to build the sacred container that allows such work to be done with respect, integrity and safety. The feather headdresses of a chief signify great learning, training, wisdom, and experience and are offered after years of service in a ceremony of great respect and honor, and now we see white faces wearing them not because they earned them from their tribe, but because they were purchased as part of their "costume." These are some of the indignant emotions arising among indigenous folks who have already had so much stripped from them, so much trampled and stomped on. When so much of one's culture is lost, you tend to be very protective of what

remains. So emotions run high, and this makes efforts at reweaving the family even more difficult as the errors of a few leave a bad taste in the mouths of many.

Although I carry Native American blood and some of the vestiges of it can still be seen in my face, there are many other nations featured there as well, and it is not prominent. To most indigenous I look like any other white face. I grew up completely isolated from Native American culture, and so I do not carry the cultural heritage that would allow me to be accepted as part of the community, and so I, too, fall into this category of "other." If I were to wear feathers in my hair, I would likely be accused of cultural appropriation by many whose ancestral anger runs deep. But the truth is, that if I wear feathers in my hair it is because the sky people have gifted me feathers, and I wish to honor them and carry their specific medicine with me as a reminder of how I wish to live. This I do, not because I was ever trained or taught this way, but because it just makes sense to me. I have had many extremely intimate and powerful experiences with birds, my other animal relations, and with grandmother trees that have offered me deep insights into the web of nature and the patterns of existence. I trust my inner guidance to tell me what is right to do with my experiences. I live inside an altar made of pieces of the natural world, which helps to calm me within the cacophony of this disposable modern lifestyle. I grew up generations away from when the indigenous children were taken from their culture and brainwashed into some other way of life. Are the children to blame for what they became? Somewhere back in our history we were all indigenous. Are we to blame for every crime our ancestors perpetrated to cause us to be born into the culture we were born into? I have carried the shame of my non-indigenous blood as my ancestral cross my entire life, burdened by the horror of what so many of my non-indigenous family did to my indigenous family until I felt like a battlefield with a war being waged within my soul. This is not useful. This shame, this guilt that does not belong to us? I don't think it is what we are or where we came from that matters, but how we feel and what we choose to do with that knowledge. Living inside this war zone of my ancestry kept me from reaching out to indigenous elders for so long, kept me from asking questions and making connections and building relations. I was so afraid of being rejected for the color of my skin, for the crimes of my ancestors, that I spent 40 years alienated from the ones who could teach me the most about the person I want to be. And my deepest sadness is that although in my heart the indigenous ways ring true, they often can't see me in this face, so in some manner I remain an outcast from my kin, in exile in my own skin.

can, even if it means forging forward alone and untrained, sometimes tripping over sacred grounds along the way, however well-meaning. Let's not let misplaced shame, guilt, fear and ignorance keep us from learning from each other. Instead, let's use that awe and fascination to inspire a collective movement.

Everywhere there is still wounding and fear, but there are also pioneers out there who carry this vision of a global family. They are reaching out to the world and all of our relations, ready for forgiveness, to clear the slate and call it good. Ready to acknowledge our strong hearts beating with one vision, united in purpose.

traditional elders share indigenous ways to ALL interested community. Couchsurfing and Project Nuevo Mundo are examples of the many tools available for exploring other cultures, widening your perspective, helping to celebrate diversity and acknowledge our sameness at the core.

We all have much to learn and much to share. Some of us have maintained the old stories and practices; kept the disciplines intact for certain spiritual and social technologies holding on to an understanding of balance with the natural world and our community. And some of us have forged into scientific realms and honed how to make use of what we have learned to help all of us to live a better life with less hardship, while still being attentive to those laws of balance and considering the consequences for the next seven generations. Humility is so important, and often learned with age. Western culture is still in its adolescence and still rebelling against many things, not the least of which is a sense of abandonment and the theft of the sacred from our lives, barely even knowing what has been stolen. This is not to be taken as an excuse, but simply to aid in understanding how it may take a while for those who grew up in its grasp to see the larger picture. All of the woundings and indignities that have been done should not be forgotten. I would never suggest that. Remembering our history helps us not make the same mistakes again. But we can remember and let it go. We can always try to be respectful of all of our fellow humans and what they hold sacred. We can always try to not offend. But we can also always choose to not take offense when the offending actions are borne of ignorance and immaturity, and instead try to keep an open mind and seek a way forward.

What are you sorry for? I'm sorry too. What good will sorry do? Let's just shake hands and agree to start right now from today and write a new story.

When considering issues of cultural appropriation I think it is so important to note the difference between: a) purposeful slander which denigrates and is demeaning, b) thoughtless appropriation which is often misguided and stemming from ignorance, and c) true mimicking which can be the highest form of flattery. A whole generation of people with indigenous spirits have taken measure of the world and see where their hearts lie. They wish to learn, but don't have the cultural tools to know how to do this well, or are afraid to try for fear of even more rejection. Like orphans seeking our parents and teachers, innately feeling called back to Gaia and balance, in awe of something beautiful but with no acceptable access to learn more. Instead we grasp at it any way we

There are groups all over who are not restricting participation to indigenous alone, but who are reaching out to ALL WHO ARE INSPIRED to come. The International Indigenous Leadership Gathering, Four Worlds International Institute, Earth Peoples United, Tribal Convergence Network, the Ojai Foundation, Unify, Tribal Alliance, the 13 Grandmothers, the Compassion Games, are just a few of the groups encouraging all aligned hearts to join so we may sit together at the table and share what knowledge we carry for the collective good of us all. The Nawtsamaat Alliance (which means: "One House, One Heart, One Prayer, United in Power to Protect the Sacred") is a movement building in the Pacific Northwest of the United States and Western Canada which, for the first time, is holding regular teaching circles where

I once heard Grandma Aggie, of the 13 Grandmothers, speaking at a benefit event to protect the salmon and the waters, and she was talking about all of her travels and speaking with non-indigenous folks. She said that people kept coming up to her and saying they were sorry. And she would say something like this: You didn't do anything to me. What are you sorry for? I'm sorry too. What good will sorry do? Let's just shake hands and agree to start right now from today and write a new story. Apologies for my paraphrase, but I heard you and I couldn't agree more, Grandma Aggie.

If we can let go of that deep-rooted and wholly understandable fear that we are going to lose something more of ourselves by building bridges with people who are different, then we might finally be able to take the first steps of reweaving our global family, rebuilding a network of trust, and re-gaining access to all that we may have lost.

There is a place for all of us at this table. We are ready to move forward as one imperfect amorphous family, and maybe together we can weave the two-heartedness back into one great heart beating in time with the earth's resonance. We ask for all aligned hearts to come join us at the table. Bring your wisdom to gently show us another way. Let us forge this new path together. Let us use our bodies as the bridges which will find our way home.

With respect, humility, and deepest love
and most of all with hope,
to all of my relations.

Aokha Skyclad

Aokha Skyclad (Betsy Cacchione) is a chameleon shapeshifter and finder of common ground. Anthropologist and linguist at heart, she's traveled the world in solo journeys, delving deep into indigenous cultures, and become welcomed family among countless diverse peoples. Adventurer extraordinaire, she devotes her life to helping us find and follow our passions, and promoting her ethic of the Nomad Academy: aiming towards the dissolution of fear, and empowering the superhero within each of us to bravely become. She is an author (Soul Laundering under the pen-name Ishtar Raizine), and professional organizer / life coach (ArtOfGettingLighter.Me). A co-Founder of Tribal Convergence Network, she focuses on Social Tech & Communications, Networking / Alliances, & Relations. Her mission: maintain clear lines of communication, promote transparency, inclusivity, and community participation, and support an holistic evolution of this multi-faceted entity so all parts stay mutually informed. An efficiency guru experienced at directing flow of operations, she brings her vision as a pattern seer and a communicator to help weave the gifts of this nexus of activated leaders into an effortlessly humming system of interlocking spheres. Always tracking ways to improve efficiency in collaboration by seeking tools / practices to help resource sharing and cooperative innovation.

http://artofgettinglighter.me

The Guardian Alliance

Stewarding Personal & Planetary Transformation

Adam Apollo

We are creators. We create every day for ourselves. From our time dreaming, to the moment we arise, and through every action, we are creating. Yet as much as we create for ourselves, our real inspiration and drive comes from creating for others. We rejoice in seeing our gifts reflected in the world. It reminds us of who we are.

It is a time of great challenges, and great changes on planet Earth. Conflicts over politics, religion, and resources are still relentlessly instigated around the world, and the environment continues to be ravished by the addictive consumption patterns of humanity. Yet as these and other atrocities are surfaced through the internet, global communications, and open media, people are beginning to understand these problems and identify their roots. With enough connections between us, and an understanding of the interconnectedness of the planetary biosphere, it becomes impossible to simply shirk away "other people's problems" and ignore the shared collective impact we have on the world.

An illuminating awareness of ourselves grows, and we are beginning to see the many dynamics behind our relationship with the world as a whole. We are no longer isolated beings, living off our immediate lands and resources, but a global organism that can both bring immense gifts to the world, and take gifts from the world and its species as well. Human progress has been a runaway train, with our mental systems and physical capacities outrunning our intuitions and emotional response systems. Many people feel stuck, stifled by the sheer magnitude of change required to bring our planet back into balance. We take as many small steps as we can, but we fear that it is not enough. This fear is something many bury down deep, filling their lives with things that distract them from this underlying pain. The reflection we face is simple: we have the power to impact the world. Yet we have to choose how we are going to use this power. We can either improve it for all life, or we can destroy it.

Need for Change

Here's the key: it's all part of being human. It's all part of experience, and we're in an experience of evolving on a planet, multiplying, and slowly coming to understand and make sense of our role in this world… and perhaps in this Universe. Along with this growing understanding comes an ability to respond to what is happening around us: our responsibility.

We are global adolescents, coming into an awareness of our responsibility for the whole world. And wow, doesn't that feel like a big job to take on!

We may run from it, we may shirk from it, but eventually we must take responsibility for our actions and our impacts. We have to learn that it's not okay to hit people. We have to come to understand ourselves enough to be kind to ourselves and each other, and to be forgiving of mistakes. Those of us who see this necessity must shift our emotional vibration from a state of anger to compassion, and begin acting like older brothers and sisters, guides for those who are just realizing that a change needs to happen in the first place.

Yes, we need change. We also need to claim our choices of what we are changing things to become. We need *Visionary Arts*, that may show us the future of where we are going. We need good science, and courageous physics that is willing to look at all the possibilities, past and future, and bring forward a Unified Physics that incorporates and accepts all of our insights. We need builders who honor the living cycle of nature, educators that empower people to be who they want to be, and leaders who provide clear templates. We need Guardians. We need people who are willing to take responsibility for themselves and their impact on the world, and begin to help others see their power and responsibility. Perhaps more importantly, we need to recognize that evolution is also about bringing forward new gifts for the world, and stewarding the world in a way that cares for all life.

To fulfill this quest, it will take all of our courage; we must face our darkest shadows. Yet through facing our own pain and fear, we also face the world's, and we begin to understand how all that struggle has given us the greatest gift of all: the opportunity, the inspiration, and the energy to make the world a better place for all. This is the initiation time.

Our True Origins

We are not taught how *alchemy* works in modern schools. There are no mainstream initiation practices in much of Western culture, and many of the fundamental insights of cultures around the world (in regards to practices for healthy living) have been regarded as relics due to their "spiritual" nature. In traditions as diverse as Kabbalah (Judaism), Taoist Alchemy (Taoism), Yoga (Hinduism, Buddhism, Jainism), and those inherent to many Native American tribes, there are simple principles for dealing with the dilemma of feeling stuck behind buried fears.

In many of these traditions, there are ceremonial practices involved with facing our fears and pain, and accepting them fully and deeply. This is an initiation process, where an individual creates space within them for a fire that consumes these latent energies, and transforms them into love, vitality, purpose, and drive.

This transmutation process is well understood. When we come to

change in the air or on the earth. Forms shift as energies change.

While that might sound quite esoteric, the energy we are discussing here is solid science, though it still may lack some theoretical understanding in many scientific circles. The energy we are referring to is what modern physicists call the "quantum vacuum energy," the "zero-point energy," or simply "spacetime" as Einstein referred to it in General Relativity. For thousands of years, physicists called it "Aether," the fundamental field whose aspects have slowly been divided into particles, forces, and wave constituents.

At this point in time, many experiments have confirmed the existence of an underlying energy field that permeates, or rather, composes all matter and the apparent "empty space" between it.

For most people, these scientific confirmations only spark a reminder of an inner knowing. We are made of energy, immersed in energy, and everything we experience is based upon this fact and the innate qual-

about by someone somewhere else. We are connected, we are psychic, and that's just part of being human …if we're willing to accept it.

As Spiritual Practice

The act of Guardianship is not merely the act of watching over and caring for another, it is an act of standing for and protecting that which will serve another's evolution, growth, and wellbeing. Guardianship requires good principles, wisdom and patterns that help people to anchor deeper into their personal Truth, their innate sense of themselves, and what they are here to do. There are many people around the world who are dedicated to this service. They know that their personal path and purpose, as dynamic and different as it may be, is inevitably focused on bringing others closer to their True Nature. Beyond any labels or dogmas, this is a pure spiritual practice.

When we acknowledge that "spiritual" is not synonymous with "religious," we begin to understand its deeper purpose. If you dig into the roots of "spirit," you find that it simply means Aether, life-force energy, consciousness, or even just energy itself. If we are all made of energy, and we have consciousness, experience life-force and live in spacetime (the Aether), then how can we not be spiritual?

In the publicly edited online collaborative resource Wikipedia, the introductory paragraph on spirituality expresses its deeper roots, and the various definitions this concept holds across a variety of sources:

The reflection we face is simple: we have the power to impact the world. Yet we have to choose how we are going to use this power. We can either improve it for all life, or we can destroy it.

understand that the fundamental element of the Universe is energy, every experience is seen as a form of energy, which will *always, inevitably,* change. Notably, it will not only change, it will shift to its opposite and completely transform in the process. We see this when we mix fire and water, or when temperatures

ities of this quantum energy field. We have already integrated it into our common language, through the "vibes" we feel when we walk in a room, the sudden "intuitions" we have before things happen, the "itch" of someone looking at us from across the room, and the warm or "burning" sensation of being talked

"Spirituality is a process of personal transformation, either in accordance with traditional religious ideals, or, increasingly, oriented on subjective experience and psychological growth independently of any specific religious context. In a more general sense, it may refer to almost any kind of meaningful activity or blissful experience. There is no single, widely-agreed definition for the concept."

Guardianship is a pure spiritual practice, because it is simply the act of facilitating and enabling the experience of personal transformation and growth. This is a Universal practice based on the simple ethics of compassion, acceptance, encouragement, understanding, and empowerment. Regardless of religious roots, corporate ideals, social indoctrination, or political beliefs, these are generally recognized as characteristics we are all striving to achieve. Initially, most of us want these qualities for ourselves, and often judge others based on their abilities to exhibit them. Yet eventually we discover that by practicing them with others we transform the way we treat ourselves, and the way we are treated by others begins to change as well. This is not magic; it is a fundamental property of the energy dynamics in the Universe, and through coming to understand these dynamics more deeply, we begin to see the keys that can transform our lives, and the world.

This is why "The Secret" by Rhonda Byrne was so successful, why Deepak Chopra and Eckhart Tolle consistently top the self-help book list, and why so many books and movies these days tell the story of people doing incredible things with their consciousness, changing the fabric of reality itself. If we want to change the world, we need to understand these principles, and the deeper secrets of the Universe offered up by our ancestors. Yet, we cannot do it alone.

A Guardian Alliance

A Guardian Alliance is forming, a community of guides and explorers, people whose lives dive into the layers of reality and our responsibility as an awakening species. This Alliance can connect global communities of Guardians and bring them together to refine and anchor the principles and projects that will lead to a peaceful, sustainable, thriving planet Earth.

As communities form, villages develop, and urban neighborhoods come into relationship, there are individuals who feel the call to help mediate conflicts, restore balance, and empower the community-building process. These people are often left without support, clamoring against the chaotic waves of human dynamics, emotions, and relationships.

In many Western countries, police forces are able to address some of the worst of these waves, but they are generally not community centric organizations, and they cannot (and probably should not) intervene in most of the more subtle community dynamics. In the light of recent events, it is clear that police reform is a serious issue, and public opinion for police departments in many locations is very low. Part of the problem is that police are often not community members, and many government protocols leave local communities feeling like their own police are an oppressive, fear-inducing force.

Most of the problems in communities need to be addressed at a deeper level, where family dynamics meet social circles. This requires that more people in general learn about principles and practices that can support community development. Techniques for conflict resolution, mediation, healing, and creating mutually beneficial agreements should be taught at the grade-school level. For that to happen, societies as a whole must begin to develop a deeper understanding of the importance of these techniques and the principles behind them.

The Guardian Alliance is a template: through self-development and personal transformation, we become capable of holding greater responsibility for ourselves and others, and begin to apply ourselves to educating and empowering others. Through assembling a curriculum of techniques and practices that can aid in this process, and delivering highly applicable skills to face the challenges of life, we can build communities rich with Guardians.

Guardians of the Earth,
Guardians of Destiny,
Guardians of Life.

It's time.

With Justice
In Honor
For the Truth
Love
With Respect
In Peace
For the Joy

I Am
Adam Apollo
Guardian.is

Adam Apollo

Adam Apollo has been leading integrative presentations, interactive workshops, and self-development training for over ten years. Though well known for hundreds of "Jedi Training Intensives" around the world, he has also offered insights on global transitions, technology, conscious leadership and the future of humanity at the White House, in a UN Summit, and at a variety of conferences and festivals.

Founder of several education and technology based companies and organizations, Adam Apollo is dedicated to achieving a sustainable and thriving interplanetary culture. His professional experience includes management in multinational corporations, business development for non-profits and strategic campaigns, and executive officer roles across a series of companies.

Adam Apollo cofounded and developed the "UNIFY" initiative through Unify.org, bringing hundreds of thousands of groups and organizations around the world into synchronized conscious connection on December 21st, 2012, and periodically each year since then. In 2013 he founded Superluminal Systems to develop new learning systems and technologies, and it launched it's first Academy through The Resonance Project Foundation on their new website that his team designed and developed, Resonance.is.

In 2015 he opened the Guardian Alliance, a global "Jedi" Training Academy for self-mastery, personal transformation, and leadership. Join at http://guardian.is

The Artemis Project

A Letter to My Daughter

Dear Daughter,

This is quite a time to be born.

Quite a time to becoming a woman.

I know you have a lot of questions.

I know you look around at other girls and sometimes make secret wishes to be them, thinking their life is so much better than yours, and that for this you feel guilty sometimes.

I know you look forward into the future and are angry at the state of the planet and how much college costs and wonder how you will ever make it on your own.

Some pretty bad things have happened to you or to people you love and you often don't know what to do.

You look to the future and wonder what will happen to you. Will you have a good life?

I wish I could tell you it will be easy but it won't.

"The most beautiful people we have known are those who have known defeat, known suffering, known struggle, known loss, and have found their way out of those depths." Elisabeth Kubler-Ross

Your life will most likely be riddled with the pain of injuries both great and small, to your body, your spirit and your mind.

You will feel alone a lot of the time and greatly misunderstood.

And whether you chose it or not, I won't always be there for you, not forever.

I think about this. Often.

"The truth will set you free. But first, it'll piss you off." Gloria Steinem

This is the world you were born into and there is nothing I can do to protect you, not really.

"You have to accept whatever comes and the only important thing is that you meet it with courage and with the best that you have to give." Eleanor Roosevelt

I have something for you though.

It is called *The Artemis Project* that I created along with some really great other women and girls

... just for you.

If you take the first step, this is what you will find:

Mentors, Mothers, Aunties, Sisters and Grannies to be there for you as you grow up, in your times of celebration and in your times of need, for when I cannot be there for you myself.

For within that community you will find the wisdom and stories of great leaders and a safe haven from the world that would otherwise consume you.

They will be there to remind you of who you truly are when you have forgotten.

Of the moment of your birth, when the world held its breath and you took your first; to remind you of the 1 in 400 quadrillion chance that you would ever be born at all, that you are nothing short of a miracle.

To model self-respect and self-love and integrity in thought, word and deed.

"When we speak we are afraid our words will not be heard or welcomed. But when we are silent, we are still afraid. So it is better to speak." Audre Lorde

To remind you that your life matters even when everything is dark and you cannot even feel the light of a future where your dreams could ever come true.

To model for you how to survive and then thrive, despite circumstances to construct success out of dust and the sky.

To give you instruction in how to care for the earth and your body with equal reverence.

To wipe away your tears and listen to your stories of victory and your night terrors, and to sing you ancient songs until you fall asleep again.

To model self-love and acceptance so that you know it is okay to feel good in your skin.

To be there to answer your questions so you can make informed and intelligent choices about your body, about having sex and about having children.

They will show you how to care for your body during your cycle, how to tap into the power of your connection to the tides and therefore, into your own power of creation.

To take you to the women who would heal you with their medicine.

To dream out loud too.

Daughter, they will show you that you are a Jedi, a Warrior, a Priestess.

To convey that there are always, always choices and that there is always, always forgiveness.

To grant you the wisdom of millions of women who have survived so that you could live.

They will show you how to access your intuition and activate your mind, body and spirit for the good of yourself and all of humanity.

They will offer to you their hearts, time, knowledge, and tools.

"Is solace anywhere more comforting than in the arms of a sister?" Alice Walker

They will bring this to you with offerings of nutritious foods, sacred objects and beautiful materials.

They will take you to nature and walk with you, sit with you, listen to you, sing with you.

They will show you how this works, knowing that you will pass it all along to your daughter, your granddaughter and great-granddaughters.

Come here, my daughter.

Rest in our arms.

Kristen M. Rivers

The Artemis Project, a non profit program of The Children Are Our Future, is located in the Bay Area and offers summer camps, after school programs, specialized empowerment curriculum and events for girls 8-18 as well as mentorship training for women 18-24. The organization acts as a portal for teen girls to gain access to the wisdom of women of all ages. We educate and uplift the next generation of leaders by providing mentorship with love.

A Voice from the Future

This is what I want to see in the world, what inspires me, as a voice from the future generations.

I want to see more equality and more resources for the people who don't have the resources to meet their basic needs like water, food, and a shelter. I want to see more action to show people who don't understand how they are using all the resources and harming others. For example, there are big companies in the world bottling up water in places that don't have enough for the things that live there. Water is one of the three basic needs. Without free water, you don't have free people.

In this world, I want to see more gardening and people helping to take care of the earth, for people to remember that they are a part of the earth. I wish to see more "green" practices, more connection with the mother earth and with the green beings, plants and trees, and animals around us. More than just human, we need to respect the animals. Since the beginning of time, animals and humans have had a strong connection. Without animals many things that exist today would not be here.

I think ecovillages are a good thing. My idea of an ecovillage is a green sustainable community, working together to create a better environment for all beings. In our future village, I see gardens, treehouses made from all natural scraps of wood, not plastic, rubber, or metal. I see rivers and valleys full of healthy beings, and smiling faces filled with joy, love, peace, and happiness. Those are the main qualities of a good community. And to have a good community, you need lots and lots of good community builders working together. You need all the voices of the village heard. If you have one representative who speaks for all then you don't hear how things might affect someone else. We need the children's voices, too. The youth have a powerful voice when they are not afraid to step up, and they can change the world. Along with the children, we need to respect our elders. When we are talking about the word elders we are not just talking about older people, but wise people and those who came before us; they have a mighty say in this world too.

Coming from one of the next generation who will inherit the world, I see potential for a better life for all the next seven generations. The concept of the seventh generation came from Indigenous people and meant we think about those who will come after us. How will the decision we make today affect them 1000 years from now? But to reach

WITHOUT FREE WATER,
YOU DON'T HAVE FREE PEOPLE.

this, we need to work our hardest, in the life that we live now, to make sure that the resources are there for the next generations.

If we can make a sustainable life for the seventh generation to come, by then the message will be passed on.

We have now a plan, but if we don't treat each other with respect then it will be hard to treat mother earth with the respect she needs to continue to be a prosperous, abundant world for us all. We need to treat each other as if we were looking in the mirror. You wouldn't go and punch yourself. You would say "Good morning, you look lovely today! I love your new haircut!" Just a few kind words can make a downpour of a day feel like you are flying on rainbows!

And the best part of all is that it doesn't just make the reflection in the mirror feel good, it makes the entire world around you feel good too. You, yourself, your neighbors and your country could feel much happier if someone just said hello and let you know they care.

What I hope for is a stronger community and all the voices to be heard in a better world. Our future is dependant on the now and the actions we take all the time. Everyone's part matters, the young, the old, the rich, the poor, the disabled, everyone. No matter what, everyone matters.

Amelia Stevens, Age 11

Amelia Stevens is the 11 year old daughter of Jamaica Stevens, who is the author of this book and project manager of ReInhabiting the Village. This article was written with Amelia's words, with mom helping to edit them to be clear, but the true wisdom and expression comes from this vibrant young woman who is quite aware of the world around her and feels passionate about the things that we must do to make a difference. With bright and curious eyes, she sees the world with innocence and a realism that compels her mom and the other adults who know and love her to do everything we can to make the changes to the way things currently are so that when she is an adult, the world is still filled with promise of her children's well-being. The village is only whole when our children are included, considered, and held in love.

Health & Healing

THE WOUND IS THE PLACE
WHERE THE LIGHT ENTERS YOU. Rumi

A foundation of health is essential for all living systems. Yet what is the true measure of health? From a holistic lense we have to consider health from a physical, emotional, spiritual, and environmental perspective.

If we are in separation in any one of those facets of our lives, then we create the conditions for dis-ease or imbalance within the ecosystem of the self, the community, and the earth.

Healing the imbalance on the planet begins with healing ourselves. The place we have the most empowerment to determine the quality of vitality is in our own bodies, minds, and lives. As we come into balance with the food we eat, the water we drink, our level of physical exercise, our quality of sleep, as we limit the toxins we ingest, the stress we take on, the constant stimulus to our nervous system, we open ourselves to the possibility to not only survive, yet to THRIVE.

Many do not have choice about their exposure to environmental toxins, to the stresses of economic hardship, ravages of political oppression and war, or are not equipped to overcome the limited access to resources as the basic means of survival, including sanitary conditions, adequate health care, nutrition, clean water, or education about lifestyle choices that contribute to the conditions of health. We will not see justice in our world unless these basic conditions of health are met for all.

Yet in any and every place that we do have a voice, that we do have a choice, that there is a solution, that we do have access to the necessary means of health, there is a compelling call to engage. As we are each like a cell within this collective organism of the earth, we must play our part to be a vital and functioning cell, contributing to the wholeness of the system we are connected to. We are a microcosm of the macrocosm.

With the miracles available through Western Medicine, brought into symbiotic partnership with the wisdom of Ancient Healing practices and the advent of innovative Alternative Medicine, we have the means to use the best of what is available to us from a place of service, in order to advocate for the healing of dis-ease in our fellow humans and uplift humanity from the conditions of suffering into the joy of wellbeing.

As humans come into balance within, we will see our choices as affecting the balance of our organism as a whole. We can ripple healing through our actions, we can restore what has been damaged, we can revitalize what has been overtaxed.

Just as the human body is masterful at resilience and regeneration, we can mirror our body's systems for healing, along with looking to the natural systems of regeneration, and apply these approaches towards the rebalancing of our planet.

Jamaica Stevens

The Oasis

Transportable Community Bathhouse

Vision

The Oasis is a pioneer in mobile bathhouse environments; with artistry, ecological responsibility, and community building as our guiding principles. The Oasis is available for events and gatherings.

Mission

To provide full service mobile bathhouse environments for events and gatherings. To create experiences that inspire well-being, transformation, and community interaction. To partner in the research, development, and application of environmentally appropriate water management technologies.

Description

The Oasis is a unique and inspiring mobile bathhouse environment that consists of a customizable arrangement of "modules." These modules currently include showers, sauna, grooming station, and tea lounge. The potential market for our service is expansive and currently includes multi-day camping festivals, retreats, and private gatherings. The Oasis can be hired at a flat rate or can be offered as a pay-per-use service.

www.oasisexpress.com

Reclaiming the I

An Exercise in Self-Awareness for Healthy Relationships

Aokha Skyclad

If you have no relationship with **Self**, how can you ever expect to relate to others? All group endeavors, collaborative projects, and community development will stumble and eventually crumble if each person is not doing their part in exploring Self.

I work a lot with **relations** in my life. I seem to have a talent for listening, being a finder of common ground, and being able to chameleon myself into other people's shoes so that I can understand their perspective. I learned early on that two people can be completely different and have very different ways of doing things, and when they try to work together from a place of disconnect, this can cause conflict and strife. But when you are able to reach a place of connection, understanding, empathy, then it is possible to realize that there is no one right or wrong way to do things--just an infinite number of DIFFERENT ways. This is easy enough when we are healthy and whole beings, able to share our thoughts and opinions with detachment. However, because we are a species that has trained itself into disempowerment--with a global epidemic of abandonment anxiety, scarcity motivation, and a surety of our own solitude and disconnectedness--we waste much of our energy in relationship trying to gain power from others around us, because we lack the ability or training to be energetically self-sustaining.

Often, I find myself in the center of conflicts, listening to all sides and then helping translate to all parties, so that they may see past their own anxieties and fear-based motivations into a measure of compassion and empathy with the other person. Once all are able to connect, and no longer see each other as enemy or "other," the problem swiftly dematerializes, and it is just a question of finding a creative win-win solution to whatever ails.

How can we start from a more enlightened perspective to avoid falling into these conflicts in the first place? I feel the solution lies in all of us continuously seeking to be that whole healthy being, self-aware, and able to live in a state of detachment, so that we do not feel a need to WIN for the sake of being right, being in control, or to have power over someone else.

The first relationship we ever have is with our Self. There is only one perspective to take into account, opinion to listen to, and set of desires to try to meet. If we remain a hermit our entire lives, then that is the only relationship we need worry about. But few practice this type of lifestyle. Most of us live in the company of others. As soon as another person enters the equation, we begin to learn the need for compromise. Now there are at least two perspectives, opinions, and desires to be taken into consideration as decisions are made. Often they are congruent, but sometimes they conflict. We have grown up in a world where we are not often taught to look inside. If we have not taken a moment to delve into the inner workings of Self, the relationships we have with others will be driven by the strongest will in the room. Our deep underlying fears of not being loved, being left out or left behind, being alienated or outcast, being publicly ridiculed or embarrassed, not measuring up--these are all driving factors for how we create relationships to please those around us. The loudest and strongest voices can easily manipulate others, even unconsciously, because no one else knows what they want, or what they really care about. So instead, they fall in line with what is easy. They do what they've been told is normal, following a path that's already been tred by others before them, fulfilling someone else's desires, because they either don't know they have any of their own, or don't believe they have the right or ability to achieve them.

So many relationships, partnerships, friendships are embroiled and muddied by these kinds of power struggles. The results are well hidden, because we are a culture that buries our truths, trained from birth to want to fit in, be like the rest, be liked by everyone. Anyone who doesn't try to follow this pattern is labeled a misfit, or unsociable. Some people spend their entire lives in what appear to be calm, happy relationships, while underneath they are slowly dying inside, watching from within a cage of their own making, as they and all they have associated themselves with slowly trample their dreams into dust.

on the shore, or the bird song of the forest or jungle, whether in real life or digital. Try to feel the rhythmic unending pulse of Nature and let your own breath fall into that rhythm. Let yourself be calm for a moment, just listening to those peaceful sounds without the clutter and clatter of everyday, and everyone else, and all of their needs all around you. We all have an energetic field that we carry around ourselves, and when we are around others, our fields overlap quite a bit. Most of us do not know how to shield them, so we are constantly pushing our thoughts and desires onto others unintentionally, and constantly receiving those of everyone around us. This clearing helps us get into a space of just our own thoughts. Our own ideas, desires, needs. This is a place many of us have never been, or even realize existed.

is all you really need to get started. Play with it. Write the story of the inner super hero you always knew you would grow up to be, when you were a kid, back before they trained you that you couldn't. What does that person look like? What do they do in their average day? Where do they live? Who do they surround themselves with? Money is no object. Time is no object. It is like the best wishlist ever and you don't have to feel guilty about asking for EVERYTHING you want. How do you dress when you can afford anything? What activities do you spend your time on when you have all the time and resources you could ever need? I would try to spend a good few hours on this, still holding the rest of your world at arm's length, giving yourself this space for ONLY YOU.

Once you have an outline of the qualities that are innately you, then you can bring back the people and obligations you currently have in

How do we escape this fate? There is really a simple answer that doesn't have to cost anything. The key to freedom has always ever been right in our own hands. We just need to take a moment for ourselves. Step away for an hour, a day, or a weekend. Go someplace alone: a mountaintop, the ocean. A place of epic beauty is helpful, but you can lock yourself in the bathroom with headphones on if that is the best you can do to get away. Then sit down with pen and paper (or some magazines to make a collage, or a voice recorder, if writing doesn't work for you) and just for one moment, allow yourself to let go of all of your obligations. Let go of your relations, your children, your parents, your significant other, your job, your everything. Forget it all for one moment. You don't have to feel guilty. It's not going anywhere. Let it drift a little away from your head in a cloud that billows up and out and leaves a little room to breathe inside your own mind.

Then sink down inside yourself. Feel into your very bones. Feel how they are grounded into the earth with gravity. Listen to the waves crashing

The optimal situation is to have a deep sense that we know and love ourselves regardless of anyone else, and can support our own needs, without requiring anyone else to agree or validate us.

When you have gained a measure of calm, the next step is easy and fun. Still holding all of your obligations at arm's length, safely in a cloud about 30 feet away from you, try to imagine, if you had NO other obligations and YOU were the only person in your world, what would be the best possible scenario that you could imagine for yourself? For all aspects of your life: work, play, education, lifestyle, relationships. I have whole exercises that outline the types of questions you can ask yourself for more directed work, but this

your life, by choice or otherwise, and match those up alongside this new map of the inner you. You can start to look at these with more clarity of mind and determine which of them are redundant or extraneous and not really pertaining to you at all. Perceived obligations which you took on mistakenly because of cultural norms or expectations, but which have no real bearing on your life, and not even any real significance, can be let go. The ones which ARE important to you, or the people to which you DO have true love and obligation, you can now address the issue of compromise and brainstorm how to weave these together in a balanced way with your own true needs and desires, so that your whole life is not about fulfilling other people's dreams and needs, but a mixture that you can be happy with, without fostering underlying resentment.

I run this exercise with people in greater depth; going through calendars to see what they spend time on, through bank accounts to see what they spend money on. Filtering through outlets of energy and reshaping existing systems so that they can become closed loop self-feeding processes. But if there is one gift I would try to give everyone as a first step forward is to take this moment for Self. Find out who you really are.

Realize that we still need to practice humility and detachment for ideal interactions to occur. Once you start to know yourself, it is easy, especially if you have been silent for so long, to go overboard and start overpowering others with YOUR desires and needs, pressuring them to agree with you so you can finally get what you want. The optimal situation is to have a deep sense that we know and love ourselves regardless of anyone else, and can support our own needs, without requiring anyone else to agree or validate us. To hold onto a sense of detachment, so that we don't need our inner agenda to be affirmed by others around us. And to remember to place ourselves in each other's shoes, so we can step away from ego and be compassionate with others when our desires and needs do not coincide. There is always a gentle way to disagree, when we are detached from needing to be "right."

For all true partnerships to work, I feel we must aspire to be fully whole, healthy beings. Not requiring anyone to "complete" us, because we are complete in and of ourselves, but who CHOOSE to be in relation with each other because it nourishes us, gives us joy, and helps us to fulfill that which we are here to do and be.

Know that this is not a static exercise. We are nothing but creatures of change and it is imperative that we reassess regularly. What an empowering gift we can give ourselves, every few months though to sit down alone and take this time to reconnect with Self and see who we are becoming. At the very least, performing this exercise lets you know yourself better, and allows you to enter into relationships with confidence and an ability to share who you are and what you need from the partnership. In a best case scenario: change all the words you wrote down into present tense, as if this is already how your life is right now, and start manifesting it today!

Aokha Skyclad

Aokha Skyclad (Betsy Cacchione) is a chameleon shapeshifter and finder of common ground. Anthropologist and linguist at heart, she's traveled the world in solo journeys, delving deep into indigenous cultures, and become welcomed family among countless diverse peoples. Adventurer extraordinaire, she devotes her life to helping us find and follow our passions, and promoting her ethic of the Nomad Academy: aiming towards the dissolution of fear, and empowering the superhero within each of us to bravely become. She is an author (Soul Laundering under the pen-name Ishtar Raizine), and professional organizer / life coach (ArtOfGettingLighter.Me). A co-Founder of Tribal Convergence Network, she focuses on Social Tech & Communications, Networking / Alliances, & Relations.

Her mission: maintain clear lines of communication, promote transparency, inclusivity, and community participation, and support an holistic evolution of this multi-faceted entity so all parts stay mutually informed. An efficiency guru experienced at directing flow of operations, she brings her vision as a pattern seer and a communicator to help weave the gifts of this nexus of activated leaders into an effortlessly humming system of interlocking spheres. Always tracking ways to improve efficiency in collaboration by seeking tools / practices to help resource sharing and cooperative innovation.

http://artofgettinglighter.me

Nature's Dreams
Our Collective Journey towards a Healed and Whole Humanity

Nature Dreamweaver

"I want to be remembered as a great healer," I told the interviewer at my Gaia University advisor meeting. I don't really remember what else I said, but this has echoed in my brain for almost ten years. Initially, growing up watching TONS of television and a ridiculous amount of movies, I wanted to be a famous altruistic film actor. I still love to act and perform and entertain, but my own bouts with health and healing, dealing with lyme disease, chronic pain from scoliosis, depression, and addiction has inspired me to become a healer in the truest sense of the word. The truth is I've dreamt of playing every heroic role in life you could possibly think of and now it's time to synthesize, integrate, and weave them all together into a beautiful mosaic tapestry. A sustainable symbiotic singular synergy. A profoundly poetic and prophetic permaculture system of the highest order.

What does that mean these days and what does it mean in the context of reinhabiting the village? I don't have all the answers, though I do believe the founder of permaculture Bill Mollison when he says, "Though the problems of the world are increasingly complex, the solutions remain embarrassingly simple."

After years of touring the country and having all sorts of enlightening and educational experiences creating sacred space through nests and art installations and being a part of festival culture, I have a different dream for myself and for humanity.

If we truly desire to reinhabit the village, then we must first reconnect within and then, perhaps, find a connection to the people and places where villages still exist. If we want to solve the world's problems, then we must first solve OUR OWN problems. We must embody the new consciousness and new paradigm in our minds, bodies, and souls. It's a

daily lifelong practice of hard work, commitment, and dedication. Individuals must use and choose the power of free will to do our own personal work to grow, change, evolve, awaken, and heal. If we can be autonomous and sovereign within our own beings, this freedom will allow and inspire us to serve our communities with our skills, gifts, and talents. Rooted and grounded in some form of spiritual practice, this will create a foundation from which a community village can grow and last.

Though the problems of the world are increasingly complex, the solutions remain embarrassingly simple. Bill Mollison

I know that when I am taking care of myself and loving myself and balancing myself on all levels, centered and grounded in my heart and doing my work, then I am able to serve the greater whole in the highest. I am able to show up and be fully present with my heart full of love. It's really that simple. When I try to overextend myself or operate only from my ego, then my

world quickly flips upside down and falls apart.

I either become sick, stressed, depleted, or even diseased.

We can look to some indigenous cultures of the past and present, and see that many of these cultures share a deep reverence for the earth, for spirit, and for the universe. They have humbled themselves to a greater power and honor the earth for providing all that they need to live and survive and thrive. These cultures are mostly based on living in harmonious balance and in right relationship with all of creation.

Some days I wonder if we might not all be happier remembering that we too can live in this way, much closer to the earth and the stars. Small groups of people all over the world working towards change and healing.

Society as a whole continues to charge forward into the future at lightspeed trying to figure life out while our ancestors figured it out a long time ago. None of our technological advances or space travel has made us any happier. It's simply inflated our egos and fooled us into thinking that we're the best thing that ever happened to the universe. Sure, we're evolving and becoming smarter and stronger, but we've been losing our humanity in the process. We don't seem to be any closer to health, happiness, God, or living our lives with deeper purpose and meaning. We've become so attached to acquiring stuff and things and possessions that we've lost touch with the simple pure magic and miracle of life itself.

That's why in the context of reinhabiting the village I've chosen to focus on the arts, mainly the healing arts in order to restore balance. If we continue to humbly come together and continue to dive deeper into the practices of permaculture, yoga, meditation, ceremony, and slow down enough to watch and observe ourselves, our patterns, our wounds, and see where we've come from and have enough foresight to see where we're going, then we can become more and more clear on what will truly bring more happiness, love, and joy into our lives.

Our story is an interconnected spiritual spiral of evolution and if we can practice sensing this inside ourselves every day, then we can start over (or continue, depending on where you are at on your soul's journey) telling stories that are not full of pain, suffering, violence, and horror, but overflowing with an abundance of truth, beauty, and love.

Life is as poetically simple as it is infinitely complex. We're born, we grow, we learn, we blossom, we live, we love, and we die. As we slowly wake up and remember the truth of who we are and why we're here, we begin to realize what it truly means to live in a village, on a planet called earth, floating in space across the universe.

Nature Dreamweaver

Nature Dreamweaver (Nate Hogen) is a multidisciplinary visionary artist, healer, yogi, poet, and student of life on a eternal quest for truth, beauty, and love. His higher self mantra to "Love, serve, heal, educate, entertain, and free all sentient beings," has led him on a journey for the better part of the last ten years touring the country creating and building sacred space art installations at eco-villages, events, festivals, retreats, parties, art galleries and weddings in the form of altars, earth mandalas, and human sized NESTs (New Earth Sacred Temples). He has also performed several cacao comedy ceremonies (sillymonies), led art workshops, and has been proactive in continuing to educate and enlighten himself on his path in order to adapt to the changes and crises of our times.

Presently and forever, Nature is practicing, studying, and training to become a yoga teacher, shaman, natural doctor, permaculturist, priest, and a leader in the movements of planetary change, healing, awakening and transformation.

http://naturedreamweaver.com

Paititi Institute
Collective Evolution through Individual Transformation

Cynthia Robinson & Roman Hanis

"Paititi," in Quechua (Andean indigenous language), is an enlightened realm manifested through the awakening of our shared human heart. The vision for our work was born through the highest aspiration we could imagine for humanity and mother earth. We believe that through the awakened spirit of the individual comes the greatest potential for the transformation of the planet. The Paititi Institute for the Preservation of Ecology and Indigenous Culture is committed to embodying this paradigm shift and demonstrating what is possible through ordinary human efforts, in service to Mother Nature and the infinite human potential.

Through all of our programs, we dedicate our efforts to supporting a society of deep, nurturing and true values. We serve as an intercultural bridge, supporting individuals and communities to live activated, responsible and joyful lives in harmony with both inner and outer landscapes. As humans, we are inexorably linked to the earth and each other. As part of this magnificent symbiotic organism, we are dedicated to co-creating the enlightened world we all have in our hearts.

Overview

The health and well-being of each individual, our communities and the global environment are interdependent and therefore it is essential that our efforts to create a harmonious life encompass a full spectrum of aspects which all support each other, including:

• Conserving and restoring the Earth's natural resources

• Preserving and integrating indigenous wisdom and culture into modern life in practical & skillful ways

• Cultivating health through natural, holistic approaches - integration of both traditional indigenous healing methods supported by modern science and technology

• Embodying true stewardship via eco-regenerative, community living

• Understanding consciousness and engaging in personal cognitive evolution and self-realization

In modern society we often look at ancestral cultures and label them as primitive, however in many ways these cultures were far more advanced than we give them credit for. Ancestral cultures carry the cumulative wisdom of humanity and through practical application of ancient wisdom within the modern world, we can build bridges for the physical, emotional, and spiritual healing of all people from all cultures and from many spiritual traditions and walks of life. In this natural way, we channel the healing and rejuvenation of the planet herself. Using holistic models based upon our points of focus listed above, we can engage as a community to improve social and environmental life via spiritual evolution and maturation on the collective and individual levels. We utilize the planetary wisdom of humanity to become fully conscious and responsible human beings in service to the shared heart of all humanity.

Intercultural Bridges

Today the modern mindset comes from a much different perspective than that of the indigenous ancestors from around the world who had a deep intuitive understanding of their interconnection to each other and nature. These cultures, who lived in deep connection with nature both internally and externally, realized deep states of interdependence with all life. They understood how life energy flows and how to work harmoniously within these flows.

In today's society there exists a conditioning of separation from everything and everyone else. This sense of separation seems to be the root of all of today's problems. Yet, even with all the conditioning accumulated today, we live in very fortunate times. There is a profound and unprecedented opportunity to access the collective wisdom of all of humanity and utilize it to benefit our global society and eco-systems around the world.

Restoration of profound human values so pertinent to ancestral traditions from around the world in a practical and non dogmatic way is the cornerstone to the regeneration of inner and outer landscapes of life on this earth. Through the formation of intercultural bridges at our Institute we make sure that these ancient insights into the very nature of existence are not only maintained in the

modern world as mere exponents in a museum but are practical and dynamic core-fusion elements of a paradigm shift in the global emergence of evolutionary consciousness.

This groundbreaking rediscovery of skillful means and cumulative millennial wisdom from around the world unravels the highest creative potential geared towards individual and communal well-being in a balanced and regenerative relationship with the ecology of the planet. The transformation of consciousness within oneself and resolution of inner conflicts inevitably leads towards healing of the organism as a reciprocate conduit of positive change in the world.

Building a Sanctuary for Humanity

Our programs involve an interwoven fabric of Permaculture projects geared towards ecological preservation, restoration and regeneration, research and implementation of the indigenous healing modalities as well as sacred and medicinal plants that are inherent in these cultures, natural medical campaigns for the native populations and international patients, consciousness transformation retreats and awareness raising global tours. These projects are all developed as a part of the shared heart vision towards a healthy and a viable example of an intentional, sustainable and a thriving community based on healthy, purposeful and joyful way of life in harmony with nature.

The Institute has been working on a 100 acre land in the lower Amazonian basin in Peru since the year 2009 both as a base for working with indigenous natural healing and transformation as well as an example of reforestation and restoration of degraded landscapes, a research facility for organic food production, and creation of fertile soil for the impoverished native population.

As of mid 2014, we at Paititi Institute with deep reverence and honor embrace our role as guardians and stewards of a pristine and virgin forest 4000 acres in the Mapacho Valley of the Peruvian Andes bordering the Manu National Reserve and area which is a key watershed for the Amazon rainforest. The land is intended for preservation and protection against exploitation by the mining and logging companies that threaten this still pristine buffer zone of the virgin rainforest in which some of the last un-contacted indigenous tribes still exist.

This sacred land is a sanctuary for self-realization, transformation and evolution. Through this land we are reawakening our interconnected and interdependent relationship with all life and are taking the steps daily to build an example of a harmonious, nurturing and regenerative community.

Our continuous focus is to form deeper relationships with like-hearted people, communities and organizations in a way that is synergistic and complementary. We look forward to the most optimal manifestation of evolution within our current human society towards the luminous network of intentional communities around the globe geared towards bio-diverse productivity of abundant well-being and natural lifestyle.

Cynthia Robinson

Working in the corporate world of New York City led Cynthia to discover a more profound passion to support herself and others in establishing deeper connection to nature, ancestral values and unlocking our highest potential as human beings. She traded in the corporate boardrooms for a life in Peru where she co-founded the Paititi Institute.

At Paititi, Cynthia works to create a platform that practically applies ancient wisdom to our modern world, building intercultural bridges for healing on physical, emotional and spiritual levels for all people from all cultures and all spiritual traditions. She currently works to develop and facilitate holistic experiential education models, weaving together our inner and outer landscapes through nature awareness, Permaculture education, health education and Indigenous Ando-Amaonian teachings and practices integrated with ancestral wisdom from around the planet.

Additionally, Cynthia is currently in the process of being initiated into Indigenous Amazonian and Andean healing practices to become a medicine woman, carrying on this ancient lineage of healing wisdom.

http://paititi-institute.org

Roman Hanis

Roman Hanis is dedicated to the study and implementation of the many deep transformative healing methods that have been in use for thousands of years, paying special focus to the construction of intercultural bridges between Amazonian, Andean & Tibetan spiritual traditions. Using these very traditions, Roman was fortunate enough to be cured of a terminal, genetic illness in 2001. The non-conceptual universal essence of these tools and the wisdom behind them are being integrated into western culture by Roman through his daily healing practices, aimed at providing maximum benefit and the profound re-emergence of ancestral values in our modern world.

In 2004, Roman was pledged as a healer-curandero by the Whitoto and Yahua tribes and has served the international community as a medicine man ever since. He is also a Traditional Chinese Medicine practitioner in Peru, studying under the director of the Open International Institute of Oriental Medicine, Myriam Hacker, in Iquitos, Peru. Additionally practicing physical trauma rehabilitation, medical massage therapy and Eastern bodywork, having earned his degree from New York's Swedish Institute of Health. With these tools and concentrated efforts, he has helped numerous individuals overcome many health issues on physical, mental and spiritual levels.

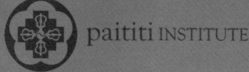

paititi INSTITUTE

Reinhabiting the Heart

Healing our Relationships to Create Synergy

Matheo James

We have set out upon a journey to rediscover a healthy way of life - one that works for plants and animals, for the soil and the waters and the air, in addition to making life enjoyable and comfortable for human beings. This requires that we learn to live well with each other, for if we are to dedicate time to balancing our way of life, we must avoid the petty, destructive squabbles that have become commonplace in the modern world. A resilient, local, interdependent network of co-creators – in a word, a village – is the most valuable resource we can cultivate to this end. In order to create healthy, long-term relationships in the context of our chosen tribe, we must first do the necessary work to create harmony within our own hearts, and extend that harmony first to our family, then our neighbors and friends, and it is only then that we can truly begin to co-create an abundant life in right relations with the world around us, together.

Fundamental to human life is this aspect of relationship. It defines all we do from the moment we are born. We are brought up in a family, where we learn to relate to the world around us based on the way those closest to us live. Each of these interactions and relationships shapes our heart, laying the broad outline for how we will show up in the world as we continue to learn and grow. Ideally, we are surrounded at this stage by many loving people who will demonstrate for us how to work, play, and communicate effectively - aunts, uncles, grandparents, friends and older cousins, who make up an extended network beyond our immediate family. Exposure to this variety of perspectives and ways of interacting with the world allows the young heart to choose from a broad palette, with which it can begin to paint an individual expression on the canvas of human life.

Many of us have grown up far more isolated than what might be considered ideal, and furthermore, within the insular conceptual framework of a culture seemingly dedicated to its own destruction. The abstractions of the modern way of living seem to encourage the development of an individualistic, mental interface with the world at large. Without the firsthand experience of interconnection provided by an open heart, people become statistics, the living world becomes numbers on a balance sheet, and the Earth beneath our feet appears as nothing more than property to be bought and sold.

In order to reverse these trends, we must learn not just to think about the world around us, but to feel it. We must learn to expand beyond the confines of ourselves, and see people, the living world, and our place within the larger web of life.

While this may seem simple (and really it is simple!), we must transcend the complexity of modern life in order to reach this place. We must heal our relationship with ourselves first, which means unweaving the tangled web we have been born into. This requires effort, often over a period of many years. By learning to see the lack of harmony in the world around us as a manifestation of an inner condition, we go from a place of powerlessness--placing the blame on something outside ourselves, something beyond our control--to a place of empowerment. In this way, we can truly own our world, by owning our experiences, seeing what we have experienced as a manifestation of our inner beliefs, and choosing our responses according to the heart's inner guidance, rather than reacting immediately and allowing our emotions to control us.

It is an odd thing, then, to recommend joining community efforts before one has accomplished total self-mastery. After all, how can we create well together without each of us demonstrating this inner balance? When we delve deeper into this, we can see that the purpose of returning to village life is not to step directly into harmony and balance, but rather to join our hearts together in a journey of creating it. There is this familiar story, which anyone who has lived in community can appreciate: a new monk, joining the monastery, was surprised to see that it was not the peaceful, blissful life he had imagined it to be. Disturbed by the infighting, gossip, and jockeying for position among his seniors, he went to ask the abbot how such behavior could possibly be part of the spiritual life.

The abbot smiled quietly, and replied,

"It is the pebbles rubbing against one another in the river that make each other smooth!"

We can, of course, rub each other the wrong way at times. We each come to community with our own beliefs and expectations, and these must be removed in order to see the path before us. This requires an attitude of humility, which can often only be acquired through experience. It is essential then to have enough cumulative experience to gain a new perspective. To reach this level, there must be a commitment, not to anything outside of ourselves, but to our own spiritual growth and character development.

It is by activating and receiving the collective intelligence of a diverse group of people that we are able to see ourselves clearly, and accept who we are, in order to change ourselves. This is where surrounding one's self with elders is of the utmost importance. All functional villages hold an honored place for the wisdom of its elders, and actively work to integrate the perspectives of those who have

Two things are of utmost importance in choosing one's allies in the journey towards a balanced way of life: sharing a vision and developing the ability to effectively pursue it together. While we all may have different missions--that is, the way we each choose to actively pursue this vision--any community composed of people working towards differing goals will constantly be at odds with itself, and never achieve the full potential it might otherwise have. Another way of putting this would be: the "hows" may differ, but the essential "why" must be the same. It is the commitment to this higher vision that binds people together despite their differences, and can inspire them to overcome differences in opinion in order to achieve positive results.

ing belief structures that are working to shape civilization, the stories begin to look more and more similar: the various cultures that compose it all share similar beliefs about the world and the place of human beings in it. All "civilized" cultures to a greater or lesser extent believe that man is the ultimate form of life on Earth, and that ownership of land, possessions, and even other forms of life gives humans the right to do with them as they wish.

Enacting these stories is systematically dismantling the web of life and destabilizing the climate at an alarming rate. To counteract these trends, the answer cannot be new government programs, or novel approaches to owning land and other forms of life. We must choose to step outside of the aforementioned assumptions, and attempt to tell a story about ourselves which replaces the abstractions of money, government and "ownership" with the

When we delve deeper into this, we can see that the purpose of returning to village life is not to step directly into harmony and balance, but rather to join our hearts together in a journey of creating it.

gone before us into daily life. It is similarly important to listen to the unfiltered reflections of children, and to take them seriously. The fresh untrained eyes of children, who have not yet learned to censor themselves in order to avoid hurting the pride of others, are often able to pick out and describe inconsistencies in our behaviour and beliefs that might otherwise go unchecked.

The stories that we tell about ourselves and our place in the world are fundamental to the act of creation, as they form the framework within which all decisions are made, and all plans are conceived and carried out. There can even be many ways of telling the same story: if we take the modern world as an example, it may seem on the surface that there are a diversity of stories being enacted.

Yet if we are able to "zoom out" to a level where we can see the underly-

immediacy of direct interaction with the resources upon which we rely. To be truly sustainable--that is, to enact a story which can be continuously passed down from generation to generation indefinitely into the future, without destroying the very systems upon which we rely for food, shelter, and warmth--we will have to find new ways of relating to the world of life which surrounds us, and of which we are a part. We must also inspire others to change their

story! While this may be one of the most daunting tasks that could be asked of a human being, the people of Earth are coming closer and closer to a choice between doing so, or setting the stage for our own extinction. Faced with this kind of stark and depressing reality, the unexpected answer is to create a way of living that is so lush, abundant, meaningful and fun that people will abandon all of the trappings of their old way of life just for the opportunity to join.

Imagine if the kinds of places which breed terrorism, drug use, neurosis and random acts of violence--places of vast economic equality, ideological domination and control, and apparently no way out of the "machine" or "the game"--were interspersed with permaculture villages, where people created abundance and a grounded sense of place, and shared the abundance with their neighbors. It wouldn't be long--perhaps the span of a generation or two--before this new way of doing things spread throughout the hungry, disenfranchised, marginalized and poor.

The point at which we find ourselves now, in 2015, is surprisingly far along this path--a massive sea change in the way humans relate to each other and the world around us is already underway, as evidenced by this volume and the support it has received. Our question is no longer what to do, or even why - these questions are answered here, and in many other forms of media circulating the globe with the lightning-fast speed of the internet. Our question now is how to do these things effectively, in inspiring and accessible groups.

This requires forms of collaboration appropriate to our ends. To those unfamiliar with the broader history of human beings on this planet, it is often a surprise to find that the answers have been with us dating back to a time preceding the dawn of modern civilization. In the vast repository of wisdom contained in tribal cultures are many paths to the same end: cooperation and mutual support, communion with the world of life, myriad forms of obtaining spiritual insight, and innumerable ways of interacting with the world which, through tens of thousands of years of trial and error, have been refined to be both sustainable and enjoyable for human beings. While it is naive to suggest a wholesale return to these ways of life, the values and wisdom residing therein can be integrated into our visions for new ways of living by following the same process we'll need to use to create incredible outcomes in the villages we choose to create together.

Matheo James

A community organizer, healer, and musician from Denver, CO, Matheo "Golden Dragon" has led non-profits, played music nationally with a touring circus, and been Mayor of the Burning Man theme camp Red Lightning. A strong advocate for tribal wisdom, spiritual healing, and gender alchemy, he works with individuals and groups to increase the impact of their world-changing efforts.

Believing that fun and inspiration are the keys that unlock our hearts for spiritual growth and community building, Matheo leads wisdom-sharing circles, facilitates team-building experiences, and brings diverse teams together to co-create immersive festival experiences, such as the CAVA visionary art gallery at the annual music festival Sonic Bloom.

As DJ Golden Dragon, he strives to bring joy and insight to people through sacred bass music. He creates playful and spiritually uplifting dance experiences that weave in elements of indigenous shamanism, eastern heart-based mysticism, and western esoteric traditions.

These endeavors are designed to build and create healthy relationships: within our communities, the web of life, with spirit, and most of all within our own hearts. With this end in mind, he practices Natural Path meditation and teaches Non-Violent Communication, as well as facilitating Restoration Council healing circles and offering personal energetic healing work.

http://facebook.com/GoldenDragonHealing

Healing of Love

*Sex, Partnership
and the Village*

There will be no
peace on Earth as
long as there is war
in love. Dieter Duhm

In the book **Sex At Dawn**, authors Christopher Ryan and Cacilda Jetha convincingly demolish most of our assumptions about love and sexuality in the dominant culture. For instance, they argue that humans are the most sexual species ever evolved, with origins closer to peace-loving bonobos than war-mongering chimpanzees. They reveal that women are biologically geared for multiple partners per sexual encounter and that monogamy is extraordinarily rare in the animal kingdom.

And finally, they reframe our assertions about "natural" human partnership, where men offer their protection and resources for the sexual fidelity and fertility of women. Instead, they suggest much of what we "know" about human partnership & sexuality is in fact an adaptation from the shift to agricultural civilization - where we abandoned the ancient structure of community for the recent nuclear family.

Let me detonate that again: all of our modern beliefs about sexuality & partnership are, in fact, born out of the trauma of losing the village.

This is why it is so difficult to talk about love and sexuality. Many of us who grew up in the dominant culture carry deep wounds inflicted by the story of separation. From the religions that tell us sex is sin unless practiced within the sanctity of marriage. From the sexual abuse that ripples through generations unobstructed in the silent shame of families "saving face." From the fairy-tale conditioning to seek "The One" upon which your innate wholeness depends.

When your only chance for companionship and intimacy is limited to a single monogamous partnership, they are the promise of a life raft in a world that seems, to the mind of separation, like a sea of chaotic, cosmic cruelty. This is why the myth of "The One" sounds like a good idea - we then manifest the structures of scarcity that already dominate the rest of our lives. We seek in our partner that which they cannot provide: a vast spectrum of sexual fulfillment, superhuman parenting without inflicting their own neuroses, solid productivity in a material economy, and a never known but deeply longed for connection to the web of life.

To be clear: I'm not attacking monogamous marriage nor am I advocating reckless polyamory.

What I'm declaring is that one cannot have authentic conversations about their own sexual and relational proclivities until we're willing to recognize the depth of our woundedness. The sexual revolution of the 60's was sincere but ultimately premature. "Free love" when practiced as freedom from responsibility left a generation of broken families and disillusionment. But the lessons were not lost. The true depth of "free love" is revealed when understood as "freedom to love without fear."

The first step toward the liberation of love is not the forms. Instead we must ask: how can we recreate communities of trust, where we can share our anger and our fear, our love and our longings, where nothing is banned from the circle? How can we recognize our kinship in our shared vulnerability?

Upon this foundation of trust can we can build the momentum necessary to call forth our highest selves. Suddenly, we are free to love as we are free to breathe the air. Jealousy dissolves when we no longer compete for intimacy. We are nourished by the community in ways our partners could never, and should never, have been asked to carry.

Our life energy can now be joyfully harnessed toward the real work: healing the damage wrought upon this wild earth.

May it be so.

Ian MacKenzie

Ian MacKenzie is an award-winning filmmaker & media activist from Vancouver, Canada. His work has appeared in The New York Times, National Geographic TV, CBC Documentary, The Globe and Mail, Adbusters, and film festivals around the world. Recent films include Occupy Love (2013), Dear Guardians (2014), and the forthcoming Amplify Her (2016). His focus covers a range of diverse topics & subjects, though all fall under his mission of uncovering and amplifying stories of the emerging paradigm.

http://ianmack.com

Stepping into Our Power

If you are reading this, you are one of the amazing people who thinks about humanity, how we live and work together and what it will take to design a world that is thriving. I acknowledge you for being part of this creation process and, I invite you to explore the topic of Power and what's possible when we embrace it.

We've all seen Power out of balance. We've seen bullies in our schoolyards taunting little kids and corporate leaders seizing resources while violating forest, fauna and humans. We've seen tyrants wielding power for personal gain, and electorates abusing it to benefit their sponsors.

Our conversations are often about these power injustices, clarifying a social disapproval that prevents each of us stepping into our own power. We end up fearing each other's corruption as well as our own. In our determination not to be those bullies, we've rejected all forms of Power, and tied money to that view.

But we are not those bullies. We have strong, deep values. Our desire is for a good world, one that's healthy on all levels and cares for everyone. Many of us have developed ourselves through yoga, meditation, non-violent communication. Our shared passion is for a common good, not personal gain. I believe that we can trust each other to become more powerful.

Power is a force in our universe like air and water. It is inherent inside of each of us. We are designed, by nature, to be radiant with Power in our fullness. Unlike other forces in our world, we don't need to mine it or gather it, all we need do is stand up taller and it rises to fill us.

Power is our ally, and if we can turn our minds around to embrace it, it could help us achieve our dreams bigger and broader than we ever could have imagined.

Just for a moment, freeze where you are. Don't move a muscle. Now, noticing where you normally are, lift your chest. Lift your chin. Feel that? Feel it fill you? Power is as natural within us as the air we breathe. Individually, where we stand is a choice. Collectively, we have access to the most powerful wealth and resources on the Planet and if we don't use them for Good, taking our work seriously, the old way will continue to dominate.

What would the world look like if we, both individually and collectively, stepped into the power and wealth of America and used the resources there for Good?

Just for a moment, try this on – "WE are the one percent." Feel the ripples it creates.

What shifts would we, would you, need to make internally, to recognize yourself as being the most powerful force on the planet, bringing new ways of living to the world, like Microsoft did?

What would a world managed and cultivated for Good be like, if we stepped into a much bigger, more powerful vision of who and what we could be?

Power is not to be feared, it's ours to re-interpret and use for Good. If we - the visionaries of a kind and caring world, don't step into an impactful role on a much grander scale, the world will continue to be what it has been.

When I look at what we've created so far, there are so many things we have embraced, fully explored, and then redesigned on our own terms, toward our new culture. Look at how we eat, how we grow food, how we address both health and illness, how we dance our relationships, how we raise children, how we work together. All of these things we have faced head on, studied every version of throughout history, and then modified to fit our values and our vision. We have re-created each of these aspects of our lives to facilitate a caring world, and each is nothing like anything done in the past.

I see great successes in all of these efforts and I believe that when we embrace Power, and unlock its mysteries, we will again succeed and create new models that will uplift the entire world.

Power is a force of nature, built right in to every layer of our existence. Power is like water, we can study its properties, and build new ways to direct it, channel it, distribute it. What would it feel like to take on a role with greatness beyond your norm, to empower a team around you, to push the edges of your potential? How would it impact our work if we built for profit, generative models and designed them for Good, with the intention of generating millions of dollars in overflow to pour back into our communities? What if we found ourselves at the helms of projects as Powerful as Microsoft, Xfinity, Exxon? What would the world be like if we ran it?

As we learn new ways of acculturating each other to stand up and be more powerful, it will mean that we have to be strong enough ourselves to stand in a roomful of empowered people and collaborate at a whole new level. I know that we look with respect to one another now, but I believe that there is a much bigger level we can go to, that will completely change the game.

Saphir Lewis

As a teenager, Saphir Lewis hitch hiked around the USA, up and down both coasts and to many midland states to brainstorm with others about cultural transformation or the re-invention of humanity on earth. How can we build a culture of caring – for each other, ourselves, our neighbors, our planet? It was a very exciting time.

From there, she began producing Transformational Festivals. Rich in education, these gatherings are collaborative open-sourced villages, devoted to experimentation and inspired co-creation of the world we want to live in. In 2013 she launched FestivalFire.com, a network of websites, to serve the rapidly growing movement.

Today, she runs Festival Fire, including a blog showcasing thought leaders, a suite of event calendars featuring transformational events around the world, a free annual Festival Guide and the Festival Fire store, an online marketplace with artisan made festival wear and costumes.

She also currently co-produces several events, coaches young upcoming producers and travels to assist with events across North America.

http://FestivalFire.com

Women's Liberation, Global Evolution

How we can create a culture that empowers women's voices and inspires true change

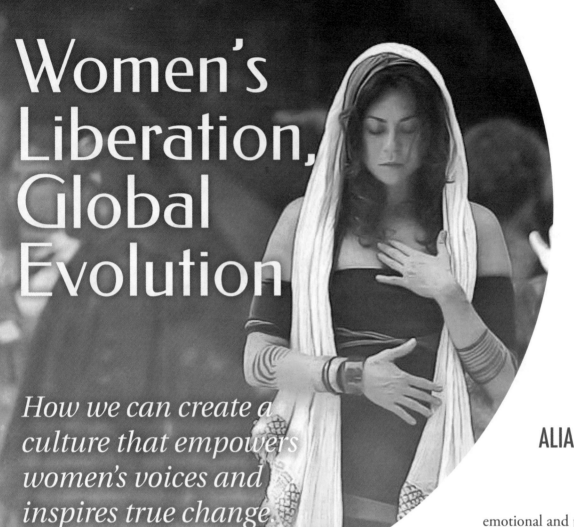

ALIA

Looking around the planet we can find examples everywhere of women being marginalized and their freedoms being stripped away. We can see the human and environmental devastation that is created when women are removed from the decision-making.

Over the last few decades more and more women have stepped in to leadership. Women entrepreneurs, politicians, authors, business owners, speakers, artists, and visionaries are sharing their voice now. As a result, we have seen women receive more freedoms.

Yet as of this writing, many women around the globe are still enslaved, mutilated, tortured, and subjugat-

ed. Even more subversively, many women that live in the "free" world do not feel free.

All of the women I have been speaking with are generally privileged and educated, they have a certain level of financial resources beyond what most of the population of the planet has, and their life is not being immediately threatened.

Yet, many of these women are struggling with their confidence, with fear, with a feeling that they are not fully self-expressed. They keep their visions and projects small. They censor themselves and they do not speak their truth.

Many of these women have been licking their wounds from real

emotional and physical traumas they have been through that have taken a toll on their self-esteem. These women often have not had an internal experience of freedom.

To create true change on the planet, we must come together in new ways to help heal this significant percentage of our population. To create a village that is a new model for the planet, we need to explore how we can inspire true women's liberation in our communities, and how the masculine and feminine can co-create and collaborate together in harmony and unity.

There is a new movement emerging and I believe the audience for this book is on the cutting-edge of it.

THE VISION

In this piece I am leading with the idea that women's liberation = men's liberation too and that if we focus on freeing the feminine, the masculine finds purpose and inspiration in supporting the feminine to find equal footing. When a culture experiences this, there is a massive rate of inspired, visionary creation from both sexes. This has been my experience in my communities.

For some of us when we look at what is happening now on the planet, trying to understand how to create this vision of a harmonious world in which both the masculine and feminine are free can feel daunting. However, we can look to examples of communities that are putting practices in place that are helping the feminine feel safe, free, and expressed and fostering better relations with the masculine that, if replicated, can help us realize this bigger vision.

I happen to have an inside look at some of these communities and inside this segment I share some valuable practices and knowledge that are working very well and can easily be modeled.

For the last 15 years that I have lived in the San Francisco Bay Area I have found myself a contributing member of a variety of communities practicing radical models of being and relating — from communication and relationship-oriented communities to entrepreneurial business-oriented communities, communities oriented around women's leadership or priestess work, spiritual communities, dance-focused communities, and neo-tribal communities that have formed around the underground electronic musical culture and festival circuit. I have worn many hats as an Executive, Entrepreneur, Coach, Course Leader, Speaker, Founder of Femvolution, and most predominantly now Musical Artist and Visionary behind the Feminine Medicine™ Project.

My experience traveling through these communities has given me the great privilege of seeing how we have been modeling a new way of being. As a result, I have personally witnessed a lot of healing between the sexes, and, perhaps not surprisingly, my communities are a hub for transformation and are producing some of the most influential Visionaries, Thought Leaders, and Artists creating impactful projects and businesses. My communities in particular have produced some extraordinary women's voices.

So what have we been doing right?

THE MARKS OF A CULTURE THAT EMPOWERS EXPRESSED WOMEN

1. We acknowledge there's an imbalance.

In my communities the women and the men have their eyes open about the fact that there are prevalent injustices against women and imbalances on the planet. We are awakening to the fact that there has existed a centuries-old patriarchal structure designed to quiet women that has made its way into every facet of our lives — it's in our collective story, our thought patterns, our religions, our media and entertainment, and our advertising feeding us with destructive stories. The men in my communities see this and want to help heal it in themselves and the women in their lives.

2. We are playing with an evolutionary Feminism that empowers women and includes men.

In response to patriarchy we saw Feminism emerge and with it women-only spaces were created. Feminism has helped create massive shifts and opened up rights for women that we never had before that have been essential for our evolution.

I've observed that our new breed of women leaders have felt something missing from the traditional Feminist discourse, something so essential to the true realization of our leadership and our ability to help facilitate change on the planet now - partnership with the masculine.

Feminism emerged because of a global imbalance. Looking around we can see that the work of Feminism is not yet done but the difference is we have a lot of men who get it now and are promoting a conscious masculine paradigm that empowers women too. The unconscious masculine has made it ok to give women fewer freedoms in our society. For a long time we have met this exclusion with exclusion. I think this has led the good men out there to think that that they don't need to help or we don't welcome their help.

The Fourth Wave of Feminism being called forth is one that chooses to include men, not fight with them, in order to invite partnership with the conscious masculine while still holding the space for women to keep coming together in gender specific circles. Bottom line the women in my communities love men, and we want to honor the good men surrounding us and co-create together.

3. We are playing with an evolutionary Conscious Masculine that inspires men and empowers women.

There are many men's circles popping up that feature a focus on reclaiming men's integrity, the conscious Warrior archetype, and brotherhood while intentionally fostering healing with women.

4. We are exploring how the masculine and feminine can experience peace with one another

In my communities we are exploring how to heal and communicate with the opposite sex. There are now multiple courses, workshops, dialogues, circles, and specialized communities offering new relational approaches to address this.

THE PRACTICES

There are some specific ideas, practices and ways of being that you can replicate and put into practice in communities anywhere that can help empower women's liberation, and therefore the liberation of the whole community.

1. CREATE WOMEN-ONLY SPACE.

Sometimes women-only space is the only way a woman can feel safe to feel everything that is there for her. Women's circles are a very powerful way for women to come together to support each other to begin to open up.

It's important to understand that many women have been abused, disrespected, dishonored, and made to be something she is not for a long time. We are experiencing tension and trauma in our nervous system. We have been scared of being in our bodies because we have been so violated. We have felt shut down and numb as a result. There is a collective wound here that feels very personal for each woman.

Having a place to cry, yell, share the places where we feel vulnerable, and have our challenges and triumphs witnessed is very healing.

Women's circles can get created simply by one woman declaring that she will and inviting other women to be a part of it. It's helpful at the beginning to create a shared intention for the circle and guidelines that describe how you want to relate with each other. The important thing is that a woman feel safe enough to be fully expressed.

You may want to play with a "closed" circle format where you circle with the same women for a period of time, or an "open" format where you allow women to come and go. In my experience, closed circle formats create way more safety and depth.

2. GIVE WOMEN STAGES AND PLATFORMS TO SHARE THEIR VOICE.

There is a powerful trend of women giving each other space to lead and speak and share their wisdom at their events and workshops. There are also multiple courses that have been designed to support women to find their voices as speakers, leaders, and artists.

Women have needed to remember that our power lies within and it cannot be taken away from us by an external force. We have needed to remember that we can speak up for what we want to see changed because for so long our voices were shut down. There is so much healing that becomes available when women find their voices again. The more we create space for this, the more we will see women step into leadership roles.

Women supporting each other to share their voice publicly is a powerful evolution of the women's empowerment paradigm demonstrating the values of collaboration and co-creation as well.

Within the electronic music industry, I saw an opportunity to empower many of my fellow women artists by creating my Feminine Medicine™ platform which is creating an album and stage show that features collaborations with almost 30 women musical and performing artists. Each of us were struggling individually to win the coveted spots at shows and festivals and I thought it could be much more powerful to bring us together so we can all benefit and expand from the increase in visibility that comes from aligning together.

The Feminine Medicine™ Kickstarter campaign was extremely successful generating support from hundreds of contributors as well as media coverage and the support of many of our male peers.

I also see a lot of women entrepreneurs becoming affiliates for one another's programs or creating programs together. When we do this more people can experience the great work. This model is what will help us all rise and it can be applied anywhere and in every context.

There is still a lot of room for more women to be given stages and platforms everywhere, in every domain and industry, every organization and boardroom.

MOVING FORWARD
LET'S HOLD A SHARED VISION

Men and women have the opportunity to come together here at this critical juncture in our history. We can envision a world in which all women feel free and in which men and women live in harmony.

There is no doubt women have been hurt by men. Many good men have been hurt by women too. We have much work to do to heal the hurt and trust again on both sides. So here is what I want to say to you:

Men: We need you. We need to feel your support. We need to hear your voice. We need you to tell your brother, friend, father, that it's not ok to beat women, it's not ok to rape that woman when you catch her alone. We need you to speak up when you see policies that strip away women's rights. We need you to speak out when you don't agree with something. We need men who are not threatened by the full expression of woman. We need the conscious masculine as a visible force in numbers.

Women: We must do the inner work to become self-resourced and heal. We can no longer fight the men in our lives making them the unwitting faces of patriarchy. We must celebrate their desire to support us, and honor their stewardship of us. Give them the same respect we request, and we will have far more stewards for our cause.

There is tremendous power in holding a shared vision. When we put that kind of collective energy behind it, our chances of it being realized are infinitely greater. It is time for the masculine and feminine to coalesce together to create the kind of critical mass we need for long-range change. To begin, every man and woman that believes this needs to speak up for this change. This kind of mutually empowered partnership is the way we will evolve humanity and the state of our planet.

ALIA

ALIA is an acclaimed San Francisco-based Electronic Music Producer, DJ, Vocalist, Author, Women's Mentor, and Founder of Femvolution.

ALIA has crossed international borders and traversed coasts to perform at many transformational festivals including Envision Festival, Lucidity Festival, Enchanted Forest Festival, Beloved Gathering, Wanderlust Festival and Burning Man. ALIA's music is featured in the acclaimed video series showcasing the transformational festival movement, "The Bloom Series," and a musical contributor for TedX.

Addressing the lack of women artists in the electronic music scene, ALIA is the visionary behind Feminine Medicine™, a series of collaborative albums, a stage show, and an educational platform that gives the feminine more of a voice. She is currently producing the first album and stage show with over 25 female musicians, vocalists, and performers.

ALIA is a long-time champion for women's liberation and a former corporate marketing executive and graduate of Brown University. ALIA created the movement Femvolution. She was named a "Woman Changing the Planet" by ORIGIN magazine.

http://aliasounds.com

http://soundcloud.com/alia_sounds

http://facebook.com/aliasounds & www.facebook.com/femvolution

http://kickstarter.com/projects/630282426/feminine-medicine-music-project

Renewing the Sun Fire Within the Heart of the Masculine

Davin Infinity

Welcome. We have crossed an incredibly important threshold, into a new era that supports the Return of the Feminine Wisdom in its many forms. This also presents the invitation to step into and embody the internal Union of Man and Woman in each of us. This energetic shift is awakening the Courageous Lion King Heart of the Man that loves powerfully and acts heroic. When a man goes on the hero's journey he eventually returns home initiated into great powers and a new awareness. A new responsibility is born within him - a total commitment to the Feminine, to Union, and to the ending of Separation. Today's global reality and all of the challenges and opportunities we are facing, deserves a new code of honor from the masculine heart of integrity. A code that will script a new heroic story projected from the light of our hearts onto the canvas of a planet in transition.

We are in the midst of a global uprise of women's empowerment unprecedented in all of recorded history. Women are encouraging men to join them – for men to awaken their hearts like a spring flower blooming. The feminine is ready and waiting to help men heal. But men must make the choice to join women by leaving the battle-field or depleting their life-force climbing a competitive corporate ladder. The time has come for men to be brave in the arena of Love, to be brave enough to trust life itself and let go of control. The control over possessions, finances, women, and the mother earth's territory and resources. To face this painful truth is a first step to heal the masculine heart which in some men can feel wounded, guarded with armor and lacking in authentic courage.

Since creativity and developing life vision are essential to my being, my life's journey has naturally led me to my life's work in the field of discovering universal consciousness, natural laws and human potential. One of the main ways consciousness takes form is in symbols and archetypes. I have created three categories for some of the great archetypes that are not only symbols but energetic frequencies of embodied life-force that give men strength, wisdom and creativity.

1.) The Ancient Deities of a World-spirit Energy (Shiva, Christ, Buddha, Hermes, Ra)

2.) The Classical Archetypes & Roles (Guardian, Magician, Sage, Lover, King, Scholar)

3.) The Neo Archetypes just emerging (Corporate Shaman, Integral Philosopher, Maverick, Visionary Entrepreneur)

An understanding of these great archetypal energies prepares men and women to come face to face with the powerful archetypal source of all co-creation: the mother and father.

The Father & Mother: Bringing Light to These Roles of Responsibility

We incarnate as human beings through the genetic alchemy of our parents. Their ancestral genes combined with the energetic imprint received when we were in our mother's womb has a degree of effect on us. As we age, we face another series of conditions such as: gender, race, national culture, and environment. Then we develop beliefs and habits, we wear masks and uniforms and we become players in the game of life and actors in the great story unfolding called life on earth.

Life is a growing process where individual beings and their experiences weave the threads of a larger interconnected tapestry of a collective humanity. This process plays out as

our self-identity and then we transition into our conditions, lessons, life experience and growth. All of these arise under the guidance of a father and mother relation to the natural world and can show us the truth of our being enacted through diverse lessons of guidance, betrayal, genetics and everything else under the sun. We are here to learn. A new world is born forth from our tragic and triumphant experiences and our ecstatic revelations as we mature into adults and enter the potential role of the father/mother ourselves.

My Own Journey into Awakening

I have found myself a contributing member of a variety of communities practicing radical models of being and relating. I have met thousands of people who honor the sacred qualities of masculine and feminine and also seek a deeper sacred union of these polarities within themselves. Since the year 2000, I have been a part of many men's rites of passage work and men's support circles that have influenced my own healing of the wounded masculine so that I can empower the divine masculine energy. I have brought this renewal into my creative career.

I have played many roles as Student, Facilitator / Teacher, Creative Director, Filmmaker, Entrepreneur and Life Coach. A big part of my mission is to work with the leaders and co-creators of this new era. We are inventing an entirely new cultural paradigm that has never existed before since it is a synthesis of every great tradition, lineage and wisdom that can be drawn from.

But most uniquely, over these last 15 years I have worked on over 70 projects with clients, teachers and artists of the Divine Feminine Wisdom. These projects have taken form as full color books, films, graphic design and marketing / branding. I have co-created books like ***Heart of the Flower: Book of Yonis, Raising a Goddess Temple and films like Divine Nectar and Birth of the New World*** which explore women's sexuality and empowerment, conscious conception, natural birth, and sacred relationships. Most recently I have just finished 55 original images of artwork to a deck of self-growth oracle cards called: "the God Cards: Empowering the Sacred Masculine.' By working on such deeply insightful and intellectual projects with women and men ushering in a new era of sacred masculine and feminine union, I have undergone rapid transformation in my psyche. Layer after layer of armor that guarded my heart and my potential has been removed.

The Ascension of the Empowered Masculine Spirit

As we have entered the twenty-first century, we face a gigantic crisis in the masculine identity of most cultures. If you look deep into today's crisis in masculine identity, you will see two major causes: the disappearance of masculine rites of passage and patriarchy itself. The most outward result is the modern day imbalance in business industrial capitalism that has destroyed the living eco-system in just 100 brief years. How is it that billions of people have to go to work everyday to participate in a system that is not providing a desirable future for them or their children? Within this patriarchal matrix is a programmed virus that completely removes the presence of love, emotions and the solar masculine strength we know as courage. Masculine energy has been that which has been credited with much of the disharmony and destruction that is still going on upon this planet. There will be an estimated 9 billion people by 2050. This means up-leveled stress on the environment and resources. It is important to realize that the masculine energy manifested upon this planet is out of balance. It needs to come into balance with the feminine. Men are in need of support and guidance at this time. Men are beginning to feel a calling within themselves that is a desire to find balance within themselves. When this balance within them is stabilized, they will not seek to create destruction and rule with competitive-based power, but seek to cultivate inner power that serves all.

What is the basic nature of the masculine energy? What are its strengths? What are its current rites of passages? The great aspect of the masculine is that it is focused, action-oriented and outwardly creative. It wishes to explore, discover and build outwardly something that is an extension of itself. So whatever is forming inside the collective masculine gets projected outward into society (whereas the feminine engages and creates more internally using the powers of the body, such as through emotions, dance and the healing arts). This masculine energy has created a great deal of society's pain and suffering due to its separation and limitation. Man has projected his shadow and fears out into the world, instead of the light of his heart.

When the masculine comes into balance within each person regardless of gender identity, there will be harmony and peace upon this planet. There will then be a balance of the masculine and feminine. These energies are meant to work together, they are meant to be in a co-creative dance. The unspoken prejudice, violence and hatred towards the feminine has corrupted the creative power of the polarities and has created unhealthy and unsafe tension. Instead of co-creative power we mostly see the power over another. How did we get this far away from the natural balancing energy of the opposites? The imbalance has come from the governing social systems removal of rites of passage, initiation and the ancient traditions that took a man on a path of discovering his inner power. Without this inner transformation and complete awareness of his virtue, the wounded, unconscious masculine has rejected his own spiritual core in anger and leashed it out on the world. The changes in the evolution of consciousness that are transpiring upon this planet now are beginning to assist those in the masculine form to seek harmony within themselves and their world. How can you serve this great awakening of balance and union?

By studying how to become a better man, artist and human being, I have discovered the daily practices that allow the balance of learning and memory formation to connect to a state of flow in the body that has the grace of water. This is a benchmark for the new human potential. It activates a key missing link that causes men's wounding: a lack of water to balance all of his fire. Modern man is "dehydrated" from the flexibility, agility, and soothing sensuality of the water element that balances out his fire. The out-of-balance masculine fire that we see globally is rage and war. The global droughts and climate change often seem like either metaphors or actual outward physical manifestations of the destructive empire of man. If the masculine can once again cultivate this inner state of creative flow, then he can return to the source of his original nature. Just like water which moves from the ocean, to the clouds, to the rain, to the mountains, to the rivers and back to the ocean, the delicate cycles of nature are within us all.

The bravest qualities of the masculine are rising up once again. I have met thousands of men who have found union within. They have become a loving being connected to earth, connected to emotions and creativity, and connected to form and structures. This full spectrum gives them the ability to build enlightened structures in business and community without controlling others. The new masculine energy becomes a guardian and a role model for the balance of sovereignty and interconnectedness. Their core values become accountability, humility, honor and being in service. Everyday I find the power of this simplicity. To wake up and through the sun in my heart wish that every man develop:

A beautiful mind

An open heart

An ecstatic body

Tuned into emotions

A creative spirit

A radiant field of life-force

Deep breathing of graceful mastery

A remembrance of who they are.

Davin Infinity Skonberg

Davin Infinity Skonberg is a Creative Director with 15 years of experience as a creative professional and teacher. His life's work, research and creative play is at the intersection of art, social psychology, human potential and culture design. He has traveled the world visiting and studying sacred sites and cultures as well as visiting over 50 eco-village and new paradigm societies like Damanhur & Auroville.

Davin has also spent the last 15 years training in several lineages of shamanism, martial arts and Mystery Schools. Davin is an experienced and powerful metaphysical guide into the Quest of human potential and transformation. This work is a major influence in his films, books, workshops and visionary art.

http://shamaneyes.net

Art &
Culture

IF MAN IS TO SURVIVE, HE WILL HAVE LEARNED TO TAKE A DELIGHT IN
THE ESSENTIAL DIFFERENCES BETWEEN MEN AND BETWEEN CULTURES.
HE WILL LEARN THAT DIFFERENCES IN IDEAS AND ATTITUDES ARE A
DELIGHT, PART OF LIFE'S EXCITING VARIETY, NOT SOMETHING TO FEAR.

GENE RODDENBERRY

So much of our lives is woven into the context of our culture. But what is culture?

Culture is the mythos we collectively share, it is the fabric of our lives, threading the experiences, people, values, beliefs, stories, symbols, conditions, and behaviours together into a tapestry that we are inextricably bound to.

Humanity is comprised of a melding pot of many cultures and our global culture is challenged with the impetus to learn how to embrace the strength in many unique expressions of people while finding the common language and set of agreements by which our societies can peaceably overlap in cooperation, not in competition with each other.

It is an unavoidable and necessary shift in humanity to find the way to overcome the barrier of separation and unify in respect of each other, even if we do not understand or agree. To come into greater understanding of a culture is to connect to its soul. And the soul of any culture is seen through its art.

As artists, we sculpt and shape the landscape of our culture to bring our visions into form through our voice, through our words, in our song, through dance, painting, building and design. Art is our culture in action.

LIFE IS ART. BINAH ZING

Just as we arose from the origins of nature's palette expressing its artistry as elegant function and wondrous beauty, humans have evolved to become co-creators, consciously shaping the natural world around us. As we developed the capacity for engagement, reflection, intention and perspective we transcended mere survival into a celebration of sound, color, light, emotion, texture, shape, movement.

In this theme we explore how creativity and expression is the framework through which we dance the edges of individuation and collectivism, of grief and praise, of disconnection and resonance. The stories we tell through our art are our inheritance and our legacy. Let us tell stories of triumph over despair, of love over fear, of peace over war, of creation over destruction.

Jamaica Stevens

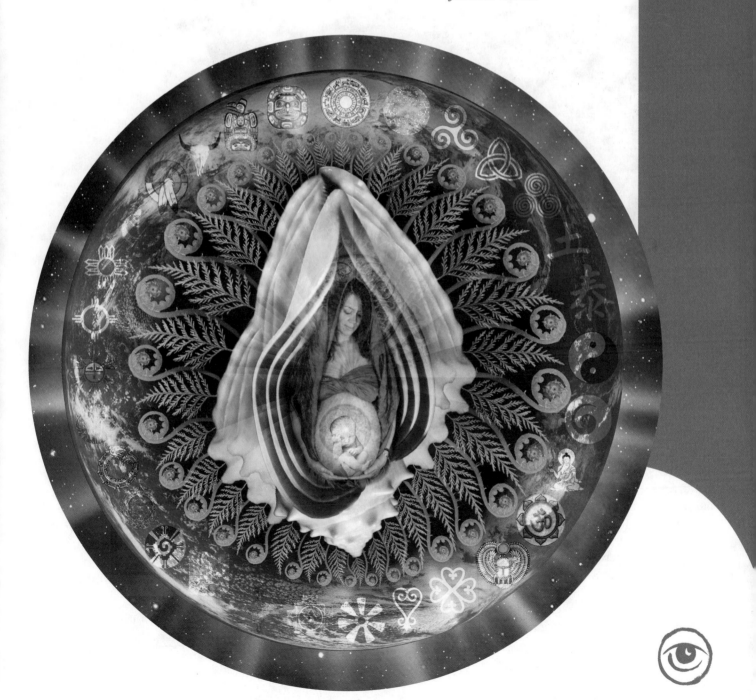

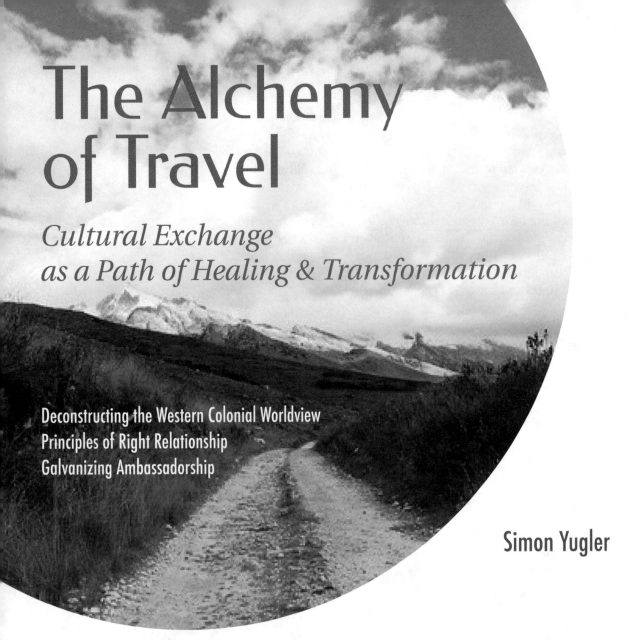

The Alchemy of Travel

Cultural Exchange as a Path of Healing & Transformation

Deconstructing the Western Colonial Worldview
Principles of Right Relationship
Galvanizing Ambassadorship

Simon Yugler

World travel is now more accessible and popular than at any other point in history. As a result, cultural exchange is an undeniable aspect of our evolving world. Yet travel is, and always has been, about more than just narcissistic consumption in exotic locales. Through travel, we are presented with life-changing opportunities to engage in authentic cultural exchange, cultivate right relationships, and become ambassadors of our communities. By integrating these practices into our lives, we can become empowered agents of intercultural healing, and transform the lead of our global legacy of colonialism and oppression into gold.

In 2009, I spent the summer traveling in Australia, and found myself living in an Aboriginal community at the farthest reaches of the Northern Territory. Receiving a grant to do ethnographic research and to deepen my own study of the didgeridoo, I embarked on a journey that would take me to the heart of an ancient people.

On my final day, I caught up with my "uncle," a local elder and community leader who I had developed a special friendship with. We came to know each other at the annual Garma Festival, a massive gathering that hosted many local tribes and families, as well as a handful of intrepid white attendees. I would bring him coffee and juice, helping

him in his old age, and showing my respect.

We would sit on the ground, often in silence, while he and his wife would watch the young people dance in their traditional style, kicking sand through the air, making war-like gestures and imitating the animals that comprise the Australian fauna. Amidst the peaceful rustling of eucalyptus trees, sounds of the didgeridoo and clap-sticks reverberated throughout the balmy night, punctuated by songs sung in an ancient language that I would never fully comprehend.

On the day I left, I caught up with him before he stepped into the local courthouse. Placing his hand on my shoulder he me bid farewell.

"I've never met a white person like you before," he said.

"You listen to us, and you understand."

In that moment, I knew that no matter what came of my research, my journey had been a success.

Cultural Exchange as a Path of Healing

Why had my uncle's words affected me so deeply? And, in turn, how had my actions brought him to convey such a powerful truth?

Years later, while traveling through Africa and the Middle East, I would experience the same depths of connection and exchange with my hosts, confirming that intercultural healing is possible when done with an open heart.

Trained as an anthropologist, and raised in the early 1990's, terms like "cultural exchange" and "diversity" filtered through my subconscious, often undefined. These ideologies of inclusivity and political correctness can very closely border on the impractical and cliché. Yet clearly something had occurred between myself and this old Aboriginal man, as well as with many others on my journey, which went beyond any academic formula or rhetoric.

Years later, I can say that what occurred between my uncle and I was a gradual process of authentic relating, sharing, and cultivating mutual respect. This moment revealed to me the profound potential for intercultural and international healing that can be achieved when we wholeheartedly engage in authentic cultural exchange.

What is Authentic Cultural Exchange? (How can Cultural Exchange heal?)

At its root, authentic cultural exchange is a process of remembrance.

Uruguayan author, Eduardo Galeano, famously states, "My greatest fear is that we are all suffering from amnesia. I write to recover the memory of the human rainbow, which is in danger of being mutilated."

In many Native American cosmologies, there are countless stories and teachings about bringing the world back into balance. The Lakota tradition calls this process, "Repairing the Sacred Hoop." Throughout indigenous cultures the world over, there is an understanding of the importance and need for healing in our world, and our communities.

Perhaps this is because it is the indigenous peoples that have bared the brunt of the abuse inflicted by what we can call the Western legacy of colonization and violence.

Dismantling the Colonial Legacy

In order to become agents of change and intercultural healing, we have to understand and be aware of this legacy of violence, which some would argue begins at the very roots of civilization itself.

Yet this dark thread of history, which weaves itself through our cultural heritage, can be best understood by looking at the legacy of colonialism, and its associated psychology of extraction and appropriation, which saturates our modern worldview.

To interact with the world from a place of healing, we have a responsibility to cultivate an awareness of this history, and to act accordingly, especially when we approach a people or a place which is not our own.

This is especially true when we travel. If we fail to do so, we become a tourist, or worse--an unconscious representative of the colonial legacy, and an unwitting agent of extraction and appropriation.

A Word on Cultural Appropriation

Cultural appropriation is a naïve, yet nonetheless living remnant of the colonial legacy and its inherent violence. Thankfully, the topic has begun to bubble up to the surface of conversation in many different communities.

Appropriation means to take something as your own without giving anything back in return to its original caretakers. Like it or not, we are literally sitting on a legacy of appropriation.

Yet guilt-tripping and shame are no longer adequate sentiments in this conversation. To bring our world back into balance, and to serve as acting agents of intercultural healing, means to come from an informed place and to act from there. The salve that heals appropriation is exchange, and to truly be in exchange means to give something back.

Author and ethnomystic, Martin Prechtel, teaches that, "You have to give a gift to that which gives you life."

Everything we have, both materially and culturally, has been given to us, willfully or not, and therefore requires something in return.

What are we giving back?

Coming into Right Relationship

In Australia, I was struck by the immense poverty and difficulty that the Aboriginal population is faced with. Like many indigenous communities here in North America, terrifyingly high rates of suicide, domestic violence, and abuse, not to mention substance addiction, depression, and rampant health problems plagued their communities. The destructive symptoms of what academics call "structural violence" were overwhelming.

At first I was paralyzed by what I saw. How could I possibly help any of these problems that were so deeply entrenched into present-day Aboriginal life? The colonial legacy of violence goes so deep, that I felt powerless against it.

As I spent more time amongst Aboriginal people, I became less fixated on the systemic problems they face, and more focused on interpersonal relationships, which felt like the only way to cut through the dense cloud of distress. In the end, meaningful interactions and relationships proved to be the only way I could "help" anyone.

Human relationships offer the best opportunities for authentic cultural exchange, and for the intercultural healing that may follow.

The reason so many NGOs and aid organizations often fail is because they lack grounding in genuine human relationships with the population they serve. These organizations often lack an understanding of what it means to be in right relationship with another person or culture, expressed by paternalistic attitudes and a neo-colonial mindset.

Though the mission of NGOs is built on a desire to help improve the lives of marginalized people, many fall short in understanding of how one can actually relate to a different culture in right relationship.

Right relationship, in a cross-cultural context, can mean many things, and is a dynamic process that often takes time.

But you can start by practicing these 5 principles:

1) Offerings of Respect

In Mali, West Africa, there is a saying that "You are no better than anyone else, except your parents, and nobody is better than you, except your parents."

Coming from a place of deep humility and respect is the cornerstone of fostering right relationship with anyone and anything. Much can be said about the importance of respect, and it may look very different in different cultures. Learn the customs of honoring and respect for a given culture, and you are well on your way to forging right relations.

A central piece in building right relations in many First Nations traditions is the act of gift giving. What gifts do you have to offer? Perhaps you have some skills or knowledge that is needed in the communities you visit on your travels. If nothing else, each of us possesses a pair of hands, which when put to use, can serve as a valuable offering of time and energy.

2) Mindful Listening

If you have nothing material to give, sometimes silence is the best offering. Deeply listening is a powerful practice that both shows respect and opens up roads of conversation and healing that could not have been revealed otherwise.

When we are truly at a loss for words is often when the deepest healing occurs.

3) Awareness of History

Do you know the story of the land on which you stand? Who walked there before you? Many people have fought and died for their land. If you are a visitor, you have a responsibility to learn the history and stories of that place and its people. Read up.

As mentioned above, the colonial legacy of violence goes deep. The reason the Spanish built churches over temples of Latin America was to suffocate any history of indigenous culture still alive in that land. There are many untold histories that inhabit the earth, some of which are still in danger of being stamped out.

Learn the stories of the land. And if you don't know the story, ask. But be prepared to really listen.

4) Love of Language

Nelson Mandela said, "If you speak to a man in a language he understands, you speak to his head. If you speak to a man in his own language, you speak to his heart."

This principle of right relationship ties all the others together, and is perhaps the most impactful. Many ancient traditions teach us that one of the greatest gifts given to humanity was language. To learn another culture's language is to learn another culture's heart. This is true for all people, but especially true for indigenous or ancestral languages.

Even if you're "not a language person," learn one word or phrase. Your efforts will be appreciated more than you know. For inspiration or guidance, check out Benny Lewis from Fluent in 3 Months (http://fluentin-3months.com).

5) Enthusiastic Sharing

When you share something you are passionate about, you share your heart. Enthusiasm is contagious, and sharing personal interests is amazingly effective at establishing common ground. I recommend sharing music—the language that transcends all borders.

I've found that oftentimes sharing our passions can open doors for mutual education as well. In both Australia and Africa, I remember sharing about Native American cultures, explaining that the original inhabitants of North America were not white Europeans.

Sharing about our family and friends is also a big step in establishing common ground, as family is a universal aspect of human life. I've found that sharing about my community has opened doors of cultural exchange that I could have never imagined.

If we take it upon ourselves to act as representatives of our communities, we take a step towards reclaiming our ambassadorship, and potentially towards more lasting levels of exchange, connection, and healing.

Reclaiming Ambassadorship: Becoming an Ambassador of your Community

Often as travelers, we have no obligations and answer to no one but ourselves.

Yet, like it or not, we always represent more than just ourselves. Everything from the color of our skin, to our clothes, gender, language, and nationality, communicates something to the wider world. It's not always positive, and it's not always under our control.

Yet every time we travel, we have an opportunity to communicate something new or different to the world around us--to break a long held stereotype or change someone's perspective--especially our own.

To reclaim our ambassadorship means to take a higher level of responsibility--to take accountability for our words and actions, and to act as representatives of our communities, our families, friends, and tribe.

Building on the 5 principles of right relationship, we can begin to see ourselves as ambassadors to something much greater than just ourselves.

We can even extend this awareness beyond the cultural sphere, the sphere of human conscious, to the sphere of the environment, and the earth itself.

Invoking the concept of right relationship as understood by indigenous cultures, David Abram, in

his book "The Spell of the Sensuous," implores us to see ourselves always relating to something greater, stating, "If the surroundings are experienced as sensate, attentive, and watchful, then I must take care that my actions are mindful and respectful, even when I am far from other humans, lest I offend the watchful land itself."

Reclaiming your ambassadorship requires commitment, impeccability, and vigilance. Yet it is an essential piece if we wish to become instruments of healing and transform our world, and ourselves.

Bringing it All Back Home: How Travel Affects our Lives and Community

Finally, we return where we started. When we return from foreign lands, we unwittingly carry with us treasures in the form of knowledge and culture, like gems ground into the soles of our shoes.

Upon coming home, ambassadorship becomes a two-way street and we often find ourselves, as anthropologist Victor Turner put it, "betwixt and between." As we come to represent our communities at home to people across the world, we also come to represent the people across the world to our communities at home.

This inescapable predicament is the true test of our ambassadorship. How can we continue to integrate, represent, and maintain connection with the people and places we have encountered? How do we continue to give back once we leave? How do we continue to honor those who have touched our lives, and whose lives we have touched?

The act of honoring lies at the very roots of right relationship. Travel forces us to forge right relationships, as so often the traveler depends upon the kindness of others for their well-being. Travel is both a practice of, and a test in, this concept. And if we practice it enough, it's easy to see how we can apply this principle to our daily lives.

Some would say that travel is about encountering the other--about finding the most outlandish customs or places on earth, and reveling in the utter strangeness of it all.

But I would say that there is no other, and that travel is about encountering our true selves. No matter how far we venture, at the end of the day, our selves are the only things that remain with us. Travel is about making the strange familiar, and the familiar strange.

By integrating these practices into our lives, we can become empowered agents of intercultural healing, and transform the lead of our global legacy of colonialism and oppression into gold.

Simon Yugler

Fascinated with ancient cultures from a young age, Simon is on a lifelong journey to bring people together through celebrating the wisdom and beauty of world cultures. With a degree in anthropology and religion, Simon has found himself studying the didgeridoo with Aboriginal Australians, living on the banks of the River Nile, traveling Africa overland, and learning from indigenous shamans in the Peruvian Amazon. He currently writes about travel and conscious living for his blog, Travel Alchemy. When not abroad, he can be found enjoying life in his native Oregon.

http://travel-alchemy.com

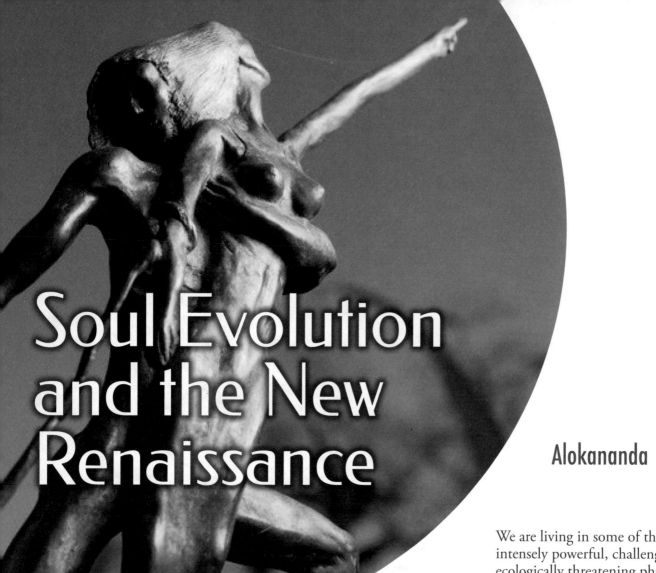

Soul Evolution and the New Renaissance

Alokananda

We are living in some of the most intensely powerful, challenging and ecologically threatening phases on our planet as a species. Thousands of experts in varying fields and paradigms all point to a similar conclusion: we need to make some serious adjustments in our choices as a planetary culture, otherwise our future does not look good. Many of our teachers, Elders and Wisdom Keepers speak of this time as an acceleration in consciousness and that we have reached the tipping scale where we can actually start evolving our consciousness as a species beyond the destructive tendencies of the last 26,000 year cycle; a cycle that according to many sacred indigenous prophecies and great minds, ended in 2012. Therefore, we are at the beginning of a new cycle, a new era where our choices have more impact on the planet, inspiring a culture that invites deeper self-responsibility. We are in a new renaissance, where the emergence of Soul-based choices

It's important for me to start by sharing that this article is but a small piece of a larger body of work and simply introduces the subject and theories of Soul Evolution and the New Renaissance. How these themes and theories relate within the context of an emerging planetary culture, festivals and community-building has been intimately connected to my personal journey and is an area of passion in my life. I don't want to paint the picture that this emerging culture or this new renaissance has all the answers; that it is by any means perfect or fully realized yet. Each part that makes up

the whole of this new renaissance, including each person's subjective experience of Soul, is completely and utterly a discovery process. Just like the word Soul poses a challenge to define, the concept of soul-evolution is equally challenging to define because at the very core it is a multi-dimensional and subjective experience, involving feeling, thought, energy and beliefs. I am not here to fully define this movement; I am simply impassioned to introduce some key themes that exist within this new renaissance. A major theme being that this emerging planetary culture includes, celebrates and honors Soul evolution as a primary focus and the driving force behind making healthier choices in one's life and for the planet.

is leading us toward inner transformation, rippling outward to create transformation in our external world. This new era cultivates the deepest empowerment at the level of personal purpose in service to the planet and how we relate to ourselves and others. Key pieces to this new renaissance are Transformational culture, community gatherings, and the celebration of Soul-based choices.

This emerging Transformational culture hosts a new breed of explorers, visionaries and innovators; who instead of sailing across the vast oceans to discover uncharted lands in the previous eras, cross an ocean of populism and set out to discover alternatives to our contemporary worldview, consumer culture, and societal norms. We are in search of the fundamental building blocks that create fulfillment, unity, direct spiritual revelation, authenticity and a sustainable way of living on the planet. This is a way of life that we are in discovery of for ourselves and how we share it with the world. This is happening now; an organic blossoming is literally unfolding as you read this. Some try to compare it to what first emerged in the 60's, yet that movement was only the first wave. The new version of this is a movement that is more integrated and spans a much vaster range of

"The requirements for our evolution have changed. Survival is no longer sufficient. Our evolution now requires us to develop spiritually - to become emotionally aware and make responsible choices. It requires us to align ourselves with the values of the soul - harmony, cooperation, sharing, and reverence for life." Gary Zukov

demographics, people and places; this Transformational culture is a global phenomenon that cannot fit into any one mold of culture that has previously existed.

This is a culture of transformation that transcends, age, gender, genre of music, religious preference, socioeconomic status or a definable social cause. This new renaissance shares a multi-dimensional culture that includes the impact and place of Soul within our choices in the world. It is here to help others discover that there IS in fact a more nourishing, supportive, and connected experience available within our planetary culture. As a community, as advocates of this planetary culture of transformation, we are here to dispel the myth that this is one group's "thing," when in reality it is a human thing. We were meant to evolve beyond what is familiar and comfortable. It is possible to make new choices outside of the default conditioning that we have all experienced and have been subjected to by society. We can evolve beyond a modern culture of separation, competition, scarcity, exploitation, un-sustainability and violence. At a Soul-level, we are meant to help each other and work together towards our common goals, towards win-win scenarios that honor our fellow man, our earth and all its life. While the explorers, innovators and artists of the earliest Renaissance were discovering third-dimensional perspective, we are now discovering fifth-dimensional perspective if you will; one that includes intuition, and a more profound understanding of how the universe works; where formlessness meets form, where spirit and matter are interconnected.

We are discovering how powerful we are as individual souls and how our consciousness co-creates our reality, influences our environments, each other and our experiences. We are discovering how effective and fulfilling it can be to transcend competition and work together for the betterment of all.

Defining Soul

The word Soul can potentially stimulate such a wide plethora of personal beliefs or concepts, sometimes over-intellectualized. Soul or soul energy is deeply connected to feeling and is beyond belief or dogma. Soul is in essence a completely subjective experience and yet it is as universal and diverse as each person's unique connection to the life force moving within them. "Soul" can come down to either being an abstraction or something tangible, depending on your own personal experience and relationship with the soul energy within yourself. Our Souls are rooted in unconditional love and they connect us to our higher-intelligence, to the source of all that is, the universe - whatever you want to call it. It is the individuated aspect of divinity, with an intelligence that is beyond the mind and holds a deep reverence for all of creation. As we cultivate a relationship with our Soul, the nature of the cosmic forces, how life functions,

and how karma works is revealed to us. Soul transcends gender, identity, and is the most profound aspect of our consciousness in its capacity to witness life and to experience all that is. Countless cultures have found ways to cultivate Soul energy, from ancient Egypt, India, the Inca, and so many more.

New Renaissance: Soul-based living vs. Ego-based living

The new renaissance acknowledges that there is something inherently divine, that carries the codes of our gifts, challenges, passions and fulfillment deep within each of us. It holds our longing to be of service to the planet and to assist in building a new planetary culture that will allow a truly nourishing experience for all beings. This motivating factor, this wellspring of inspiration, this aspect of us that is rooted in love we can call the Soul. We acknowledge that it is different than our personalities and our ego-based safety mechanisms. The ego-defense structures or survival-based aspects of consciousness are the aspects of self that defend against life to feel safe. They seek to separate, control, blame, destroy, act out or check out in connection if there is a perceived threat in the other or the environment. It is what keeps us in separation and protecting a false sense of self or identity. In my opinion, this ego-based consciousness is the premier causality of dis-ease in the world. As

a community, we are learning how to assist in creating safe places, or "containers" where the Soul of each individual can emerge as the primary aspect of being in relationship with self, other and everything around us, as opposed to the primary ego-defenses that are rampant in the world and popular culture. We are discovering how the Soul-based aspect of our being can infuse the personality, can utilize the ego, but it is not the primary focus anymore. It's not as if we have one soul-revelation and are functioning fully as Soul 100% of the time, all the time. It is rather that the Soul of the individual "comes online" more and more in percentages as the primary operating system. As Souls inhabiting bodies on Earth, we are here to grow, to learn how to flourish, thrive and end suffering within our own lives and the lives of others.

This process of rapid awakening of Soul-consciousness, of accelerated Soul-learning, heightened creativity, communication, inspiration, innovation and intuition I call the New Renaissance; others describe this as the process of Ascension. It is where we are learning to embody completely as Souls, beyond survival consciousness, beyond projecting and acting out our wounding onto others and into a totally sovereign interdependent network of beings choosing to co-create with one another for the highest good possible. This co-creation inherently includes our own fulfillment and transcends self-sacrifice, which is just another expression of ego-based consciousness. It is a Soul-culture that is discovering how to deeply nourish both its humanity and its divinity.

One of the main ways we evolve as Souls is by learning to love something in ourselves that was previously perceived as unlovable. Another way that I understand it is that the Soul evolves when the personality is willing to face and claim any of its unconsciousness, especially and including suppressed emotions. We all carry pockets of emotion in our DNA and within our chakra system (ie. nervous system), and the function of the Soul in the age of ascension, or in this case, the New Renaissance, is to fully embody as love. What that means is that we need to get support in accessing and unfreezing the stuck emotional energy in the DNA, sub-conscious, and nervous system. This process is fundamentally energetic and is literally bringing one into direct experience of initiation. A version of a mid-life crisis, a loss of a loved one, or a dark night of the soul are all examples of initiations, in which one is thrust into a situation of being confronted with their identity, fulfillment, choices and how they relate to life itself. The Soul evolves as we move out of mental energy and concept into direct experience of feeling, of relating, and of discovering ourselves and the universe. It is the journey of the head to the heart. It is the journey from protection to vulnerability. The Soul evolves when we claim aspects of ourselves that have been living in shadow and bring them into the light of consciousness to choose how we want to engage with ourselves as whole beings. In the larger body of work I expand in much more detail the different ways the Soul evolves through community-building and Transformational culture.

In conclusion, the most practical, linear, and challenging aspects of our modern world and the most profound potential for planetary healing and unification can only come about fully through a cultural paradigm that includes Soul. When we live from Soul we include others and have an inherent reverence towards all of creation and the web of life. To me, the culture we identify with is the single-most powerful influence on our choices. In developing a conscious connection to the aspect of our Soul-self, which functions as empowered love, we bring those aspects to the forefront of our choice-making. If we do this, the world can change for the better. May we all discover the deepest harmony we have ever known within ourselves and with each other, for all our relations.

Alokanada

Alokananda is a modern mystic offering many gifts and passions where they can be integrated in the heart, grounded in life and applied to the process of self-mastery and planetary service. His primary service consists of his healing, Soul coaching and consulting practice working with individuals, organizations, C.E.O's, entrepreneurs, high impact leaders and healers that are transitioning into a more holistic paradigm and empowered life path.

Alokananda has mentored in Bioenergetics therapy for the last 7 years, apprenticed under master shamans, healers, and teachers in other modalities and lineages undergoing consistent initiations and tutelage since 2004. Over the last year he transitioned out of 6 years of holistic event production experience and creative direction through large scale festivals, international retreats and locally in community building endeavors to focus primarily on his private practice and creative work. He continues to support and assist the building of vibrant and empowered community as a visionary, writer and a catalyst for multiple projects. His mission is to support the individual and collective ascension process and ground heaven on earth.

http://alokananda.com
http://tribalallianceretreat.com
http://sunsoftheearth.com

Dancing the Village

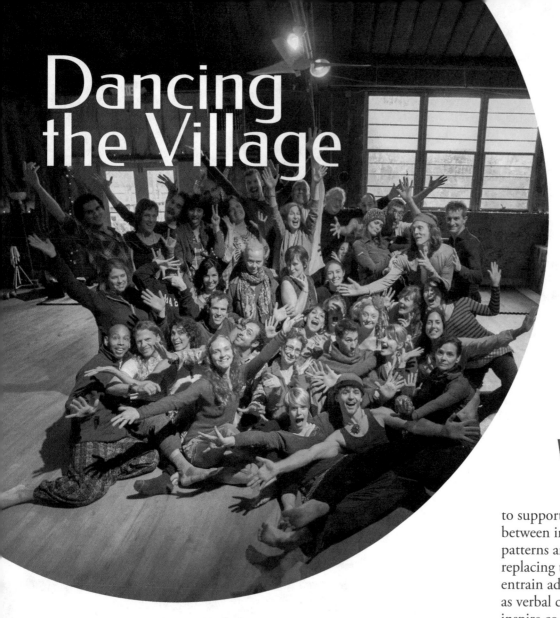

Wren LaFeet

Historically, social partner dance has been a social activity; something that people come together to do for fun, or to find connection, and always to revel in the joys of being intensely physical creatures (try as we might to repress it). From swing to tango, salsa to waltz, and my personal favorite blues and the emerging style of fusion, humans have created these art forms that are absolutely dependent on the cooperation of two or more people and an ability to physically express the complex languages of music. In all of these forms, there is a vocabulary which, when both people have an intimate knowledge

thereof, can result in an absolutely stunning example of communicative mastery.

Intimate knowledge or no, social partner dance is a proven tool for developing authentic connection, presence, and effective, compassionate communication within a community, as well as a social diversion, full body workout, and intelligence-improving activity. It provides a fully integrated experience of mind, body and spirit through which, when practiced with integrity, we can develop resilient group bonds and come into a functional, and deeply gratifying symbiosis with one another. Bringing a practice of partner dance into one's life and community holds immense power

to support connection, develop trust between individuals, heal wound patterns around physical touch by replacing them with healthy ones, entrain adept psycho-somatic as well as verbal communication skills, and inspire co-creative action vital for the healthy functioning of a whole organism.

On dancefloors around the world, I notice the same thing. If I show up and give myself permission to fully express my feelings through movement I inevitably leave feeling more alive and connected to my life. When I bring this same willingness into a partnered context with those who are aware of - or even just receptive to - the vocabulary of structured partner dances (such as swing or salsa), the feeling of connection can expand to include belonging, vulnerability, acceptance, intimacy, and trust. Presence and intention are the essential elements for using dance as a means for connection rather than a mere diversion or entertainment. These practices of presence and intention pulse at the center of implementing social partner dance as a tool for creating relational stability and harmony as well as transformation throughout the diversity of systems with which we interact.

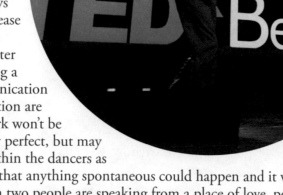

Like anything, learning the basics and practicing will create better dances, just as learning the ways of a romantic partner will increase pleasure for both, or those of a business partner will create better collaborations; just like learning a verbal language makes communication possible. If presence and intention are absent however, the magic spark won't be there; it might look technically perfect, but may not have the same response within the dancers as two who are so totally present that anything spontaneous could happen and it would still seem to flow beautifully. When two people are speaking from a place of love, poetry is the result. The same is true of a dance, and the same is true of community.

Connection and Collaboration

I believe effective collaboration on and off the dance floor requires a five-fold connection; Connection to Self (Center), to Partner, to the Earth, to the Whole (Community), and to the Music. Feel free to substitute "Music" with Source, or Spirit, and allow yourself to see that in the context of the dance floor, Music can be synonymous with these things. It is descending into us and inspiring us to move our bodies in ways that free us and feel good. Connecting to our Center, our deepest place of Self, allows us to stay balanced in our sovereignty. Coming into a knowing of our Self, where our internal compass and balance always brings us home, we can avoid forming co-dependency on others, or external stimulus. With this kind of awareness, connection to our Partner begins to create intimacy. It is this polarity between two clearly defined individuals that we require to develop real trust, and therefore, real collaboration. It is an interesting catch-22 of course, since we need to first trust our partner before we can experience the depth of connection we are longing for, both to our own center, and to others. Partner dance can be the parachute for such leaps of faith that working in teams often demand.

Introducing non-verbal partner dance practice into the operational procedures of your organization and relationships holds the power of being a balancing element, bringing the different parts into equilibrium with each other, and making people more effective learners and doers. Science behind dancing shows that the "split second, rapid fire decision-making" required by this art form acts on the plasticity of the brain, making intellect more agile and actually increasing our IQ. There is a set vocabulary of movements with the freedom to improvise; a shared set of agreements about how to initiate and respond to ideas, how to share those roles within an exchange of energy, and the permission to express and co-create within those boundaries. There is built-in accountability to honor and respect one another as we are engaging with each other's sensitive bodies, and there is an imperative for each to ensure the safety of all in the environment. Integrating verbal communication practices - such as reflective listening and non-violent communication - into a regular partner dance practice for your village, your organization, and your relationships, can provide a sensitive, responsive, intuitive, and relaxed environment from which to expand.

Healthy Living Systems

The added benefit of dancing together is the release of all those good chemicals like dopamine, oxytocin, and endorphins, making it one of the most effective forms of healthy touch out there. A great way to run preventative maintenance on your group is to gather once or twice (or more) a week for an hour of social dance. Put on your favorite music, come into connection with one

direction or another. There is also a chance that we'll both relax into the feeling of being held by one another in just the right way to be suspended in mutual support. A balance will be struck. More often than not this awareness does not exist right away, but given time, a partnership will succeed in dancing their unique homeostasis. We become two cells in equilibrium in the greater organ of the dance floor.

as I passed behind her in our crowded café. Touch comes with a host of issues resulting from repression and trauma from our past: the abuses of women's bodies by men, enslavement and control exerted through pain, and the psychological torment around our bodies imposed upon us by religious and societal institutions the world over. Yet in North America as well as Europe, Australia, and parts of Central and South America, partner dances are enormously positive cultural practices. They allow us to heal the wound between the masculine and feminine (or commonly known as "the war of the sexes") by entering into consensual, collaborative, and most importantly, non-sexual physical interaction. Once again those magical elements of presence and intention enter into the picture. Imagine this scenario: a man and woman who don't know each other approach on the dance floor. The man asks the woman to dance with his hand outstretched, she says yes and places her hand in his, he immediately pulls her toward him and wraps his free arm around her pulling her in close to his body. What do you think will likely be the woman's reaction? Now imagine the same scenario, but instead of the man pulling the woman close, he simply opens his right arm and creates a space for her to step into; a consensual, safe space where they

Dancing in partnership is all about striking balance.

another, and at the very least, just move together. No one is looking for or needing to be Fred Astaire and Ginger Rogers, and there are plenty of videos out there for inspiration and instruction, not to mention top notch instructors in most places you'd expect to find them.

Dancing in partnership is all about striking balance. A sharing of weight must be achieved in order to create the dynamic movement we see in the most extraordinary of dances. When I stand across from someone whom I have never met and hold hands, and we both sit back equally as if into a chair, there is a chance the partnership will fall in one

Making Contact and Consent

One of the dynamic and sometimes edgy things about partner dance, and group dance in general, is the element of physical contact. In acknowledging dance as a vehicle for restructuring our consciousness around touch we find a critical ally. Touch is a complex aspect of cultures throughout the world. We've only in the last six decades in America experienced a loosening of the grip around our societal freedom in regards to how we're able to express ourselves physically with other people. Even so, touching strangers accidentally can come with serious consequences. I once got written up at a job because a customer complained about me touching her on her back can both meet. The difference is the intention enacted with presence tending to the other person's comfort and safety. Now imagine it doesn't matter who asks.

Partner dance as a conscious dance practice entails witnessing and nurturing an emergent model of a powerful, joyful and gentle masculine man. Women are being offered, and encouraged to use, tactics to empower themselves in navigating away from unwanted physical advances so they can be authentic in who they are without having to worry about being preyed upon or concerned with feeling validated or accepted. As they develop, some communities are empowering themselves to support one another in facing the challenges these changes present in our cultural norms. In communities adopting partner dance as their mode of expression, the gender binary of male/female and the normative view of sexual expression and gender identity commonly promoted both in society and within the historical context of partner dance is being questioned. When we take gender out of the picture, and can simply view each other as people, a whole new world of possibility opens up before us. This inquiry presents many inroads for conversations around how to invite touch and dance in ways that set both parties at ease and make them feel understood and respected. There can be support of strong boundaries at dance events and a web of accountability created by the community attending them that empowers individuals to speak, act and be in their authentic truth.

Functional Integration

How does this create a good village? Imagine that at least once a week, every member of the community you identify with came together with this tool of partner dance and practiced it together. You see where you are attracted, where you are repelled; where you have natural, strong chemistry with some, where you have work to do with others. You feel the floor beneath your feet more acutely, and it ripples out to the ground you walk upon daily. Your body becomes more sensitive and attuned to the touch of others to the point where you can understand another's intention just by standing close to them. Your intuition grows so powerful that you no longer think when you dance, the MUSIC moves YOU; information held within moments no longer needs to be decoded through the processes of the mind into words you hope your partner will understand; inspiration becomes knowing and is communicated effortlessly, confidently, and immediately. And this transfers to every aspect of your life. Your body, your village, our world, balances itself.

Wren LaFeet

Wren's journey into service and earth stewardship has led him through theater, Vaudeville, the circus, and festival culture. Through it all, his love of dance has guided his evolution and rests at the core of his passion for life. His work with Stacey Printz at The Branson School gave him a diverse foundational training. Apprenticing with Actors Theater of Louisville from 2003-04 showed him what goes into running a successful repertory theater. In 2007, following completion of his degree in Drama and Dance at the University of Washington, he created The Harlequin Hipsters, a Vaudevillian partner dance troupe culminating in the ensemble dance theater variety production, "Passion, or Death!" Turning towards teaching full time in 2011 answering a call to service, Wren created Nomad Dance envisioning a sustainable, collaborative global culture through dancing together. At present he co-creates Nomad Dance with his partner, Antje, as they synthesize authentic communication and relating practices with mindfulness, and partnered movement into a unique living teaching called Cocréa. He participates with organizations such as Tribal Convergence Network (http://tribalconvergence.com), Oregon Country Fair, GeneKeys.net, is a celebrated TEDx speaker and continues to bridge communities internationally. His work has been called "the next evolution of conscious dance."

http://nomaddance.com

Anything but Temporary

*Building Blocks
toward a New World*

A reflection on our collective architectural roots

Luke Holden

Join me now as I explore the intersection between ephemeral and traditional architecture and how temporality in form translates and reflects to our culture.

What is Temporary and Site Specific Architecture

The concept which underpins all temporary structures relates to the creation of infrastructure, which when in place creates the most space with the least mass and material, and is designed with modularity and mobility in mind. Architects, engineers, industrial designers and artists create this infrastructure by first studying the minimal surfaces, membranous materials and fluids, structural integrity, efficiency and mobility found in nature. The spider's web is a beautiful example of temporary architecture, as is a badger's dam, or a bird's stick and mud nests. The influence of nature can be found in the designs of our more mobile architectural achievements and certainly in their purposes. These structures aim to offer refuge, shade and safe shelter quickly and efficiently, and in their efficiency emerges an elegance and a seeming simplicity as the mathematics and physics lend themselves to co-existing.

Theory, Romance Art and Buildings

Temporary Architecture is to traditional construction as drawings are to fully-realized paintings. A sketch is the essence of a fleeting moment captured in a gestural form while a finished painting is the result of all things decided. Imagine a sandcastle being ever built by children who grow up pouring sand, creating form, experimenting only to grow up and pass the bucket to the next child to pour and try themselves.

The world of *Temporary Architecture* champions things and people that allow for motion and does not create blockades in the overarching process. It may be said that freedom can be best expressed through objects, actions, and people that allow life to remain a process towards betterment. As we all co-create this movement together, I believe we are playing with the building blocks of any fully-realized society.

Similarly, the essence of the collective movement chronicled in this book is movement and mobility itself. If any change is truly to be holistic or sustainable it is essential to contribute and evolve the existing paradigms of society by innovating the practices of education, the functions of politics, and concepts of environmental and architectural design. Proposed to stay poised in an ever-kinetic relationship to what is needed in the world in an ever-flowing moment we experience as "now," the Transformative Culture's Movement embraces continual change.

EVERY BUILDING IS UNSTABLE: ALL ARCHITECTURE TRIES TO DO IS TO TEMPORARILY MAKE STABLE WHAT IN PRINCIPLE IS UNSTABLE. WHILE A BUILDING EXISTS, IT MUST BE STABLE: IT MUST BE TEMPORARILY STABLE. IF IT HAS COLLAPSED, IT'S ALREADY PERMANENTLY STABLE. WHEN IT'S LYING ON THE GROUND, NOTHING ELSE CAN HAPPEN, BUT THIS TYPE OF IDEA HAS TO BE ADDRESSED BY PHILOSOPHY. Frei Otto

Consider the density of bird bones which have provided inspiration to bridge builders, among others, about how buildings' productions and designs can be both light and extremely robust and strong. These are nature's innovations built for the purposes of peak functionality, survival and shelter. The animal kingdom shows us the best examples of the variety and function of temporary architecture. Animals are creating structures expediently and when complete are meant to exist for a finite duration before the animal or insect moves and builds again, leaving the previous structure to return to homeostasis with the environment.

It is important that we not conflate efficiency in nature with simplicity; generations of study find nature, while exhibiting simple elegance, is infinitely more complex than it is simplistic. Instead what researchers find is that what evolves naturally often designs in the interest of minimal waste and maximum strength. It is these types of natural structures which have informed our technological advancements. One example to consider is the load-bearing strength of bamboo, which retains flexibility or the integrity found in hexagonal shapes we see everywhere in nature. The repeating hexagonal shapes fitting into a repeating triangle have become well-known building blocks of natural forms and also have influenced many of our advancements. We apply what we've learned from these sacred geometrical shapes to everything from fabric to concrete.

THE MOTHER ART IS ARCHITECTURE. WITHOUT AN ARCHITECTURE OF OUR OWN WE HAVE NO SOUL OF OUR OWN CIVILIZATION. Frank Lloyd Wright

One of the most recognized temporary structures today is the camping tent, and the farmers-market style 10'x10' pop-up tent. Though these pop-up tents are dreadfully boring due to their repetitive use and lack of stylistic variation, they serve a great function to many in need of easy-to-create shade and also provide needed environmental context. The camping tent, however, has no shortage of stylistic variation and I am in constant awe by the sheer diversity of design and innovation going into these spectacular examples of tension structures. As your awareness of this design sensibly grows, you will begin to see the theories are being put into physical form all around you.

Keep your eyes out and let your imagination and curiosity become newly inspired by balloons, bubbles, foams, insect wings, umbrellas, stage canopies, shade structures, bamboo structures, circus tents, camping gear, hammocks, and rubber tires. Even the human body is an example of a temporary tension structure. I encourage us all to use our kid eyes and take another look at the world around us: it is staggeringly inspired for those who have interest.

INTEGRITY IS THE ESSENCE OF EVERYTHING SUCCESSFUL. Buckminster Fuller

A Brief History of Temporary, Site Responsive and Ephemeral Structures

We live in a single, ever-prolonged *NOW*. It is this moment which is the most exciting time in history, and architecture is among the reasons why that is. Architecture has been temporary, site-responsive and ephemeral from the first cave painting, hut and thatch roof through global cultural reinvention of yurts, castles, temples, coliseums to skyscrapers. We have evolved into the ultimate nomad: as we move through space and time we leave a trail of society rooted deep, and all who remain attempt to stave off time co-creating an illusion of permanence through mass, occupation and connection.

Cultures in central Asia have been traveling with their homes for over three thousand years. Ancient nomadic Mongolian communities created the yurt and ironically named it after the tamped, circular mark that they would leave in the ground after removed. It seems that they too were interested in noticing the legacy and mark they would leave behind.

During and also in response to WWII (1939-45), architectural innovators like Fuller and Frei Otto began to consider the occupation of space with new perspective. Otto, who was drafted into the German Air Force during the last years of WWII, dedicated his architectural work, in stark rebellion to the Third Reich's attempts dominate the world with structures that are intended to leave little to no lasting mark on the earth. Fuller, never actually trained in architecture, expanded the realms of what was possible at the time by what he considered "finally thinking for himself." In essence, he was a visionary inventor who decided to help the people of the world wake up to the finite resources on earth and our collective responsibility to house and protect all people, while also protecting the delicate balancing of ecological resources.

In *"Operating Manual for Spaceship Earth,"* Buckminster Fuller parallels our collective relationship to architecture with the metaphor of shipwreck survivors clinging to a piano top as makeshift floatation device. What he meant was that the permanent structures we live in and around for the most part today echo a past desperation for housing solutions; they are not the result of holistic design consideration based on caring for the earth and a constantly growing population. Fuller spoke of architecture as a fundamental means of survival and suggested, given time to reassess, we will likely abandon our first attempts to fend for ourselves and develop better ways through trial and and examination. He and his contemporaries looked for these new methodologies by studying nature for its answers. We, the festival designers, builders and volunteers, are among the family tree of his contemporaries, and our charge is of the highest import: to continue his work and design for a new world approaching. Fuller's belief in the way to arrive at new solutions resonates with the same ethic I spoke of regarding our Transformational movement: in the examining of conventional ideas with fresh eyes and trusting and thinking for one's self.

Razor's Edge between Risk and Solution

While our festivals and private land build opportunities can offer prime testing grounds for experimentation and development, this is not to say you are free to build without testing, modeling and rating. Festivals can provide the necessity that will fuel and fund ongoing research and development pursuits to a large degree, yet if we fail to create successfully we stand to weaken the public view of lightweight and non-traditional architecture. If we create poorly or unsafely we will

... the permanent structures we live in and around for the most part today echo a past desperation for housing solutions; they are not the result of holistic design consideration based on caring for the earth and a constantly growing population.

not be of any positive assistance to our generation or to future generations, such as if our designs prove to be unsafe due to poor testing and/or execution. It is our responsibility to honor the work that has been done by our predecessors, work to cultivate more trust and understanding regarding new building methodologies and materials, and to provide use cases for our state and county officials by exhibiting consistently well-planned and constructed structures.

We are on the cutting edge of architectural knowledge and capability, on the frontlines of new techniques, and because of this there will be many times that there is no law in place permitting our design. We will likely find ourselves pushing the boundaries of our county's permitting ability and comfort. Building codes are created to ensure design safety and compatibility with the industry standards, which means sticking closely to the traditional methods. While the aim is true, it is limiting to what is possible. If our generation of visionaries are to effectively evolve the status quo, our teams will have to be a credible line-up of creative designers, mathematicians, 3D modelers, engineers, and builders. It will be this team that will create a salient plan that existing governments will have no issue working with.

Even still, our work will continually push the envelope and will thus feel like pushing uphill. If we work together though we will write new laws which permit and include more materials and techniques.

In Conclusion

This architectural movement is led by the imagination and vision of a better future for the earth and the people living in it by continuing to innovate to ways of living lighter and with greater mobility. Our collective environmental impact and the Earth crisis we have helped create will be cured in time by the way we think, act, design and live.

We have new theories to expand on and currently have the unique ability to draw inspiration from collected wisdom of the ages made readily available by the advent of the internet.

Today we join the eternal pursuit toward betterment, preparation through examination and willingness to look with new eyes.

Luke Holden

Born in Santa Barbara, CA, 1982, Luke studied fine art and music from an early age, had parents who believed in crafting an education with Holden's passion and interests prioritized first. Luke was accepted into an advanced student placement program and attended Santa Barbara City College beginning at age ten. While attending SBCC Holden pursued majors in Fine Art, Theatrical Design, Multimedia Design, Drafting, 3D Modeling and Film. Graduated with an AA in Fine Arts and Humanities from SBCC in 2009.

After a stint away from the arts to undergo an apprenticeship with the United Brotherhood of Carpenters Union in 2010 Holden chose to dive back into fine arts bringing along a developed skillset from construction trade experiences.

Driven by purpose, Luke lives to leave a legacy of peace, compassion and creativity, committing his life to co-create better practices for a healthier planet and people. Holden feels a deep responsibility to contribute, aligning himself with especially proactive, loving, mindful and empathetic people. Through collaboration, passion and creativity, Luke aims to serve and uplift our planet and is always working to help ensure a more sustainable future for us all.

Luke is the Creative Director & Co-Founder of Lucidity Festival, and Founder & Operator of Branches Mobile Gallery.

http://lucidityfestival.com

http://branchesmobilegallery.com

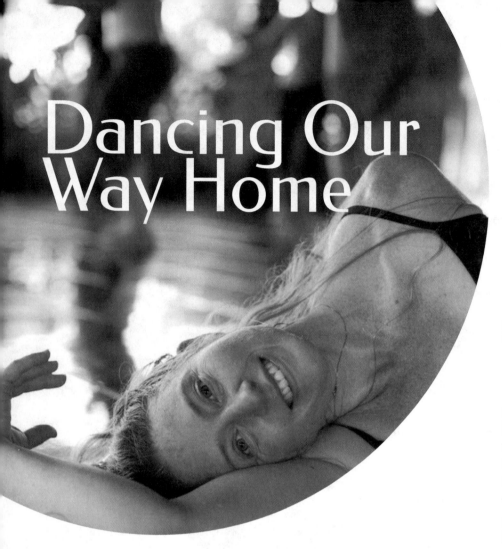

Dancing Our Way Home

Samantha Sweetwater

Do you dance? Quite likely, you do. Our movement is led by movers.

Those of us who dance know who we are. We have experienced Awakening(s) on the dance floor. We know our freedom in every cell. And, as we wake up, we find ourselves in membership - belonging and beloved in the communities we call home.

Because it leads us home within our bodies, on the earth and in relationship, dance is a foundational practice for *regenerative culture.* The simple act of moving in space and time continuously rebirths us as the people we are seeking to become. It tunes us with our souls - a universally available, uniquely affordable and efficient medicine for whatever ails you. Dance is so potent, in fact,

that it's worth unpacking the why and the how of Movement Medicine so that we can better engage it in our lives, communities, healing and leadership as we co-create the world we want and need.

Why Dance?

Creation Dances. The universe dances. When existence was born out of the great, undifferentiated field of the One, the dance was born. Since the beginning of time, being has danced with essence, Spirit with form - and so we exist. Chaos dances with order in systems evolution. Life dances with life to create more life. Our cells are in multiple and constant dances: blood rushes; muscles contract and release; mitochondria eat sugars and pump out ATP; the dance goes on. Rock dances sky as sky kisses rock. An-

cestors twirl in celebration of unborn children. Prayers braid unseen wholeness across space-time. Our souls are dancing the space where spirit and body meet.

To dance is to love our way into alignment with the motion that is life. Dance is a gateway into mastering participation and a pathway to engage our true nature as souls within the holographic universe. It's a physical main-line to the Force, a harmonization with the Way.

We are moved to dance because the universe moves. As we move, we lift up, we attune. We get really honest. We stop trying so hard to "be right." We let go. As a sacred practice, free-form dance is a training ground for aligning with the constant of change, for learning how to lead one's self, for embodying surrender, and for understanding relationships - with self, other, earth and the divine. It's a powerful, joyful way of connecting with personal presence, collective resonance and the emergent reality of a universe in constant flux at all scales. And, it's a way to pray. The human body becomes the vehicle of our prayers, anchoring and activating our cell-ves between Heaven and Earth. Good medicine.

Dancing In Community

Dance is the great unifier. Every person has a body, every heart a rhyme. Everybody wants the same thing - space and permission to be fully alive.

We see each other when we dance. A body in motion cannot lie. Dance provides a social/spiritual space where each person can go deeply within yet where no person can hide. Desire and emotion become transparent. Authentic community rises in soul and sweat.

Dance is social bonding glue that never gets us stuck. It reveals our true selves as it connects us nakedly with each other. It elevates us into ecstasy as it carries us lovingly through shadow and pain. It creates a safe space for inner process as it offers permission to be sexy, wild, and free. Animals, angels, vixens, devils, queens, kings, beggars, bitches… everyone of your inner and outer selves is welcome here. It all happens in the same space. And so, community is born. To dance is to celebrate the perfect imperfection of being a real, live human being together with other real, live, fallible, magnificent and ever-evolving human beings. In other words, it's the perfect foundational practice for a lasting culture of regenesis and peace.

How Do You Come to the Dance?

Here's my story.
I was born pretty awake. I often got lost. Dance was where I found my home.

At the age of 13, locked in the wrestling room of my high school that doubled as a movement space, dissolved in music and married to the sweat-stinking floor, I found my first true spiritual practice: Dance. I went there to get lost. I went there to get found. I went there to unbind myself from teenage projections, body obsessions, and the towering impositions of mother culture. I went there to move rage and to create beauty out of feelings too big for words. I went there listen. I went there to love. I found I could speak directly with the universe in a way that made sense to both body and soul. Over the years, through improvisation and choreography, I found myself again and again. I became a professional dancer and choreographer. But, this wasn't enough. Performance wasn't essential. What was essential were the ongoing, human experiences of connecting, expressing, giving and receiving love, and clear, courageous vision that was easy to bridge into life and form. So, I became obsessed with sharing my insights and tools with others. I wanted to share the home and liberation I continuously found. Eventually, I created Dancing Freedom, a dance practice for coming home. Now there are nearly 200 facilitators sharing that work around the globe. We are part of a larger emergence of embodied, regenerative culture rooting an ancient|future spiritual paradigm for our needs now.

You can begin anywhere. Just find a song and a floor.

Dancing helps. Life is motion. So moving your body consciously in space, time and relationship means getting more of what you want. Let your dance practice be a fun, uncomplicated and healthy way to engage consciousness and play in your daily life. Go to dance as you be and become more free and fully expressed as a human being. Use dance if you want more life.

Spirituality for All

Dance is not a religion. It has no dogmas, no boundaries. It welcomes all bodies, all gestures, all gods. Unconcerned with creed, color, size, shape, age or ability, it invites all. Equal yet different, unity flowers in diversity on our dance floors.

Thich Nhat Hanh said, "You have to do it by yourself; you cannot do it alone." We live in turbulent times. It's increasingly critical to have practices we can lean on when we need support in the context of radical chaos and change. As social organisms, we need a place to go for connection, reflection, elevation and mirroring. We need a safe space to

fall down. A good dance container can be all of this and more.

Dance is a Way of Authenticity and Joy. The Hopi Nation prophesy of the Eagle and the Condor states that when the masculine Mind of the North meets and harmonizes with the feminine Heart of the South, then will the Hummingbird medicine of Pure Joy be born. The balance of these three energies, says the prophesy, is what we require to save ourselves and our beautiful earth. When we dance, we are that Hummingbird in Tribe. We are that earth and soul- healing medicine of Pure Joy. When we put dance at the spiritual heart of the Village, we create the possibility that the improbable beauty of our co-creative, ever-evolving souls will, in fact, regenerate the world. Living the Way of Joy means reclaiming our innocent perception, our wonder, our passion, and the resiliency born of unedited love for everything and everyone. *This is the kind of energy we want as the foundation of our leadership* - a lightness that can carry the heaviness of our responsibility to and for All of Life.

And We Shall Be Free

I believe the dance can set us free. Dance is a medicine that invites us to simply be with the vulnerable reality of being a human animal. There's nowhere to hide. Letting go of hiding = opening to really living. In an embodied understanding of spirituality, "up and out" isn't meaningful. Everything meaningful, worthy, *life-y*, happens *here* - in our bodies and relationships. We are indeed the imperfect, always growing, earthbound people we suspect ourselves to be. We arrive more here, more whole, more real. This is courage. *We dance yes to life.*

When we move and just keep moving, we become like the water. We resist less. We unravel masks, armors and habituated responses and relax a little more into life. This makes us more available to love for other humans, the earth and All of Our Relations. Letting go of hiding = opening to feeling and connecting = being free.

No true freedom can emerge from resistance. *What we resist persists.* Dance teaches us how to love what is, how to embrace. I call this divine OKness. It's not about changing or fixing. It's about letting the full spectrum of life experience show you the way. Softening and opening to life, we dance in the fire of love that is at the center of our hearts. Freedom means embracing that fire.

We are fierce love warriors, yet we are soft. Those of us creating and leading the New Village are not here to create through old paradigms of power. Our power is like the dance. It is generative, flexible, responsive. It is collaborative, sensitive, connected. It is bold, daring and feisty. Our power is a power of we.

Living as Dancers

What kind of people do we want to be in the New Village? It seems we want to live as dancers. Healthy, evolving people dance with life. We respond. We take the changes. Sometimes we follow. Sometimes we lead. (We can feel what's right and when.) We "go with the flow." We know how to stay grounded. We learn quickly and integrate what we learn into our next steps. We dance with vision, challenge, opportunity and the transpersonal calling to serve while innovating new relationship paradigms, simultaneously integrating ancient wisdom with the best of new technologies. We celebrate our lives in time and tune with the greater whole. This is how we are weaving a new cultural blueprint in integrity with the knowledge that is already within our souls.

Dance is the next yoga - a spiritual practice for off-the-mat times. We can put dance at the heart of our village and know it will always bring us higher, and it will always help us find our way home.

Samantha Sweetwater

Samantha Sweetwater: Founder & Director of Dancing Freedom, priestess, coach, speaker, author & permaculturalist. Samantha is a much loved and sought after transformational guide who has touched tens of thousands of people on five continents with embodied facilitation at every order of scale - from the main stages of festivals and conferences to one-on-one healing sessions. Her meta-purpose: to ignite the evolutionary intelligence within humanity and the unique soul purpose of every human being towards a thriving future for All of Life. Her expertise and toolkit bridges somatics, yoga, dance, earth wisdom, whole-systems design science, ancient/ future spirituality and soul empowerment toward practical, poignant healing, leadership development and re-story-ing. She is the visionary creatrix of Dancing Freedom and The Peacebody School, Japan and author of the upcoming book, More Life, Please. Anchoring her life and leadership in the eco-centric rhythms of seeds, soil, water, seasons and the body, she lives and works at Seven Seeds Farm in Southern Oregon and is the Program Co-chair of The Global Summit.

http://SamanthaSweetwater.com
http://DancingFreedom.com

Onedoorland

SYNERGENIUS STEW + DIOS

OneDoorLand is an alchemical art sanctuary in Portland Oregon, home to a small community of musicians, artists, and tea monks. Over the past few years, community members have helped reshape two urban properties into thriving temples of *Life As Art*. Spanning almost 4 acres between two residential zones, these homes were once run-down rentals and now feature gardens, outdoor tea temples, meditation circles, a permaculture pond and waterfall (replacing the chlorine pool), yurt, and a geodesic dome "Holographic Light Sound Sauna."

As a media team, OneDoorLand creates inspirational videos, events, and websites to illuminate the artists and projects currently bringing more beauty, integrity, and magic to the world. The community continues to be a sanctuary for musicians and transmission artists, offering collaborative recording time in the home music studio, video artist portraits and documentary filmmaking, and occasional short-term artist residencies. The local community continues to gather for ritual music and art "vigils," using intermodal expressive arts to deepen psychological and spiritual growth as individuals and as a community. Every wall, door, and stairwell has been infused with art. Each mural is a ceremony of creativity, using the power of light, sound, music, tea, and presence to invoke a communal intention into the space itself. This process called "Portal Smiths" has been taken to other community temples, schools, and businesses in Portland as a way to anchor that particular group's intention into a creative project.

OneDoorLand is also a temple of self-cultivation featuring Heaven's Tea School of Sacred Tea Arts and the Gene Keys as tools for personal growth and transformation. Tea Monk Po, headmaster of the school, has transformed the upstairs of One-DoorLand Sanctuary into a profound library of aged Puehr tea, ancient sculptures and sacred art from China and Tibet, and a living temple for Tea as Vibrational Medicine. The tea served in this school comes from old arbor, wild leaves still charged with living vibrant energy, and create profound shifts in the human body. Heaven's Tea School of Sacred Tea Arts teaches the art of tea for inner alchemy, meditation, healing, con-

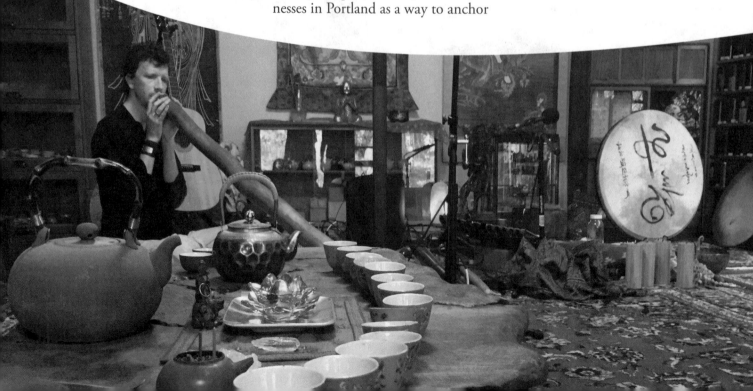

templation, and for use in art and music. Each week groups of people gather to journey with sacred tea, through both subtle energetic shifts and profoundly vivid openings, while always remaining balanced and integrated.

The Gene Keys Transmission is a contemplation tool, archetypal language, and engages a worldwide community in deep introspection, creative contemplation, and communal synergy. OneDoorLand uses the Gene Keys for personal and community emotional and spiritual processing, using this transmission of living wisdom to aid in illuminating the gift hidden within every shadow. Used as Archetypal Acupuncture, this book and its accompanied teachings provide individuals, relationships, and communities with profoundly practical and magical tools for identifying victim states and their potential higher essence.

OneDoorLand is now the media team for Gene Keys Network (http://genekeys.net): the Living Library and Community Garden for those exploring the Gene Keys around the world. OneDoorLand primarily focuses on video, music, and event production to illuminate the creative embodiment of this spiritual teaching in practical, inspirational, and magical ways. OneDoorLand hosts both local and virtual events and personal sessions for the Gene Keys.

OneDoorLand Synergenius Stew+dios was founded by Binah Zing, community leader and visionary artist. The dream will continue to live on in her honor, and is progressing towards becoming a non-profit educational center for Self-Cultivation through Tea, Qi Gong, and the Gene Keys.

http://heavenstea.com
http://onedoorland.com/

"Tea is a gateway to inner growth, healing, the subtle mysteries of existence and is a gift from nature to help us learn the flow of nature and the language of life."

Learning & Education

IF YOU THINK IN TERMS OF A YEAR, PLANT A SEED;

IF IN TERMS OF TEN YEARS, PLANT TREES;

IF IN TERMS OF 100 YEARS, TEACH THE PEOPLE. CONFUCIUS

Throughout our lives we are on a journey of discovery. From the moment we are born until the day we pass on, we are taking in information, trying to place it in context, always refining, always growing, always building on our experiences. If we are lucky, we are always connected to the experience of wonder and have access to education that cultivates wonder as a means to understanding the world around us.

Modern technology is changing the landscape of how humans learn. We are becoming more sensorial, more inter-relational, our complexity broadening as access to information is available at the push of one button. And yet, we still have much to learn in how to teach a whole being, how to honor the many ways a person learns, how to help humans learn what technology could never teach, what can only be known through observation, interaction, and through our intuitive connection to the natural world around us.

In the West, there is a movement to decolonize our education systems, bringing people back to their innate gifts, to their true potential which is not focused on output or test scores, yet on revealing the genius within each one of us.

Focusing on the development of holistic skills rather than measure of memorizing repeatable facts, whole being learning cultivates curiosity and imagination as the ingredients for success.

An integral approach to learning considers the many developmental lines in a human being such as cognitive, emotional, moral, artistic, interpersonal, spiritual, and works with the understanding of the many aspects of the "intelligence" of a person. As we honor the whole being as an individual, we foster creativity as a means to create the conditions for a lifelong love affair with learning.

An example of this kind of educational approach would be to start with the foundation of values for our students. Values-based education is a framework for modeling cultural values through the entire spectrum of the approach to education. This model of education does not insist on the "right viewpoint" yet engages inquiry as a means to greater understanding.

Also through methods like immersive education, hands-on learning, sensorial and experiential based practices, a person makes the shift from knowledge to wisdom, from conceptual understanding into embodied practice.

With the rapid shifts of a Global Village and the changing landscape of our times, it is vital to engage the voice of our youth, to bring them into the discussion and empower them at a young age to be proactive participants in creating the future they wish to see. In Village ethics however it isn't just about the youth, it's about intergenerational co-learn-

ing. The need to compel our youth to be involved in social change, to bring their innovative perspectives forward, supported by the wisdom of the elders and time honored understanding of what has already been learned, is crucial to all cultures and to our planet. Each generation has much to share. All of us have a purpose in the community.

Access to education, to mentors and teachers, to elders is fundamental in the development of a Global Citizen, capable of bringing one's unique gifts forward to participate in shaping our collective future. Regardless of the culture and its values for WHAT kinds of wisdom or skills to teach its youth, teaching Humanity to know itself, to be adaptable, and to recognize its place in the web of life is inherent to the continuation of our Species.

Jamaica Stevens

THE MIND IS NOT A VESSEL TO BE FILLED, BUT A FIRE TO BE KINDLED.

Plutarch

Metapoetics

Words are spells of magic cast to transform our World.
Words are pathways leading us home to the Village.

Through the Looking Glass

Language is one of the keys we can use to unlock creation and our creativity. And it can instead, become a force in service to limitation. In becoming conscious of our language, we take a step closer to mastering the human experience. The ancients have understood this kind of magic: the power to speak Reality into Being. It is important to remember that all language has been made up by people like ourselves in order to communicate and to express reality. This includes the naming of all plants and creatures as well as human-made creations, thoughts, feelings, needs, and discoveries... everything we humanly seek to share. Most of us filter our life experience through language, beginning in the form of thought and even when unspoken.

More than ever, we are in need of upgrading our use of language to increase our effectiveness and to become empowered as co-creators. Regardless of origins, language has a malleability that follows cognitive awareness. Metalinguistics can be classified as the ability to consciously reflect on the nature of language, our meaning making and creative choices. New modalities emerge in the realms of healing and personal growth through language. Compassionate Communication, "Conscious Language and the Logos of Now" and Neuro Linguistic Programming are a few of the fields of deep discovery into the power and transformational capacity of language. These are templates for healing unhealthy communication patterns, for upgrading all of our relationships through the artistry of our spoken and unspoken language.

The French word "portmanteau" means a suitcase with two compartments and was first used in this context by Lewis Carroll in his book "Through The Looking Glass." A portmanteau word is a synthesis of two or more words. An example of a portmanteau word is: MIRROR-ACLE. The mirror reflects the oracle and out flows a miracle! In the "***Bhagavad Gita,***" there is a Sanskrit word "samasika" that refers to the placing together of two or more words to form something new. Another Sanskrit word, "dvandhva," means a certain type of compound word in which each of the two parts are given equal significance and are unified by their relationship... including both the difference between the two parts as well as the unity: an example of this is UNIQUE AND EQUAL which synthesize as UNIQUAL and reflects a definable upgrade from unequal.

Re:Framing

Moving forward as an evolving culture depends in part on our ability to speak new realizations into form. We refer to coining new words as "word magic" and "metapoetics" as an expression of metalinguistics. There are many words being born into this fresh cultural milieu. These words speak what has often been unspeakable. Some words come complete with progressive meanings; these new and improved words construct a playful reality, that is ethical, flexible and continuously emergent.

Some of our old culture promotes individual ownership that may carry a sense of separation through our use of language. We have grown accustomed to words indicating possession such as My Partner or My Family. In fact, as we know in past times, wives were considered property of their husbands. When speaking about My Car or My Piano, as well as our relationship to the thousands of things in our lives, there may be unconscious clinging to the objects and the defined sense of self that ensues. In this case, I am suggesting that it is more important to be aware of the feelings around the words, than in actually changing them. On another level, terms such as My Land or My Property, equate to ownership that is merely a social contract but not necessarily a true assessment of our place on Earth. Let us strive to be Land Stewards rather than Property Owners-Honoring ourselves as We-Source.

GAIALOGUE

ALCHEMYSTERY

GALACTIVATION

INDIGENUITY

La Laurrien
with A. Keala Young

NOMADNESS

ECOCENTRIC

INTEGRATEFUL

KINFORMATION

Pattern Matters

There are hundreds of human spoken languages amidst the myriad languages in the patterns of nature and music creating a rich and stunning world when we take the time to listen. While pattern language has come to mean, "... a method of describing good design practices within a field of expertise," the term coined by architect Christopher Alexander and popularized by his book *A Pattern Language*, can benefit from a more expansive definition. To say it in a slightly different way, good design practice is expressed and can be learned through a language of patterns. Alexander's book inspired many other projects which utilize the living systems principle of pattern to illustrate the complexity of specific fields of expertise addressed through their own distinct language. Examples include Groupworks, supporting the practice of healthy and effective human group culture and the Food Forest Pattern Language, articulating patterns for designing and maintaining edible forest gardens.

Reclaiming

Without empowerment, language keeps us functioning within the old operating system distancing us from the freedom to express our unique gifts. By honing and refining our words, by becoming more intentional in what we say, we become active participants in human evolution, which is revolutionary~! Language is a powerful tool - let us wield it with mindfulness.

This is an invitation to be creative with the words you use. Imagine that there may be more than one message to hear in any given moment, and listen to the secret language of music and art. Read the landscape of our natural world and let the stories of the elemental forces in-form your profound relationship with the mystery.

EXAMPLES

Memes are the conceptual culture counterpart of genes. They are discreet units of cultural information: ideas and symbols that we pass around through talking, writing, gestures, and rituals.

Here are a few memes that change the game:

"Seeing is believing" has always been said to establish a rational. For something to be "true" it must first be seen.

When we shift into *"believing is seeing,"* we are affirming that our belief is a creative element that affects what we see.

Freesponsibility reframes responsibility, a word often heavy with implied obligation. By adding "Free" we gain "the ability to respond freely," a total upshift in our consciousness!

Illuminature reminds us to illuminate nature and to be illuminated in our inner nature.

Integrateful By blending gratitude with integrity, we receive a more potent meaning…to be fully integrated with gratitude and grateful for our integrity.

http://mirroracle.com

Tending the Millenial Garden

Environmental Education in the 21st Century

I could write about the real world challenges of connecting a generation of digital natives to the environment.

I could write about the desperation of seven billion souls at a crossroads; weighing choices that will forever determine the fate of humanity and the struggles we suffer on a daily basis because of it.

I could write about what it is like to be an Environmental Science and Education teacher at the turn of the Millennia implementing nature-based programming during a time of pseudoscience and a religious right backlash and the effect it had on both pedagogy and the profession.

But I am not going to do that.

Instead I will tell you that despite all the challenges that seem insurmountable, despite years of policies, procedures and priorities that shun real science and hands-on learning, I maintain my faith in our planet and in the creative potential of humanity.

After 20 years of doing this work, I actually have more hope than ever.

And it is because I am an Environmental Education teacher.

Growing up in the corporate nuclear family of the 1970's and 80's I moved 19 times in 18 years.

Mostly raised in suburbs of the great cities of the East and Midwest, I turned to the abandoned lots, stands of woods, city parks or creeks within walking or biking distance for some consistency in a world that made no sense to me.

In the dim afternoon light of a Massachusetts autumn I would lie amidst the leaf litter, the earth against my cheek and watch the birds above. When I was sad I would run to the nearest tree and climb into its arms and let my tears fall as the wind sang through the leaves.

I spent hours on end with my dog exploring woods and neighborhoods.

The quiet, the coolness, the trickling of a persistent and forgotten brook behind an apartment complex… were soothing in an ever-changing scenery.

It didn't matter what that "nature" was, how big or small… it all made me feel the same: connected and safe. These memories became me and were the first threads of a life of service to the planet.

The skill and adeptness I learned outside gave me confidence and a connection to wild places.

It was the environmental disasters of my generation that spurred me to true activism and to do as my first mentor told me: ***to find a place on the planet to love, to learn everything I could about it, teach everyone that would listen its secrets and never leave it.***

I have spent the last 25 years in the Santa Cruz Mountains doing just that and this is the essence of what I want to say:

Environmental Education is perhaps the single most important work of our times.

I believe it changes lives and therefore will save us all. I believe that connecting to nature, in however great or small a manner, is our collective responsibility and something absolutely achievable for every person. Attending the Bioneers conference in the Bay Area over the last few years, the oft-repeated message from the world's leading scientists, governmental leaders and NGOs is this:

We have all the money, all the technology and all the intellectual capital we need to turn this beautiful little ship around. However we only have THIS generation to do it. Therefore to me, this is the only work that matters. Where we begin is in turning children on to nature, sharing in the wonder of its systems, discovering together its resiliency and fragility and the inherent wisdom of the plants and animals.

It is in creating a "Sense of Place" at an early age that the seeds for global responsibility are planted. As children grow, exposure to the complexities of modern day problems will come. However especially at an early age, they must come to love their planet and connect to it in such a way that the very fabric of who they are is always in relation to their environment.

I have taught long enough to know that this is true. It paves the path for responsible global citizenship and leadership as adults. It is by design

that I focus on the children. I believe it is unjust to place such worldly problems they had no part in making upon their shoulders without infusing them with some semblance of faith and hope first. There is no way they can cope with the weight of solving the problems of energy, food and water without feeling empowered.

The first steps on the path to an environmentally connected and empowered generation begins in their homeplace and with their families, amidst a caring community of aware adults.

With each witnessing of the flight of an egret, or release of native steelhead into the river, with each track we discover, the earth is healed and the children are more prepared for what is to come. As the seasons change they witness the wondrous regenerative capacity of nature and her systems to adapt, to heal and to persist in life despite ever-changing circumstances. They observe, then immerse themselves and take these memories forward into their futures.

So I call you to action. Get outside. Look around.

What could you do?

If you have no space for a garden, plant a tomato on the balcony, or some herbs in a pot on the windowsill, or a fruit tree on the porch. Marvel with your child at the spider's web.

Feed hummingbirds. Make a bee garden. Plant flowers in a box. Figure out where the ants are coming from and why.

Open your heart to a local school garden and volunteer to tend it. Take the neighborhood kids to the riverside and poke around in the water… watch leaves float by. Look for fish in the water and animal tracks on the bank.

You don't need to do this full time, you know that.

It only takes a few minutes a day, or a Saturday afternoon.

It starts with your own children, your nephew and niece, your grandchild or the neighbor kids in your building. These experiences open your heart and theirs, empowering minds and spirits for the journey ahead; for all the overwhelming information that will come.

When we are inspired and we share that inspiration, the Earth is healed. We are healed. See yourself as an Environmental Educator and share yourself with the kids.

From my perch on the riverbanks of the mighty San Lorenzo River where I watch my sons playing, I know that there is no greater work, nor more rewarding profession, no greater path to a future for us all.

Please join me. I will be waiting at the head of the trail.

Kristen M. Rivers

I am a woman who is wise in years yet a child at heart. A Lifelong Learner and Educator, Permaculture Designer, Watershed Ecologist, Environmental Activist, Consultant, Entrepreneur and Mother.

True to my vision to be in service to the children and the planet I have taught Elementary and High School Environmental Science and Outdoor Environmental Education, Gardening and Permaculture Design with children and families for over 20 years. I had one of the first nature based Charter schools in California from 1997-2010.

I started my non profit The Artemis Project in 2012 in response to my community's request and was grateful to respond.

We collaborate with other organizations in providing opportunities for the development of feminine community mentorship and empowerment during our summer camps, after school programs and special events.

I live a joyful life of gratitude, purpose and meaning while raising my children in the Redwood forests of the Santa Cruz Mountains, along the dreamy and dynamic banks of the mighty San Lorenzo River. I dance every day.

http://artemis-project.com
http://facebook.com/ArtemisGET

Empowering Youth through Garden Education

Many social and environmental problems are caused by flawed education systems. Our education system encourages students to be competitive and value material wealth, which has created a huge disconnection in our society. We have forgotten ourselves, our place in the environment, and the things that really matter. We have forgotten what is really valuable and what is true wealth.

The next generation is inheriting many challenges and educators need to provide the knowledge and tools in a holistic education system to empower youth to make a positive impact in the world.

Teaching children about nature and to value the environment gives me hope for the future. I am the Sustainability and Garden Educator at Alta Vista School in San Francisco. I feel privileged to teach the subjects that I feel are very important and pertinent to this time. My program covers everything from soil science, garden ecology, and life cycles, to environmental issues, food crafting, and sustainable design. We teach permaculture concepts and their practical applications in a kid-friendly way. I create a dynamic experience-based curriculum that educates and inspires students to care for themselves, their community, and the environment.

In order for a child to develop an idea or value, they need to have a meaningful educational experience that invokes feelings and is engaging and participatory. Children need to feel invested and proud of their work. Education can be exciting and is not only about memorizing and routinely answering worksheets. The garden is a living laboratory that sparks curiosity to explore, discover and use critical thinking skills. The garden is abundant with interactive learning potentials for the sciences, math, social studies, art, humanities, health and nutrition, and more.

Our garden was transformed by students from an abandoned basketball court into a beautiful, diverse garden ecosystem. The students were involved in the entire process from inception, and I am continuously reminded of their experience because they speak proudly about this. Additionally, my students have designed and built a greenhouse, a water catchment system, living roofs, cob benches, ecologically designed playhouses, an herb spiral, and tipi gardens, while stewarding the garden ecosystem. Not only do they understand the purpose of these structures and their value in the garden, but they also learn that it takes teamwork to turn these ideas into reality. Students demonstrate an understanding of concepts and values through application in the garden.

Growing one's own food allows a child to understand their place in the world. This is a foundational piece that is missing in the mainstream education system. Children begin to understand this concept by simply caring for a plant from seed to plate. After they learn how to plant a seed, they care for the young seedling and watch it with fascination as it grows and changes. They transplant their young plant into the garden, where they continue to tend to it, and observe how the plants interact in the garden ecosystem. Students learn how to appropriately harvest, cook, and celebrate the delicious flavors and abundance. Once they have gone through this process, they understand where food comes from and increase their appreciation for food, while showing compassion for living things.

Children learn self-awareness tools to understand their place in the ecosystem. In every class, students take a few minutes to observe and interact with the elements of the garden. Their senses come alive as they smell, touch, listen, taste, and observe their environment. Students come together and share what they have learned directly from our environment. As a gardener myself, I have learned that the garden is my greatest teacher and my students are learning that through this exercise. By consistently practicing observation, students become aware of their place in the ecosystem through witnessing the cycles and relationships contained within the immediate ecosystem.

Youth feel empowered during garden class because they are entrusted with responsibilities, while working collaboratively on projects. They learn how to take care of themselves by growing food and preparing healthy meals. Children are empowered to choose which garden activities they would like to participate in, and this gives them a sense of responsibility and freedom. My students are enthusiastic to tend the compost pile, weed, prune, collect seeds and make a salad! During these activities, students learn life skills such as communication, flexibility, and patience. Additionally, the garden provides a safe space for making mistakes and problem solving. By making mistakes and working through them, youth understand that they are growing and learning. These tools and skills are essential for a successful, happy, and healthy life.

Garden education helps children connect to their core values and how to make a positive difference in the world. It can be overwhelming, depressing, and terrifying to learn about the current state of the environment. However, children can be leaders of change when they have a strong foundation of values and ideas, and feel empowered. Through service and stewardship, my students have not only transformed our school garden, but also served as leaders in designing and building a community garden at the local library. This community service project shows that our garden education program has clearly influenced their decision to create more gardens for others to enjoy. They understand the impact humans have on the environment and are creating the change they want to see in their neighborhoods and communities.

It has been a joy to work with children in the garden and facilitate their growth and understanding. I am constantly inspired by my students and the garden, and I channel this inspiration to provide holistic education opportunities that interweave academic subjects with life skills. Garden education is not the complete antidote to a flawed education system, but can be a part of an integrated educational structure. In fact, garden education programs are being integrated into schools nationally and internationally. The Edible Schoolyard Project has created a network of garden classrooms and resources for schools and educators. There are also useful, accessible, and widely accepted curriculum resources, such as LifeLab and Education Outside. The garden education movement is spreading and is positively influencing future generations to create a regenerative and socially just world.

Marissa Weitzman

Marissa has been working as a Sustainability and Garden Educator for four years and is currently the Sustainability Director at Alta Vista School in San Francisco. She develops and teaches an interdisciplinary curriculum with hands-on educational activities. While attending San Francisco State University, she learned about the food system, worked at farmers markets, and volunteered at Haye's Valley Farm, which made her realize that edible education is essential for a happy and healthy life. Marissa is passionate about restoration ecology, food systems, community, and permaculture and is particularly interested in where they all intersect. While in Guatemala, she received her Permaculture Certificate at the Mesoamerican Permaculture Institute. She has a Bachelor of Science Degree in Environmental Science with an emphasis in Natural Resource Management and Conservation and minors in Biology and Geography from San Francisco State University.

http://altavistaschoolsfsustainability.weebly.com

http://altavistaschoolsf.org

Generation Waking Up

I.Am.Awake

"It's (3:33) in the morning
and I'm awake
because my great great grandchildren
won't let me sleep.
My great great grandchildren
ask me in dreams
what did you do while the planet was
plundered?
What did you do when the earth was
unraveling?
Surely you did something
when the seasons started failing?
As the mammals, reptiles, and birds
were all dying?
Did you fill the streets with protest
when democracy was stolen?
What did you do
once you knew?"

Drew Dellinger

Rising from the ashes of a torn-up world, alive, alert, awake, and enthusiastic, an emerging generation is taking the stage with a bold vibrancy unlike anything previously imagined. This rising tide of conscious beings knows there is a better way to live, a brighter picture to paint, a happier song to sing, a more beautiful story to tell, and a more magnificent dream to weave, than the one currently imprinted in our collective framework. We have many names, and the one that seems to most accurately describe us is *Generation Waking Up (GenUp)*. We are awakening on planet Earth at this time to co-create a world that is "Thriving, Just, and Sustainable," that works for all, no compromise.

We are alive at a great time of transition, known to some as the *Great Turning*, the *Shifting of the Ages*, the *Collective Awakening*, the *Quickening*, and the *Age of Aquarius*. This unique moment provides countless opportunities to genuinely confront the multitudinous challenges we face, and to sublimate and alchemize them into solutions amidst the turmoil. I call this "Paradigm Composting," transforming what already exists into fertile soil from which an ideal alternative may bloom.

Through these last few decades, we have seen Generation X, Y (why?), and the Baby Boomers, amongst others. What defines the current earth generation is our easy access to catalysts of collective consciousness expansion. We are so connected! Through the internet, community gatherings, and access to more global information than ever before, we have the tools necessary to spread messages globally, to unify, and to become part of massive collective movements. Catalyzing collective awakening to our current global situation, empowering youth to become global citizens, leaders, and change agents, and mobilizing the people across issues, geography, and all lines of difference, is where "GenUp" comes into play.

GenUp is a global campaign that awakens, empowers, and mobilizes youth to generate solutions for our **current** crisis. GenUp was founded by Joshua Gorman, who, as he explains, grew up "in the heart of mainstream, middle-class American culture," strongly aware of the challenges present on planet earth. After finding a deeper connection to his life purpose while in Hawaii, he began to see that another way is possible. Starting small and expanding, Gorman began to create a curriculum for the "WakeUp" experience, which then became the foundation for *Generation Waking Up*. He and his beloved partner Cherine Badawi, an amazing Egyptian revolutionary, are working on launching a campaign called "ReGeneration," which will offer the *Generation Waking Up* curriculum as a full-time class in schools around the world. GenUp was derived from the Pachamama Alliance's "Awakening the Dreamer" symposium, and is an active organization that trains people, especially youth, to become empowered leaders and facilitators of a presentation called the "WakeUp." **The "WakeUp" experience has been offered everywhere from high schools to universities, transformational festivals (Lucidity, Enchanted Forest, Envision) and the national Rainbow Gathering, to the People's Climate Change Conference in Lima, Peru, in December of 2014.** The "WakeUp" has been presented in over 13 different countries, including the United States, China, Mexico, India, Kenya, Australia, Egypt, Germany, Brazil, and Peru. Leadership trainings occur frequently all over the USA and beyond. Thousands have experienced the "WakeUp," and hundreds are trained as facilitators.

The "WakeUp" experience is an interactive, inter-generational transformational discussion council. It includes a multimedia presentation with videos, informational slides, and a lot of group activity and conversation. The experience is adaptable depending on the audience. Some audiences really need the Wake Up Shake Up treatment, being explicitly shown the devastating information of our real world situation. Other audiences already know what's going on, and are much more interested in jumping into **solutionary** council.

A partner organization that is facilitating Wake Ups is *I.Am.Life*. *I.Am.Life* is a tribal collective based on inter-generational connection, youth empowerment, and connecting North and South American communities, with a special focus on Indigenous tribes and connection with Indigenous youth. *I.Am.Life* hosts outstanding music and art focused community events, creating spaces for inter-generational, intercultural melding of the practical and the spiritual. Everything about *I.Am. Life* focuses on transforming the individual, embracing community, and transforming the planet in the most respectful, responsible and healthy ways possible. *I.Am. Life* has been presenting the "WakeUp" with a solutionary focus, discussing ways to implement and actualize solutions to the world's cry for help. *I.Am.Life*, beyond presenting "WakeUps" at transformational festivals and elsewhere, is working on a youth empowerment council at Esalen Institute. These are transformational journeys that take youth into the Amazon with the Pachamama Alliance, to build a temple in a newly-emerging ecovillage north of Los Angeles.

Generation Waking Up and *I.Am.Life* are just two examples of amazing organizations that are actively engaged in creating a better world for all beings. The more we present these "WakeUp" experiences, the more our vision of what's possible gets painted. This is not just a dream, but a reality we can create together. We are the ones we've been waiting for, and now, there is no more time to wait. We are awake, and it's time to bloom. As they say in the Lakota nation and in sacred ceremonies worldwide, Aho Mitakuye Oyasin, to all my relations.

As I like to close each "WakeUp" with personal mission statements, I invite you, dear reader, to ask yourself, write down, and share with another, this question: *"What action will you take, in the next hour, day, week, month, and year, to help bring about a world that is Thriving, Just and Sustainable, the better world that your heart knows is possible?"* If you need help, take a piece of paper and make a list on one side titled: "What Makes Me Come Alive," and on the other side, "What the World Needs," and draw lines to find where they connect.

Onwards, to a thriving, just and sustainable world that works for everyone.

Robin Leipman

Bloom (Robin Liepman) is a bright budding beam of luminosity dedicated to dream-weaving, co-creating and pollinating a vibrant world that is thriving, just, sustainable, harmonious, and equitable. Bloom grew up in San Luis Obispo, California, graduated from UC Santa Cruz in 2012, and began exploring North, Central and South America. He is half of a musical activism project called "Compersion," and is an ambassador of I.Am.Life, Generation Waking Up, Project Nuevo Mundo, Quetzal Shipping and Trading Co., and Green New World. He taught three classes as a student at UCSC (including one he created, called "A Sustainable World: Where We Are, Where We're Going, and What We Can Do"), and continues to facilitate workshops and play music at transformational festivals, conferences, and beyond. He is expressive, musical, and can often be found freestyling and generating group singing experiences around the sacred fire.

http://iamlifeproject.org
http://generationwakingup.org
http://compersionmusic.com

Ethical Foundations of Permaculture

Permaculture (permanent culture) is a whole systems approach to conscious design that creates abundant, low input, high output lifestyles while considering optimum health for the earth, its people and future generations to come.

At the basis of the Permaculture design methodology is an ethical foundation that guides the process of conscious design decisions. This ethos guides us to the creation of a regenerative society rather than the degenerative one that we inherited.

The Permaculture ethics are Earth Care, People Care, and Fair Shares (traditionally known as Fair Share and set limits to consumption and population) and Transition (inspired by the *Transition movement* and added by the west coast in the 2000's).

Earth Care

How can we consciously work with the earth, focussing on regenerative rather than degenerative resource use.

- Respect the Value and function of all Earth's Creatures
- Relearn nature's language
- Reconnect with the environment

People Care

In what ways can we support ethical treatment of all people on the planet and also consider the care of ourselves so we can be of deepest service in our communities?

- Respect that all people have value
- Recognize the collective wisdom
- Remember to nurture yourself and others

Future Care

What can we do to leave a regenerative, fruitful planet to those who will come after us?

- Respect and generate abundance
- Reinvest into Earth care and People Care
- Rethink Consumption patterns

Transition

How can we use the remaining of our unsustainable resources to create a sustainable or regenerative society? (for ex. what better use of the last bits of oil that to create thriving permaculture systems that do not rely on such resources?)

- Respect that everything is constantly in transition
- Realign your perspective
- Realize the impact of your actions. (Be a model for change)

The ethics can seem general and abstract, which is something I always struggled with.

When I first came to the path and these were introduced to me, it was hard to relate to as they felt intangible and so broad. We all relate differently to these ideas as we all have unique situations, stories and perspectives. They can be dissected in many ways and be taken on in many forms.

As an artist, and someone who is very interested in symbolism, I created symbols for the ethics, making them more tangible for me. My friend Shannon Reinholdt took my rough drawings and made graphics with them, which are the ones you see here.

At the time, there were images for the ethics floating around, but I didn't resonate with them. Having the right representation of a concept is important in connecting the viewer with the message in an effective way. Sometimes, these symbolic portrayals are known as sigils. You may have seen a number of these throughout the book so far. They are a powerful way of directly linking

with an idea or concept, especially for more visual, artistic and existential learners. The Ethics sigils were created with hope to emanate the ideas effectively and resonate more universally with people.

Upon the completion of the Permaculture design certificate course, I knew that I wanted to share what I had learned. Already working as a creative facilitator in other realms and spending time studying with a teacher who had a dynamic style, I decided I wanted to start creating learning tools that could be used to assist people in grasping permaculture and support me in gaining confidence as an educator.

At this time, my friend Heather Lippold, came up with ideas for using the graphics Shannon made and created prototypes.

She ended up making sets of 4" tokens using a C & C machine. These were formed from a recycled dresser which I loved because it really embodied what we were teaching. We painted the tokens together and had text placed on the back which described them. in classes they are used to play games, form dynamic discussions or for integrative activities. Sets of 8" tokens were also made to be used as visuals for larger groups and using the scraps and 1" tokens created that would be used as give-aways to students.

Spending some time trying to figure out how to make these available for anyone, we worked with Shannon to make graphic representations of the tokens that can be downloaded for free or sent to an on demand printer for a minimal fee. We worked with a local *maker space* and small local business to have the small tokens laser cut. To date over 500 sets have been given out in the community, helping people to be reminded of conscious decision making in consideration of the earth, its people and the future generations.

Being raised in the traditional education system I struggled because only basic styles of learning were represented. When I was unable to learn in the rigid way being presented to me, I felt disempowered and unintelligent. Once introduced to a variety of other styles of learning, I started to realize that I had something greater to offer the world, and now have become a teacher, something that I would have never thought was possible. From my personal experience, I recognized the importance of education being accessible and able to reach a broad audience.

Making the ethics tangible in this way reaches many learning styles; visual, linguistic, interpersonal, intrapersonal, physical and existential to name a few. The more approaches to learning we can incorporate in the way we share and express, the more information can be accessed and received by the greater whole. The unfolding process of completing this project has formed an intimate relationship with the ethics in my life, facilitated the formation of multiple creative and collaborative community minded relationships, and enabled the greater community to access learning in a unique way. The journey has been a true embodiment of the Permaculture ethics in action.

Kym Chi
Vision, Creation & Text

Kym Chi is a dedicated advocate of Earth stewardship, people care and regenerative action for future resiliency. Her main efforts are as a creative facilitator, artist, gardener, healer and community builder. To date; Kym has received 3 certificates in Permaculture Design, and completed a variety of advanced trainings. She is currently completing a Diploma through the Permaculture Institute.

http://gigglingchitree.com

Heather Lippold
Vision & Creation

Heather Lippold is a visual artist and graduate of Emily Carr University. She is a maker, working primarily in ceramics and often finding her practice in constant flux. Her work is inspired by connection within her environment.

Shannon Reinholdt
Graphic Designer

A connoisseur of life, Shannon has a true flair for visual design and aesthetics, dedicated to everything that she's involved in. As a graphic/ motion designer and a lover of nature, Shannon is ever-expanding her awareness and insight into herself and the world around her.

Visionary Permaculture

by Gaiacraft Stewards

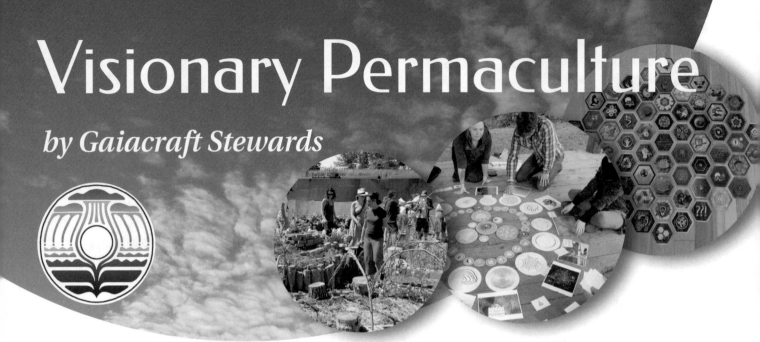

The most important application of permaculture ethics and principles is to the self, through a process of self audit of our needs, wants, dependencies, creative and productive outputs and byproducts of our very existence. Getting grounded in this way is the start of a personal retrofit or redesign process which does not require that we wait until we own land or are with the right crowd.   David Holmgren

Visionary Permaculture

Visionary Permaculture is a post paradigm, adapting and evolving to the ever-changing times. It speaks to the living and leading edge and expresses in multidimensional ways that are accessible, relatable and tangible. We sow seeds for harmonic abundance, imagining a future humanity reintegrated into the global ecosystem and being in balance with balance. Potentized by the possibilities, the present moment offers opportunities for the redesigning of our world. Permaculture prescribes a deep mapping of the elements and relationships already in place. This leads to an integrative redesign focused on regenerating ecological communities.

This starts with you. How can you get conscious about your body's strengths and weaknesses?

Your ideal diet? Exercise routine? Continuing education? Ways to cultivate emotional health?

Your home life is next. Can you redesign your home to support your diet? Exercise regimen? Education and reskilling? Conscious relating with family, friends and community?

How can you create and maintain positive relationships with others? With your paid work? With your creative contributions to the world?

Can you supply any of your needs by growing plants or raising small animals in your home, or the area immediately around it?

Once you have anchored your inner zones, it's time to play a more integrative and contributing role in the outer world.

Where are the nearest parks and public spaces, farms and farmers markets, or family-run shops stocking ethical local supplies?

Where can you access community groups, allied organizations, educational facilities and government bodies?

How can you make a conscious contribution to your neighborhood and community?

Are there ways you can become a more intentional consumer who supports local ethical food, products, and services?

Could you start, or join, any movements for Placemaking, Permaculture or Transition in your area?

With an empowered community you can begin to move out into the wild naturescapes. How can you and a group, or community of others, explore options for the healing, regeneration, protection and conservation of the remaining forests, plants and animals that live in the wild spaces closest to where you live?

Integral Permaculture

Taking the time to answer these questions and expanding your thoughts, feelings and curiosity is an opportunity to create a new map. Applying the concept of Integral Permaculture you can map the relationships of your inquiry into the four quadrants of the Integral framework. How does your visionary permaculture life manifest on the interior and exterior, from the perspectives of yourself as an individual and also for your communities, and the world?

INDIVIDUAL

I
Subjective

thoughts, emotions, memories, states of mind, perceptions, and immediate sensations

IT
Objective

material body (including brain) and anything that you can see or touch (or observe scientifically) in time and space

INTERIOR

EXTERIOR

WE
Intersubjective

shared values, meanings, language, relationships, and cultural background

ITS
Interobjective

systems, networks, technology, government, and the natural environment

COLLECTIVE

Many people think of Permaculture as it relates primarily to the lower right quadrant and the world of systems such as food, water, housing, and recycling waste streams. While those are the highest profile areas of engagement and opportunity for redesign and retrofit in our world, being whole systems designers we need to design holistically for all quadrants, levels and beyond.

Gaiacraft Education

A conscious permaculture network is forming which freely shares a number of open source learning and teaching tools as well as ongoing experiential opportunities for personal and professional development. Gaiacraft education is an expression of an emergent synthesis in learning. Representing the foundational lineages of permaculture while opening the way for a 3rd generation of teachers, Gaiacraft creates a space for novel modes of learning which open up the potentials for dynamic models to emerge. This is part of permaculture's new edge!

Gaiacraft is a planetary collective of Permaculture teachers and plant people facilitating cultural transformation. Beyond a plethora of free high resolution educational materials, Gaiacraft models the new invisible college model of participant driven, self-directed learning pathways. In addition to teaching and consulting, our collective teams up to offer ecological design and consulting, event production, educational installations and all manner of media.

Delvin Solkinson

A plant path poet and gardener from a tiny coastal village the Elphinstone rainforest, Delvin is a student of permaculture. Volunteer service supports Gaiacraft, the Galactic Trading Cards, Projections, and CoSM Journal of Visionary Culture.

A. Keala Young

A whole systems designer and healing artist with roots and wings in the bioregion of Cascadia, Keala is a world server and learning consultant serving individuals and communities for transitioning into a thriving Planetary Era.

CultureSeed

Thriving Global Communities

Synergistic community represents the fruition of human achievement, affirming our sense of purpose on this planet. Brent Cameron

CultureSeed is an educational non-profit and network, born of powerful vision and the urge to assist in birthing a New Reality. Many of us easily recognize what is not working in our world. Now is the time to turn our eyes, hearts and minds toward the magnificent future we can be creating. In reviving ancient Earth technologies, as well as developing new strategies for thriving, the seeds of our New Culture are being planted and nourished.

Mission

CultureSeed education is Visionary and Evolutionary. Our Vision and Mission support Thriving Global Communities. This refers to humanity, ecology, regional resilience as a path of healing, and conscious creative culture as a means for transformation. We are committed to strengthening existing networks of community, while teaching skill sets for thriving. Our CutureSeed mission is built upon an ethical foundation honoring interconnectedness in all our relations.

A Living Network

CultureSeed is initiating and prototyping programs in the Atlan Ecovillage Community, near White Salmon WA, in the Columbia Gorge. We share this vision with other like-minded ecovillages: to form a network with the potential to offer innovative and solution-based learning that can be applied to communities around the world. Our learning events and programs are adaptable while providing replicable modeling of new practices as they are emerging. We honor the preservation of healthy existing cultural values as we forge ahead to develop new culture and we encourage the spirit of exchange and experiential diversity in the various community centers.

Mentors

CultureSeed is growing a network of mentors, teachers and facilitators as well as life-long learners of all ages. We support learning, based on mentoring relationships and our philosophy is deeply inspired by Brent Cameron's offerings to alternative and self-directed learning through SelfDesign.

Community

CultureSeed is about fostering Community and is dedicated to including group aspects into every learning opportunity. Healthy home-cooked meals, gratitude circles, attunement practices, yoga, meditation and dance may be part of any Culture-Seed experience. Whether the course is Wilderness First Aid or Partner Dance, we look for ways to weave in personal growth and group interaction.

124

Invitation

CultureSeed invites you to share your vision of Thriving Global Communities. Which elements of our culture are thriving and which are in need of improving? By becoming active participants in a process of discernment, our future success is insured. Additionally we are learning how the practice of collaboration has more solutionary capacity than what we can accomplish individually.

When we apply this to diverse communities of the natural world, we see an enhanced concept of what is possible on this planet. By collaborating with nature, we can solve problems that threaten our existence. Respecting indigenous wisdom, we learn from cultures that have a track record of beneficial relationship with the natural world. We seek to make this wisdom visible through practice, and living life as art.

Our invitation is to live self-empowered lives as lifelong self-directed learners, sharing the vision of supporting thriving global communities. We are all in this together. The more we learn, and the more we share, the more capable we become as a global community. It is time to put our best ideas forward and allow them to evolve in the mix. Let's figure out together, what really works!

Our purpose is to support the success of your programs, events and centers and we encourage you to make suggestions through the form below for learning you would like to lead or see happen at any location where you are building community and culture - join us in cocreating the future!

Propose learning programs, events and venues at:

http://cultureseed.org/seedthefuture

CultureSeed is organized as a 501(c)3 Non-Profit Charity (pending IRS approval)

Regional Resilience

IF OUR GOAL IS A PEACEFUL, JUST SOCIETY, SELF-RELIANCE AT THE HOME AND COMMUNITY LEVELS MUST BE A CENTRAL FOCUS OF OUR LIVES. Ben Falk

Regional Resilience means different things to different people. To some it is based on a region's economic adaptability, or whether its systems of infrastructure can attend to the changing needs of a populous, its biological capacity to adapt and thrive under adverse environmental conditions, or perhaps a community's preparedness to respond to a crisis.

One way to approach resilience is to consider whether the needs of people local to an area can be met without dependence on outside resources, regardless of the conditions of economic hardship, environmental crisis, or infrastructure collapse. Resilience indicates that a community's process of rebound, adaptation, and recovery from any systemic or sudden shift can sustain the means of survival for its inhabitants.

Consider your region, your city or town. Would you be able to say that you feel your area to be resilient? Do you feel secure in your knowing of how to navigate a crisis or large-scale shift? Do you feel that the system you live in has considerations to provide services and resources regardless of challenges it may face? Does your local government

have a strategy in place to attend to a disaster? How do we fortify our communities to be sustaining regardless of shifting times and the access to the comforts we have all grown so accustomed to? Without living in fear, yet coming from empowerment, it is vital that each one of us consider these questions.

Resilience can only really come from localized, community-led initiatives that include multi-stakeholder perspectives, linking grassroots with civic leadership to implement strategies that can attend to the particularities of that region, of its unique environmental and economic considerations. This requires inclusion of all peoples, not just those of privilege and also of the bioregion itself, not only human life.

Communities need secure food systems and watersheds, strategy for meeting energy, infrastructure, transportation and housing needs, access to medical care, and most importantly strong leadership that can attend to an unpredictable shift. Strong leadership bolstered by a populous who is interconnected and participatory in the well being of its neighborhoods, its village, its ecosystem with access to information, to solutions, to strategies and to the resources to make changes through the existing infrastructure are all essential.

One step further would be to pre-emptively avert a negative reaction to crisis by motivating a populous to choose to simplify and make changes that create thriving conditions for a community regardless of calamity. How can we use the same principles of resilience and make choices to create positive impact, to vision and dream together in order to become inspired by what is possible? What can we learn from the threat of crisis that can inform us of pathways to become more engaged, more em-

powered, more thoughtful of our systems and strengthen where we see weakness? Really the question is, how do you WANT to live?

We can choose now to use the systems that are dependent on finite resources in order to intentionally create THRIVING local areas that bring the wisdom of time-tested living principles together with the best of technological advancement and propel us as a species into a future that is more harmonious with our natural environment, more interconnected with each other, built to last for all generations, not only a few.

Concepts like reskilling, neighbor to neighbor trade of services, community gardens and tool sheds, food banks, protecting local watersheds, supporting farmer's markets and co-ops, urban permaculture, upcycling reused materials, bike friendly transportation routes, shared housing and intentional living communities, shopping for locally sourced goods, as well as implementing green energy strategies that reduce our dependency on current resources, preserving green spaces and ecological diversification, creating public convergence points to share information, build relationships, and strengthen mutuality are some approaches to building resilience.

A dream some in my community have is to open land-based "retraining centers" focused on intergenerational, immersive whole systems education where one could learn how to grow food, work with tools, build infrastructure and housing, apply innovative science to energy or water needs, be trained in basic first aid and holistic health, along with leadership and communication skills, practices for personal balance, and a deep ecocentric reconnection.

What if the threat of effects from peak oil, natural disaster, environmental refugees from drought, economic collapse, pollution of water sources, threatened food security, and more all became fuel for humanity CHOOSING to change the way we live on this planet? What if in our moment of crisis we saw a widespread movement of efforts made to create balance, abundance, and connection in local regions? What if those local regions shared what they were learning, their successes, approaches, best practices and mentored other areas to support more communities engaging with direct action? What if we looked these challenges right in the eye and unified across dividing social lines to come together and create the conditions for a renaissance HUMAN FLOURISHING? Now is the moment where possibility can become realized through those who open their hearts, use their intelligence, eliminate separation, and choose to make it so.

Jamaica Stevens

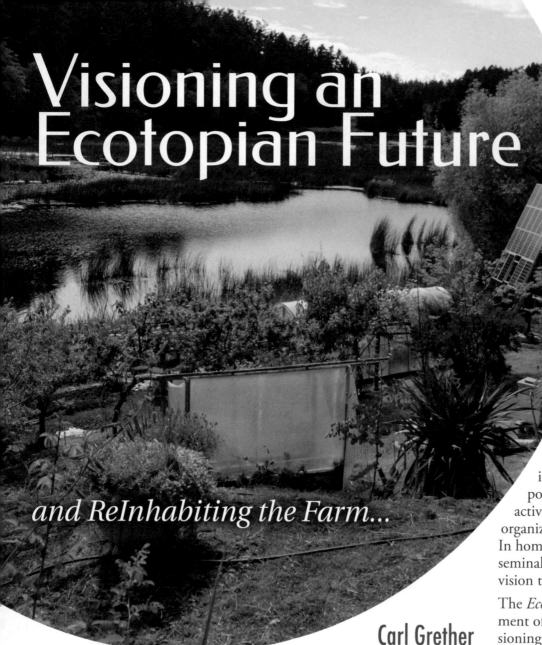

Visioning an Ecotopian Future

and ReInhabiting the Farm...

Carl Grether

by cultivating a complimentary civilization of edible cities and eco-villages; restoring urban environments and responsibly stewarding what are now huge swaths of industrial monoculture.

Over the course of consulting with countless scholars, policy experts and frontline activists, farmers and community organizers, a vision has taken shape. In homage to Ernest Callenbach's seminal novel **Ecotopia**, I call this vision the *Ecotopia Project*.

The *Ecotopia Project* is an experiment of the imagination; a re-envisioning of modern civilization based on the principles of **Permaculture** and **sustainable living**. It integrates emerging technologies and social innovation with ancient indigenous practices and agro-ecological research to build a regenerative future for humanity.

Imagine a planetary network of evolutionary eco-villages distributed across the barren tundra of industrial farmland, fusing the neo-tribal organizational structure found in Burning Man camps and the Transformational community with the ethics and methodology of Permaculture. Its core principles include radical freedom, self-expression and self-reliance in the context of a long-term, year-round permaculture civilization.

Over the course of making two feature documentaries about transforming our food system – *Edible City* and *Occupy the Farm* – I was compelled to delve deep into the big picture of modern agriculture, and it's not pretty. Multinational corporations have taken over the vast majority of agricultural land in the United States (and throughout much of the world), and are implementing incredibly destructive practices that are based on short-term profit margins and which ignore the long term health of the biosphere--leading to devastating environmental consequences.

Within our lifetimes may will see a total collapse of our agricultural foundation--and the surrounding ecosystems--unless we can make change.

Starting within our cities and moving out onto the land, we are being called upon to grow a symbiotic civilization rooted in the healthy substrate of *regenerative agriculture* and responsible environmental stewardship. Learning to live better

Villages could be created in a matter of days with shipping container homes, geodesic domes, and innovative eco-pods that include all the necessary tools, equipment and supplies to create a self-sustaining community: bicycle and wind generators, solar panels, solar ovens, shovels, seeds, staple foods, etc.

Communities can self-select themselves into autonomous tribes where a unique culture can be cultivated, with environmental stewardship as the core ethic. This project provides a refreshing openness to radical autonomy and individual self-expression beyond what is possible within the modern urban context. Evolving culture through art, music, ritual and the myriad neo-tribal social innovations that make a week on the playa feel so magical and so much like home.

Supporting these communities would be a network of traveling performers, artists and healers; a corps of Permaculture Jedi to help build, maintain and guide the settlements, along with a robust manufacturing base of the sustainable technologies needed to build and maintain the network.

This is a massive leverage point that could address a host of problems that plague our culture.

Implementing the Ecotopia Project on a massive scale would simultaneously create millions of jobs and avert environmental disaster, while creating a vastly more fulfilling and sustainable complimentary civilization–and we could have a roaring good time while we do it.

Because it all starts with groups of friends joining together to create their own permanent Burning Man camp in the real world landscape that most closely resembles the Playa; the flat depleted soil of industrial farmland. A blank canvas for human imagination and ingenuity. Regenerate the land and practice Burning Man year-round with a healthy dose of Permaculture to balance out the partying. We could call it Permaculture Burning Man. Sounds pretty good to me.

Exploring this imaginary realm is a lot like putting together a puzzle, identifying the different elements and putting them together into larger and larger pieces. The big pieces that my research and imagination has aggregated together are as follows.

THE 5 PRINCIPLES (a beginning...)

A hybrid of the 12 principles of Permaculture with the 10 principles of Burning Man

1. *Land Stewardship:* The health of the Biosphere is the top priority
2. *Reliance on Local Resources:* Minimizing the distance which goods travel
3. *Interdependence and Cooperation:* Developing systems that recognize life as an interconnected whole. Where human communities work cooperatively within their community, with other human communities and the many non-human communities to which they are connected
4. *Freedom of Expression:* Within the context of serving as Environmental Stewards, and providing for themselves, humans are allowed the space for the fullest possible expression of their true nature and creative passion
5. *No Waste:* Everything is recycled, re-used or re-purposed. Nothing is built that cannot be recycled or composted

The Pieces

Land Reform and Population Redistribution: Developing a framework and infrastructure to shift the use of agricultural land away from large-scale industrial enterprises towards small-scale sustainable eco-village settlements. Thereby decreasing the reliance on fossil fuels, developing regenerative practices of land stewardship and shifting population density away from overcrowded urban centers. This could be accomplished through a combination of NGOs, cooperatives, charities, non-profit and for-profit enterprise and government policy.

Sustainable Technology: Developing and promoting the use of innovative sustainable technologies that decrease our reliance on fossil fuels and provide the tools for a regenerative relationship to our land base (i.e. solar power, efficient building design, integrated recycling systems, etc.). Drawing on the work of Buckminster Fuller and others.

Building a Template: Drawing on the principles of Permaculture, Agro-Ecology, Biodynamics, Economics and whole systems design to build a template for eco-village design based on a relatively normative landscape (i.e. relatively flat agricultural land with access to irrigation water and/or abundant rainfall) that is both specific, replicable and adaptable to different climates, growing conditions and supply/demand fluctuations.

Education/Mentoring: Providing comprehensive instructions and mentoring for transitioning eco-villagers to facilitate their construction of sustainable land-based settlements. Including the broad scale training of permaculture stewards (skilled in sustainable agricultural practices, along with conflict resolution, social organization, basic healing modalities, and safety/ first responder) who can oversee the construction and operation of settlements.

Networking the Communities: Building connections and relationships between the relatively autonomous eco-village settlements and the existing urban populations. Thus allowing for highly efficient production of agricultural products based on supply and demand and efficient distribution of these resources through an innovative CSA model-- where specific urban neighborhoods are linked to specific settlements to facilitate the efficiency of transporting agricultural products, minimizing the distance these products must travel. Strong networks also allow for movement of population between urban centers and eco-villages, providing those in the city with chances to take breaks in the country, and providing those in the country with easy access to the community and excitement of the city. This model also allows for the possibility of "dual residency" where some may divide their time in relatively equal proportion between the settlements and the city.

The Culture: Recognizing and expanding upon the new economic, social and creative opportunities that emerge with the creation of a new cultural landscape. Including the radical autonomy and self-expression of Burning Man; opportunities for traveling artists, healers, musicians, performers; technological innovation within this new "laboratory" of distributed settlements as humans adapt to new conditions (see "The Farm" and the invention of the Geiger counter); diversity of spiritual practice; all within the context of a culture that is still fully wired to the World Wide Web and thus able to share innovations and creations with incredible ease, while still maintaining an incredible degree of sovereignty and autonomy.

Media and Outreach: Spreading the constellation of ideas, techniques, practices and lifestyle choices associated with The Ecotopia Project through mainstream and social media.

The Website(s): Building a web portal through which all of the above elements of The Ecotopia Project can be accessed, utilized and expanded upon (i.e. purchase of land, technology and materials, social network to link up with your eco-village community, access to educational materials and mentoring, monitoring of supply/demand for agricultural products, joining/creating a CSA, etc.)

Indigenous cultures are experts at biological mutualism. By comparison, we in the Western industrialized world are acting like sociopathic four-year olds. The very foundation of modern civilization – agriculture – is based on a paradigm of parasitism. We are being called to grow out of our parasitic cultural infancy and undergo a metamorphosis into the symbiotic stewards our planet so desperately needs.

In order for this vision to succeed, it must be massive. Building a sustainable future for humanity means involving the masses of humanity in the transformation. The Ecotopia Project will open up a new playing field for a multitude of new industries dedicated to developing and maintaining this emerging mycelial network of micro-cultures populated by evolutionary pioneers.

Will you join me in imagining – and implementing – this Ecotopian future?

Carl Grether

Carl sprouted from the fertile soil of Berkeley, CA. He transplanted himself to Oberlin College where he received a BA in Psychology, and then to the University of Southern California for graduate studies in Film Production. He worked as an editor on the environmental documentary "A Snowmobile For George" and for MTV before leaving Hollywood to start a garden and refocus his attention on permaculture, yoga and integral wellness. He worked as a counselor, teacher, gardener and food activist before returning to filmmaking in 2008 to amplify the message of the sustainable food movement. He has since produced and edited two feature documentaries: **Edible City** and **Occupy The Farm** – as well as recently directed the video that launched The Polish Ambassador's Permaculture Action Tour. Carl also serves on the stewardship council for the Gill Tract Community Farm in Albany, CA and is on the Board of Directors for the Bay Area based nonprofit ERIE (Entheogenic Research Integration and Education). Carl still lives in Berkeley and continues to garden, make movies, practice yoga and provide consultation for individuals and organizations on a variety of subjects.

http://ediblecitythemovie.com
http://occupythefarmfilm.com

The HeART of the Village

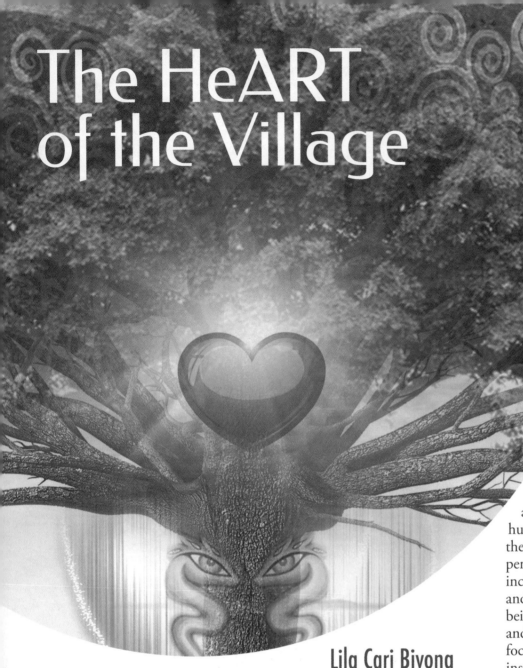

Lila Cari Bivona

dwelling human a productive method to find meaning and expression in both the complexities and mysteries, and simple commonalities that is life and the human experience.

Imagine a community in which each life is viewed as a masterpiece creation, and art as the expression and gift of a human life. Inside each of us dwells the essence of creation, and the experience of emotions, desires, and an inclination towards greater purpose and service. Uniquely designed, each being is truly a gift to be experienced and cherished. A village culture focused on artful expression, both inspires and empowers each member's special gift and contribution towards the greater prosperity, joy, and health of the community as a living, breathing entity, and thereby gift to the sustainability of a greater planetary whole.

How do we as community choose to honor and nourish the spark of love and creation that lives within each of us? Have we tired of forgetting that life is truly art? Are we happy with the legacy of the lifestyle our children are inheriting, and the culture we live, work in, and create together? What energizes and beats the heart of a community that we feel inspired to serve, to participate with, and raise our children in?

"Time is Money" is a common phrase that has driven the motivations and achievements of an industrial paradigm, now beginning to see the errors of its ways. If "time is money" is the ideology that marks our past, have we embraced the alternative, life-affirming "Time is Art" belief of our future? And how does this define our culture and our ability to thrive as a human family?

Life-sustaining and life-giving, the spirit that beats the heart of a village is essential and core to community thrivability. As we take a closer look at the word, "heart," we find the word "art" hiding inside. A spelling fun to use, heart and art feel too good to separate. What if we begin to tune ourselves to the awareness that art and culture is the palpable essence and felt energy that beats the heart of a village? Focusing on a culture of artful expression emphasizes and gives life to a creative, unique, and innovative aspect of community that satisfies a deep calling within the fabric of our being to appreciate and experience life as a creative force of good, interconnectedness, and oneness. Art and the celebration of cultural diversity provide the in-

Together we can do what we can never do alone.

How do you feel as you drive past yet another chain store corporate conglomerate? Is there something that feels lost or forgotten in a rising culture of text message kids and cookie cutter track homes? Is it art, soul, perhaps culture? Has our community design, technological trends, and architecture neglected to consider the value of personal connection and the vast wisdom and healing contained in honoring nature, and the empowerment and fostering of environments that inspire a creative, artful spirit? What kind of humanity are we growing when culture is defined by a robotic supply and demand in a television-prompted economy? Perhaps one in which our well-being is left in the hands of profit-driven corporations that practice a culture of financial boons over people and the environment. In this kind of "time equals money" culture, people are taught to believe that their worth and success is determined by their material possessions, and the promise of them to provide comfort and recreation, even happiness.

Where we sit today as a culture makes sense. Nothing humanity has created has been a mistake. We have discovered and exploited, created, and destroyed, for a desire both strong and fearless born of free will would only naturally desire such acts of fervent

What if we begin to tune ourselves to the awareness that art and culture is the palpable essence and felt energy that beats the heart of a village?

servitude and exploration. We desired to grow more food, and scaled our agricultural pursuits to immense proportions, for we could feed the world! We focused heavily on a knowledge-driven culture and assembly line production to find more efficiency, even if this meant we would leave the heart and art behind.

Press "present moment." All wounds are forgiven. With our current awareness and understanding, where and how are we choosing to focus our time, energy and money? Has the glory and pursuit of arts been overshadowed by an achievement and success-driven culture? And what if art sourced from the heart was actually the greatest gift a community could be in service to?

Reinhabiting the Village is the antidote to such a deleterious fate, by empowering heart-driven lifestyles that produce locally-sourced artisan goods that feel better, taste better, are kind to the planet, and serve the wellspring of love within the creator and the consumer, that is ultimately the core of humanity, and the heart of community. There is no doubt that a new culture of earth stewards and artisans is rising and communing in a shared vision of a peaceful and harmonious planet and her people. You many have felt the stirrings of a new time in your bones or a feeling that a transformational era is birthing. What if our time and energy was refocused towards empowering impassioned grassroots action that assigns power into the hands of the ordinary human as a creator? What might we begin to dream awaken, be inspired to create, to heal, and architect? Do our hopes and dreams ignite in the dreamscape of an artful and culturally diverse village centered community? What might you be impassioned to give your love, your energy and your time towards? Could there be an artist inside just waiting for the permission to discover and explore life through the hands, voice and eyes of an artist?

Reinhabiting the Village empowers the creation artist within to play, to create, and to be expressed. Undeniably, there is a ground swell gaining momentum, and it is breeding a new culture of humanity. The shared mythos of this culture is just that, shared, and it is sourced by the power of a collective heart field, rooted in goodness and expression. It is a way of being in the world that seeks harmony, and is motivated by a shared common purpose for the good of the individual as paramount to the success of the whole. In perfect timing, as the vestiges of an outdated design methodology fall away, a new village culture is on the rise and it must be perceived with compassion and a long view

of understanding, as it is discovering itself as a teenager would, for it has chosen to pioneer a new way of living and being in the world. Are you a part of this shift? Might you feel the tide rising to commit your hours and minutes towards a thriving planet Earth and her people?

One way a village culture can be defined is by the way it views time and creation. Time as expendable resource, as set in motion by the industrial revolution, birthed from the sentiments that time equals money, was the driving force in the processing and exploitation of resources for human consumption. On the contrary, time as art is a force that asks the individual and community to mindfully steward our planet's gifts, and celebrate the diversity and expression of the human spirit and the human experience. Perceiving life as art facilitates a culture of shared understanding, co-creation, and a feeling that we truly are all connected. The practice of art gives us a transformational medium by which to contemplate and accept the inevitable cycles of the human experience, of joy and grievance that is the beauty, gift, and mythos of an impermanent life of love and loss.

Time as seen through the lens of art in a village culture is perceived as the moments by which one is able to ask questions, receive answers, create, explore, to fuse ideas for innovation, and to produce art. Art that is a manifestation of one's contribution to life, soul calling and expression, is inevitably imbued with quality

and love. What if a village culture's success and characteristic style is determined by the art produced by its villagers, and the heart and soul that contributes to the quality of the goods and services provided? Might you be more apt to live and work here, or visit to experience the felt presence of heart and a culture that feels like family or perhaps even home?

The felt attraction and economic success of a village community is rooted in the diversity and expression of its inhabitants, whereby visitors access the culture of the community through the artful way it approaches everything. From community meal creation, celebratory events, to rites of passage, products for purchase, and even the way a guest is greeted upon arrival into the community. Each step along the way, the village experience is embedded with art and heart. Might the goal of a village community be that each being is in service to their soul's calling? And thereby contributes to the village culture with her gifts and unique way of intentionally interacting and discovering the moment?

In order for a village culture to thrive, each being is seen as equal, as valuable, and empowered to contribute their unique and diverse offering as a gift. Some may call this following one's heart, or serving one's soul purpose. Art provides the transformational medium by which one may experience and discover their place, personal development, fulfillment, and continual evolution. The success of a village depends on the freedom and sovereignty of each contributing member, as well as the mutual core belief system, that one intends their actions and their art to serve both oneself and the good of the whole.

This mutually beneficial purposeful service lends itself to a culture of creative expression that is ultimately guided by the heart. A spiritual component exists here for in the old paradigm, one's mind and ego run the show. For the belief that life is tough and everyone is in competition led to separation, anxiety, comparison, hoarding and a host of fear-based actions. In a heart-based village environment, one holds the belief system that s/he is supported, loved, and that one's peers are routing for each other's success. One being's success is thereby celebrated as the success of the whole.

The word "*compersion,*" to celebrate another's success, and then too, to feel it as your own, is a great example of the kind of vocabulary that a culture of heart and art is governed by. The cultural shift between an old separatist paradigm and a new emerging paradigm of co-creation is that the success of the village is the sum of the happiness and fulfillment of its members, and each person's unique and diverse contribution. In conclusion, a village culture that celebrates a time is art lifestyle, characterized by understanding, empowerment, interconnectedness, and creative contributions of expression and innovation breeds a high-level, solution-oriented community of inspired artists who pursue love and life, and produce superior quality goods and services that enlighten every aspect of the world.

Lila Cari Bivona

Cari Lila Bivona, visionary of The Divine Playground, is on a mission to ask important questions and tell the stories necessary to dream awake heaven on Earth lifestyles in thriving communities. She works towards this vision by producing transformational events that inspire and empower the embodiment of one's dream "divine playground" to shine and serve the world. From holistic health and women focused retreats, to performance camps, and life alchemy workshops, Cari applies her psychology degree, Masters in Leadership for Sustainability, lifelong yoga training, and permaculture design awareness to produce holistic, integrative events that feel like home and family. With a playful expressive spirit, she invites participation in transformational experiences that empower each student to act as the powerful and loving creator that is one's true nature and gift. Sharing her passionate philosophy that "Mother Earth knows best" and "Love Wins" with the consciousness of the collective, Cari also expresses her message through co-creative multidimensional "Play-Pray" art in devotion and activation of her mission in dance, song, writing, and video content. This characteristic Divine Playground style can be experienced in the retreats and events that are produced internationally "Yes! Let's Enlighten-up Together."

Village Repair

by Gaiacraft Stewards

Our people will remember a time when we had to re-learn and activate through permaculture, the movement itself will simply become the art of living in the most worthy, spectacularly evolving way that honors our ancestors while setting up ever-most fruitful futures for our descendants.

Mark Lakeman

In this time of mounting global environmental, economic, political and social crisis, we have a unique opportunity as individuals and communities to redesign our public places. This placemaking initiative emerged from Portland and its globally celebrated 'City Repair' movement with its many 'Intersection Repairs' thanks to Mark Lakeman and a whole community of other inspirators, visionaries and activists.

Heart Gardens

The Village Repair branch of Gaiacraft began as a local grassroots initiative growing in the mossy cedar rainforests of Mt. Elphinstone, British Columbia. In the Heart Gardens and surrounding downtown core of our coastal village, a unique community comes together to build foundations for a regenerative future. Here we can see social permaculture vignettes intended to inspire creative placemaking in small towns and villages around the world.

This creative media platform shares templates for a conscious, relocalized and resilient community including; community gardens, native plant food micro-forests, community mandala, eco-education center, permaculture demonstration sites, food micro-forest, bike paths, car co-op, electric car charging station, farmers market, free market, and sitting areas located in gardens on the public right of way. Our placemaking efforts includes intentional businesses dedicated to health and healing such as a health food store, massage studio, yoga school and crystal shop.

The Forest Atrium

Observation and interactions with community over the years has led to further engagement with land 1 km from the Heart Gardens. The Forest Atrium is a larger territory, blessed with second growth forest, indigenous plants and trees with many different guilds and microclimates. Getting to know what grows where and when while developing awareness of the native food and medicines of this thriving ecosystem has been a large part of the journey.

Though still in the budding phases of planning and development, its human habitats include tiny homes and sleeping units clustered around a planned central common space similar to Michael Reynold's Earthship design. The dwellings' small ecological footprint provide affordable housing and preserves nature's sanctum while considerable land remains for farming and food forests. The Forest Atrium is also used to demonstrate permaculture systems and is evolving as a model for local food resiliency, community building and experiential education.

Atlan

The Village Repair *meme* continues to thrive and is fruiting near the White Salmon River in the Columbia Gorge at Atlan Ecovillage. Drawing from a strong connection to Mark Lakeman and the City Repair Village Building Convergence roots, and carrying forward the inspiration of Gaiacraft for Village Repair, the community at Atlan is applying the pattern in a less developed and more forested area. This affords the opportunity to focus on and plan for the diversity of human,

> Build, plant and act as if this is your only chance, and you will live a life with many rich chances for the choosing. Where you are right now, with the information you have, is the time and place to start.
> Michael Becker

plant and animal communities as well as to draw the focus toward *ecovillage* design and right livelihood in direct relation to place through building regenerative economy.

Thriving community gardens, forest hiking trails, the Artifactory community multimedia arts making space, the mixed used Collaboratory Design studio, yoga decks and a camping retreat space in the forest are among the growing presence of community integration with the land. Opportunities for enriching *intersection repair* to support local area residents, rafters and boaters enjoying the wild and scenic White Salmon River, and outdoor enthusiasts of all walks are further supported by our educational nonprofit learning center CultureSeed.

Each year since 2013, Atlan has hosted an annual VillageBuilding Convergence in late September coming togetherin celebration and learning how to continue repairing our relations with the land and one another. Through our work we build together the visible and invisible structures of a thriving culture that honors all of life, celebrates diversity and fosters abundance through regenerative design.

In the Global Village

As an activated Social Permaculture program and Transition Initiative, Village Repair aims to inspire communities to redesign their own neighborhoods into thriving, abundant and diverse ecosystems which support the web of life for all living and non-living things.

At the heart of Village Repair is building a conscious community. The first step to this is getting to know your neighbors including all the people, schools, local government representatives, non-profits, community groups and all the birds, mammals, reptiles, insects, mushrooms, plants and other living and nonliving things that surround you for some or all of the year. Together we can redesign our local communities from the ground up, building a planetary network of decentralized, but effectively networked, communities that together form the basis of a harmonic civilization that will carry us into a new era of peace and prosperity.

Delvin Solkinson, Dana Wilson and A. Keala Young

A. Keala Young, Dana Wilson and Delvin Solkinson are part of the Gaiacraft Collective, a visionary permaculture design guild. They are dedicated to infusing innovation and creativity into the design process, opening up new applications of permaculture beyond the garden and farm. With an eye to rural, urban and suburban scale properties, public spaces, neighbourhoods and communities, they see the principles of Eco-Village design applying at all scales in a redesign process that provides food, shelter and community gathering places for people, plants and animals everywhere.

http://gaiacraft.com

A Community-Based Resilience Framework

Readying ourselves and our communities for economic, energy and ecological crises

The Framework shown on the following page was designed to be something that humanists, sharing economy advocates, Transitioners, activists, neo-tribalists, intentional communitarians, existential collapsniks, and even near-term human extinctionists can agree upon -- a process to increase our personal and our communities' resilience no matter what the future holds in store. It's been refined over the past year through discussions with people in all of these groups.

The Framework consists of five "stages," from the inner work of knowing ourselves better, to the collective work of building community and preparing communities for dramatic and permanent change. It has been published on both Shareable.net and Resilience.org.

A poster-sized PDF of the Framework can be downloaded at

http://tinyurl.com/resilienceframework.

Dave Pollard

Dave Pollard retired from paid work in 2010, after 35 years as an advisor to small enterprises, with a focus on sustainability, innovation, and understanding complexity. He is a long-time student of our culture and its systems, of history and of how the world really works, and has authored the blog How to Save the World for over twelve years. His book **Finding the Sweet Spot: The Natural Entrepreneur's Guide to Responsible, Sustainable, Joyful Work**, was published by Chelsea Green in 2008. He is one of the authors of **Group Works: A Pattern Language for Bringing Life to Meetings and Other Gatherings**, published in 2012. He is a member of the international Transition movement, the Communities movement and the Sharing Economy movement, and is a regular writer for the deep ecology magazine Shift. He is working on a collection of short stories about the world two millennia from now. He lives on Bowen Island, Canada.

Transition Initiatives are based on four key assumptions:

1. That life with dramatically lower energy consumption is inevitable, and that it's better to plan for it than to be taken by surprise.

2. That our settlements and communities presently lack the resilience to enable them to weather the severe energy shocks that will accompany peak oil.

3. That we have to act collectively, and we have to act now.

4. That by unleashing the collective genius of those around us to creatively and proactively design our energy descent, we can build ways of living that are more connected, more enriching and that recognize the biological limits of our planet.

Rob Hopkins,
The Transition Handbook:
From oil dependency to local resilience

A Community-Based Resilience Framework

Readying ourselves and our communities for economic, energy and ecological crises

Dave Pollard

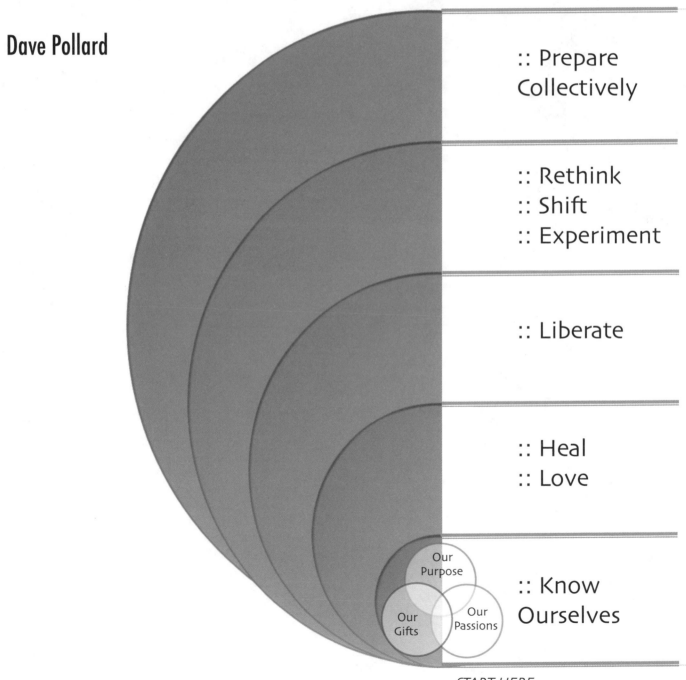

:: Prepare Collectively

:: Rethink
:: Shift
:: Experiment

:: Liberate

:: Heal
:: Love

Our Purpose

Our Gifts

Our Passions

:: Know Ourselves

START HERE

- assess and build our community's self-sufficiency, resilience and mobility
- discover what we collectively already know, have, can do and can't do
- learn what we need and don't need to live full, joyful lives
- learn how other cultures have coped with crisis and collapse
- source locally | build collective community capacity
- rehearse crisis response in our community

- learn how our complex world really works
- find people who share our passions and purpose
- rethink how, where and with whom we live and make a living
- instead of a job, find and fill real local needs
- shift to the sharing/gift economy

- strive to realize the illusion of self, ego, control, separateness & time
- self-assess and increase personal independence from centralized systems
- need less | learn continuously | facilitate and mentor others
- help liberate others by modeling equanimity, presence, generosity,
  gratitude, curiosity, creativity, adaptability & appreciation
- engage with the fearful and with deniers

- self-assess our physical and emotional health
- empathize | reconnect | give | forgive | ask for help
- learn new ways to heal and help others heal
- appreciate nature, and our true nature | insist on joy in spite of everything
- love unreservedly, even those we don't like

- know our personal capacities, limitations, blind spots, wants and needs
  and what we really care about
- know what brings us joy, fearfulness, anger and sorrow
- learn and practice self-awareness
- discover where we belong and what we're meant to do

http://tinyurl.com/resilienceframework.

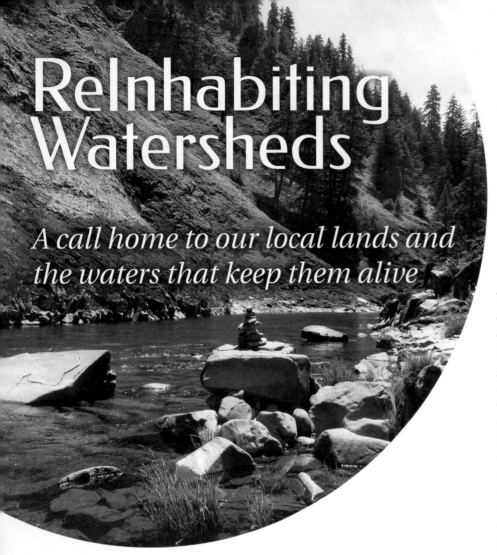

ReInhabiting Watersheds

A call home to our local lands and the waters that keep them alive

Heather Beckett and Chelsea Estep Armstrong

Reinhabitation means learning to live in place, in an area that has been disrupted and injured through past exploitation. It involves becoming native to a place through becoming aware of the particular ecological relationships that operate within and around it. It means undertaking activities and evolving social behavior that will enrich the life of that place, restore its life supporting systems, and establish an ecologically and socially sustainable pattern of existence within it. Simply stated it involves becoming fully alive in and with a place. It involves applying for membership in a biotic community and ceasing to be its exploiter.

Peter Berg, Reinhabiting California

Water is the single most common currency of life. Coursing from glacier peaks through underwater tributaries, springing up into constantly gushing headwaters, veined and linked through countless systems of interdependence, salty and fresh, clean and polluted, our waters organize our world. As climate change, drought and pollution remind us of the preciousness of this resource that is not some externalized convenience but the very liquid essence of our beings and all earthly beings, we reawaken to a call to **reinhabit** our watersheds.

Every single place on Earth--the deep forests, the rolling plains, the mountain tundras, the sagebrush meadows, and the rich sands of the deserts--exists inside a watershed. Each watershed is a unique and interlacing systems of rivers, streams, lakes, weather systems, and groundwaters that come into a cohesive body through their impassioned pursuit of returning to sea. Call it gravity or a great mythic mystery animated by some unfathomably beautiful urgent longing towards life, every drop of water journeys from heights to the depths into the mantle of the earth, and up again, joining on its way with geologies, plants, animals and humans in a breathing, pulsing, walking, winding many bodied organism of ever-changing shape.

If you are blessed enough to still live in communities that source their

WATER CAN HEAR. THANK THE WATERS! WATER IS SACRED! WATER IS LIFE! WATER CAN HEAR! WALK AWAY WITH THE THOUGHT THAT YOU CAN BLESS THE WATER. Grandma Aggie

drinking waters from within the watersheds you dwell in, you can know that the trees that grow around you, the flowers that bloom, the birds that fly above, the deer that run, and the water that flows through your own blood are connected through the same precious body of a watered collectivity.

By learning where our waters come from, what headwaters feed the rivers that feed our communities, we embark, we begin the journey back to the true story of our interconnectedness. Towards the possibility that we might "become native" to our watersheds, we turn our awareness towards a remembrance, a study, a devotion to understand the ecological kinship we share with the places that make us who we are. When we look past political boundaries that draw lines of division across inextricably unified ecological systems, we discover that we are already part of an existent community of life, defined and designed by the course of waters over and above, under and in between every earthly form.

From the grand abstract hunger and thirst for a return to relations with our bioregional watersheds, how might we begin to make this impulse into a tangible and vital catalyst for change?

Do the research. Meet your lands. Know where your water source comes from.

ROUTES TOWARDS MAKING RELATIONS WITH YOUR WATERSHED

- **Learn the basics.** What kind of a watershed do you live in? Does it flow to the sea? Or does it remain inland and plummet to the depths through underground passageways? Where are the headwaters of the little stream that runs through your town? Where are the confluences of tributaries that lead to your local waters? How many headwaters are in your watershed? Are there dams in your watershed? As you flip on your lights, are you drawing on river generated hydroelectric power from near or far? Where is the source of the waters that flow through your tap? How many miles has that water traveled? Study the maps. Draw the maps. Notice where your attention is drawn to on the maps. What might be happening there?

- **Begin to read the landscape.** Wherever you sit or stand ask, what are the geological features behind me and in front of me? What is the nearest body of moving water? What direction does it flow? Where does it course to and from? What flora and fauna are part of this community? What interdependencies sustain the health of these ecologies? What environmental threats compromise the health of the land to your left and right?

- **Study the intersections of past and future.** Ancestrally, very few of us have long held lineages in the lands we call home. What cycles of displacement made way for the cities, towns, industries and families who currently occupy your watershed to be there? Who are the Native Americans that have inhabited your region? Where are their descendants now? What were their names for the mountains, rivers, plants and animals of your bioregion? How has the landscape transformed, what used to be there? Where were the sacred sites in your region? What is there now? Where are industries located in your region? What was there before? What displacements might be occurring there currently and in the coming years?

- **Learn from watershed wisdomkeepers.** Every bioregion already has them, and they are usually thrilled to meet individuals and communities who care. Who is already stewarding the rivers, headwaters, lakes and estuaries? What are the histories and stories of plants, animals, peoples, and geologies that shape your locality? How do they echo or reveal aspects of the human story happening there around you now? What rhymes do you sense between ancient and recent histories and current events playing out in your bioregion?

- **Study the evolution of your watershed over time**. What native species have been destroyed or eradicated in your region? What were they supplanted by? Where did non-native species come from? What factors or relationships brought them to your bioregion?

- **Witness your growing knowledge of your watershed transforming you and your way of life.** As our moment to moment navigation of place is restored with a robust knowing of where we are in relation to the geologies, species, and elements of our watershed, the inner landscape of our own identities begin, at last, to arrive somewhere closer to home. A spontaneous transformation of consciousness reveals cues and clues for how our own lives might give to and receive from the life-supporting systems of our ecologies. As we approach our watersheds with a longing to return into relationship with their myriad lives, what is revealed is so much more than a list of scintillating trivia and depressing environmental crises. The mythic nature of the lands and waters, all the elements that feed them, begins to show its face.

THE CULTURE SHIFT: Towards a Watershed Culture

By beginning to learn about our bioregions, by changing our ways of cooking, eating, buying, selling, connecting and knowing within our watersheds, we set in motion a series of courageous actions that could take form in many aspects of our lives. The beneficial ramifications are as many and as varied as the people who bring them into being.

Luminaries, such as Peter Berg, Gary Snyder, and many others have advocated for decades for new forms of bioregional and watershed governance wherein each watershed, communities collectively steer where, when, and how the preciousness of their waters are stewarded. While such notions of reorganizing our communities in resonance and attunement to the uniqueness of our ecological landscapes offers promising alternatives--which are gaining steam in Canada and various communities in North America--a simultaneous and keystone movement is ready to spring to life, catalyzing emergent watershed consciousness into a thriving Watershed Culture.

Where our individual efforts give way to a collective knowing and revaluing of our togetherness, culture begins to transform. If we want to turn our abundant energies towards narratives that hold us in reciprocity with the very organisms and beings who sustain our lives, we must first examine what inherited cultural pathways caused us to forget them.

Much of the natural world has been culturally objectified as exploitable, marketable resource, bought, sold, studied, and advertised, primarily in the context of how we might make it grow, live, and die when and how we want it for our own personal or collective gain. How might we choose to dissolve these obsolete cultural ways of seeing the non-human members of our watersheds and fortify cultural patterns that hold us in remembrance of our beyond-human relationships?

In a time when we have enough music, media, and story about human hopes and heartbreak and existential crises to keep our cultural jukebox playing morning and night, on hundreds of channels, through diverse media, the time is ripe for a newfold way of uniting around songs, stories, and media that guide us home into relationship with our local lands and favored distant climes.

At one point, in lands and generations, near and far, our ancestral songs were played for the stars, the waters, the mountains, the winds, and the forests. Stories told of conflicts, triumphs and loves of the four-leggeds, the winged ones, and the swimming ones. All these were known to be ancestors in their own rights. Pictographs showed the fearsome power and the enticing beauty of earth forces. They documented histories through earth changes and evolutions in human ingenuity. In vibrant modes of cultural memory, kept alive by fireside, on migrational journeys and through every form of oral tradition, secrets of medicines, maps of the land, and symbols for spiritual teaching and interrelations were encoded. These vestiges of earth-centric cultures possess the seeds of love that held our ancestors in right-relations with our parent planet. They vividly animated meaningful connections to the magic of the many ones who bring such wild magnificence to our world.

Part of the loss that occurred with every generation of displacement was the loss of this cultural knowing. How might we convert our hunger for these ways into something that might lend breath to their return?

We see ours as a time when community organizers, artists, musicians, performers, writers, farmers, media-makers, spiritual leaders, political organizers, educators, changemakers, social visionaries might join in a devotional zeitgeist, an impassioned cultural awakening that puts our local earth stories back at the center of our circles.

We send out the call for all lovers of life, lovers of earth to turn their own natural creative urges towards making art and offerings: rituals, ceremonies, songs, dances, poems, hand-crafts, gatherings, stories, documentaries, inquiries, and expressions that renew relationship to the local watershed that keeps them alive.

For all the waters that animate every living thing, the waters who grow tall and skyward through the trees whose woods hold up our houses and fuel our fires, the waters who hydrate every cell in our bodies, the waters who quenched the creatures who became the reservoirs of oil that become the blood of our machines and manufacturings, the subterranean waters who boiled to a temperature that urged rock from below the ocean floor to rise above the sea, making mountains, the glacial floods whose torrents shaped the lands the we walk upon, the waters who land in the morning blooms of every opening flower, and the waters that sprout and nurture every uprising and down-rooting seed, to all the waters, named and unnamed, we give thanks.

Heather Beckett & Chelsea Estep-Armstrong

Heather Beckett and Chelsea Estep-Armstrong have been brought together by their shared passion for community, their awe and love for the lands, and their devotion to the waters. Their organization, Atria, gathers people to revive authentic, vulnerable and bold cultural remembrance of shared stories based in connection to place. Atria connects diverse local artists, youth cultures, conservation efforts, social visionaries, and traditional cultures to renew relationships around central stories of their watersheds through research, music, art, story, and councilship. Chelsea holds a Masters in Cultural Anthropology and Critical Theory from the New School for Social Research and brings years of ethnographic and research experience to the historic, mythic and ecological riddles that weave people and place. Heather, in a lifelong devotion to reviving myth and ritual as a vital modes of cultural memory, has worked on page, on stage and in film as a performer, writer, researcher, director and producer. Chelsea and Heather met conducting historical research through twenty-six states for a work of historical fiction in development. They are both fledglings in an ongoing, humbling, and nourishing study in Bolad's Kitchen with Martin Prechtel, whose teachings profoundly influence their work and to whom they are gratefully indebted.

http://atriarevival.com

Growing Together

The Story of the Neighborhoods Community Garden

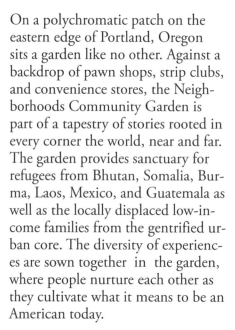

Our Mission, plain and simple, is to get healthy food into the mouths of hungry people.

Our Vision is the transformation of unused private, public, and institutional land into Neighborhood Gardens, where healthy food, resilient community, and economic opportunity spring up together.

On a polychromatic patch on the eastern edge of Portland, Oregon sits a garden like no other. Against a backdrop of pawn shops, strip clubs, and convenience stores, the Neighborhoods Community Garden is part of a tapestry of stories rooted in every corner the world, near and far. The garden provides sanctuary for refugees from Bhutan, Somalia, Burma, Laos, Mexico, and Guatemala as well as the locally displaced low-income families from the gentrified urban core. The diversity of experiences are sown together in the garden, where people nurture each other as they cultivate what it means to be an American today.

The People Who Share the Garden

- Families from Bhutan, Burma, Vietnam, Laos, Guatemala, Oaxaca, Somalia, and the United States

- The first generation immigrant children who straddle cultures

- The Outgrowing Hunger team

The Refugee Garden Cookbook

Through a focus on the gardeners and the foods they grow, eat, and share, we will explore and present a complete story in an unconventional cookbook format. The narrative will take us from how it is that these boundlessly beautiful people were brought together with disparate stories of difficulty and loss, to finding community and themselves in this one place, to an invitation to appreciate, to learn, to eat, to exalt and to enjoy their reservoir of cultural and culinary knowledge with stories of resilience lying unseen in our midst. This book is the American Story from a myriad of diverse refugees perspectives in the common language of food.

- Spotlights on 12 families that include snapshot personal and family stories, foods from throughout their experience, and recipes for a full meal that highlights how the family diet has adapted to reflect the changes in their story, while maintaining its cultural integrity.

- The children's perspective expressed through photography.

We will distribute disposable film cameras to children in order to better see what the children see in their position of straddling cultures. The focus will be meals with family, meals, at home, snacks, and time in the garden. These photos will be woven throughout the book and will communicate an intimacy of experience that likely no outsider would be able to capture.

- Gardens for mental, physical, and social health. For example, over lunch in early December, Dilhi, a leader in the Bhutanese Nepali community, shared his stories regarding how the Neighborhoods Community Garden has helped relieve the painful depression of diaspora that has resulted in many suicides.

Empowerment Through Food Story

- Mitigate the assimilation pressures endemic to living on the margins of society by bringing more focus to and appreciation for this multicultural community

- This Book will help to build confidence in the garden community as well as specifically help support the growth of a cooking class series led by refugee women as well as related economic empowerment projects

How to Support this Project

OUTGROWING HUNGER is a registered 501(c)(3) nonprofit based in Portland, OR. Financial data is available on Guidestar.org, EIN# 45-2380984. 100% of net proceeds from this book will be used for our programs serving minority and disenfranchised peoples through garden-based projects.

ONE SPIRIT PRESS (Portland, OR) has agreed to provide publishing and editing services for this project at a greatly reduced fee. You can find out more about One Spirit Press at http://onespiritpress.com

Francesco Tripoli is the Project Manger for **OUTGROWING HUNGER**. You can find out more about his work here:

http://outgrowinghunger.org

Inhabiting the Urban Village

WE CAN BEGIN BY DOING SMALL THINGS AT THE LOCAL LEVEL, LIKE PLANTING COMMUNITY GARDENS OR LOOKING OUT FOR OUR NEIGHBORS. THAT IS HOW CHANGE TAKES PLACE IN LIVING SYSTEMS, NOT FROM ABOVE BUT FROM WITHIN. Grace Lee Boggs

Many have dreamed of a place called Utopia. Since first written about in 1516 by Sir Thomas More, idealists and visionaries have waxed poetic about such an existence, the perfect world, a social and political system at peace, a life of abundance for all.

With more humans than ever living in urban environments, where resources are dwindling, existing infrastructure is already taxed and population is rising, some would argue it is an impossible fantasy, a dangerous distraction from dealing with everyday challenges that we face. And some would say their version of a utopic scenario is already a lived experience, and that while perhaps not perfect, it is possible to create harmonious and robust living scenarios. Somewhere in between lies a potential that by applying visionary imagination, successful models from antiquity, and achievable practices to our current scenario we could see positive improvements through our efforts, creating livable and vibrant cities, even with the unprecedented challenges we face.

Over 54% of the world's current populous lives in an urban environment. It is projected by 2050 that number will climb to 66%. "Managing urban areas has become one of the most important development challenges of the 21st century. Our success or failure in building sustainable cities will be a major factor in the success of the post-2015 UN development agenda," said John Wilmoth, Director of UN DESA's Population Division.

"We have to adopt a post-2015 development agenda that is holistic in nature. The agenda should put the well-being of both humankind and our planet at the centre of our sustainable development efforts.

Through these efforts, we should bear in mind that profound changes in attitudes, behaviours and policies will be required to create a world in which human beings live in harmony with nature," states Sam Kutesa, President of the UN General Assembly.

With our mazes of metal and concrete we must remember that everything we have built, every city, every machine, every freeway has come from the substance of Earth. Connecting to the living organism of Gaia all around us, we see that underneath the concrete there is a life, a heartbeat, a pulse, that is louder than

any street noise, car, or the hum of electricity. Since we built the walls of our current world we can also take them down.

All of our life is lived at the crossroads, the intersections where your stories rubs against another's, the common ground that connects us as we, of course, are not islands. By tearing down fences and linking backyards, creating gardens in empty parking lots, reclaiming intersections as public and pedestrian space, sharing our homes, our skills and resources, sharing food, sharing ideas, sharing hearts, rolling up our sleeves

and rebuilding our neighborhoods, we can recreate our Urban centers to be thriving Villages once again.

Piazzas, a term originating in Italian culture, Agoras in Greek culture, and many other names for the shared public square, have long been the center of any bustling village. This was the community's meeting grounds and marketplace for exchanging goods, stories, and ideas. This is where celebrations were held, news was shared, and also where unrest was expressed, where protesters gathered, where revolutions began.

There are so many modern-day examples of successful and essential public space in our Urban environments. With the rise of the place-making movement, we are seeing a

renaissance in the intentional design, management, and use of public spaces. It means designing for people, not for cars. It means that a neighborhood has claimed public space as a central hub for its inhabitants, promoting interconnection, placed-based knowledge, familiarity, a sense of safety, well-being and belonging. It takes into consideration the needs of the local populous, the environmental factors of the place, a nexus informed by the unique reflection of a culture.

The most stable ecological systems are ones where the diversity of life intersects and interrelates to create connection points of coherence. Applying permaculture principles in an environmental and social context to our urban landscape, we can adapt our concrete grids into flourishing centers for the full expression of our art, our culture, our prosperity, and for celebration.

Jamaica Stevens

OUR CHOICES AT ALL LEVELS — INDIVIDUAL, COMMUNITY, CORPORATE AND GOVERNMENT — AFFECT NATURE. AND THEY AFFECT US. David Suzuki

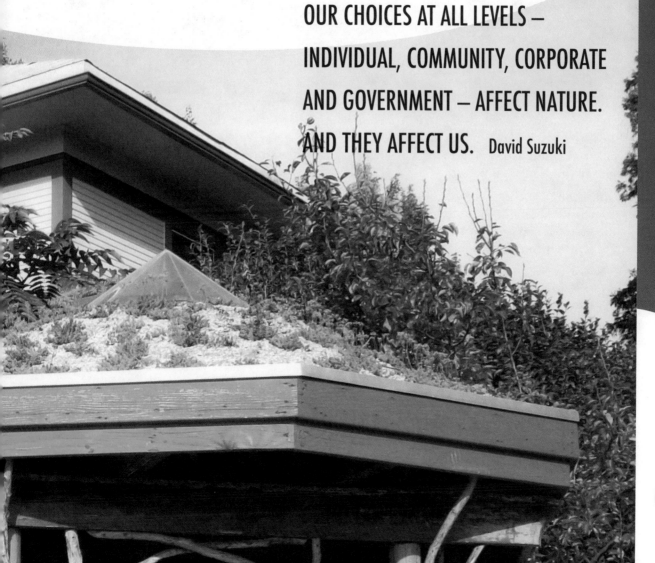

Inhabiting the Urban Village

The urban village begins with you and your relationship with its streets, parks, apartment rooftops, porches and stoops. It asks that we engage with the environment of our homes with more appreciation, responsibility, creative energy, and ownership. The urban village truly begins in our hearts.

A village implies community. An urban village calls out to each member of its city walls to participate in creating a breathing, living environment in which children play safely, elders are honored and cared for, and our businesses flourish providing services and goods to benefit our community.

The strength of our communities begins where we gather, where we meet, where we play, and where we choose to find a little more beauty and meaning in the everyday environment we inhabit. What is the course of your day? Where do you walk, drive, park and buy your groceries? How might your interaction with the urban landscape shift if it belonged to you? How might you share your special brand of energy, of perception, of love?

The nature of a village culture is that the community has the right and co-creative power to shape the energy and activity of the landscape. What do you cherish about your neighborhood? The first strategy of any engagement is to honor and express gratitude for what is working. By doing so, a culture of acceptance and positivity is grown. As we look through the lens of what we find beautiful, interesting, or even nostalgic, we express our appreciation for the generations before us who had a vision for something meaningful, artful, and celebratory.

Lila Cari Bivona

As we widen our lens, we may begin to see areas that have been neglected, perhaps forgotten or deemed unworthy of our attention. Might we focus our efforts on reclaiming these places? Might we find common ground in the rehabilitation of that which has seen hardship, perhaps death and decay? It is in the cycle of life that all things shift and change, die and are reborn. Where might we lend our energy, our time, our art, and our compassion towards rebirth?

In our love of the city nightlife, Saturday street markets, or public transport, we attune our senses to ways in which more health, vitality, and community service may find a voice. Whether through your child's school cafeteria, the flowering tree-lined sidewalk of your walk to the grocers, or the street lights that walk you home, it is our duty as village dwellers to take notice, to appreciate that which is guiding our way home, and to ask how we might make this playground, this community space, or this apartment complex feel more like home, and bring more joy, peace, and happiness into our lives.

ReInhabit the urban village. Find your passion. Ignite your community in empowered action, and plant seeds in abandoned lots to grow and nourish thriving city lifestyles.

Growing Edible Cities

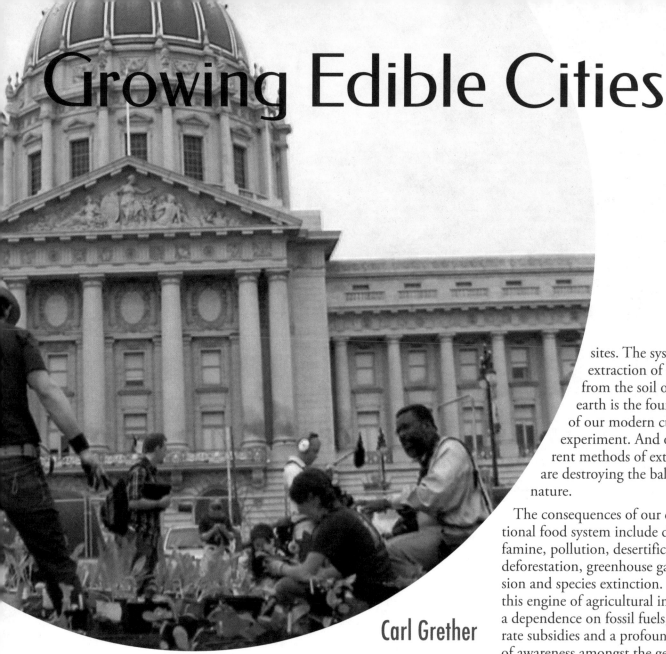

Carl Grether

Seven years ago I was getting my hands dirty digging trenches for the Permaculture (r)evolution, having abandoned filmmaking for the honest work of urban farming. I was drawn to urban farming as a means of reconnecting with the web of life through direct interaction with the plants and animals that nourish me, and as a means of understanding the concept of symbiosis and applying it to my immediate environment.

Symbiosis is how organisms live together, spanning the range from *mutualism* to *parasitism*. The garden is an ideal laboratory in which to observe and understand the varying relationships between plants, pollinators, decomposers, predators and parasites that make up a complex living system. These relationships within the garden are governed by the same **principles** of symbiosis and interdependence that regulate our entire living biosphere.

Taking a bird's-eye view of the current planetary crisis, industrial agriculture looms large as one of the planet's most nefarious para-sites. The systematic extraction of calories from the soil of mother earth is the foundation of our modern cultural experiment. And our current methods of extraction are destroying the balance of nature.

The consequences of our dysfunctional food system include disease, famine, pollution, desertification, deforestation, greenhouse gas emission and species extinction. Driving this engine of agricultural insanity is a dependence on fossil fuels, corporate subsidies and a profound lack of awareness amongst the general population about what our food system is and how it is affecting life on earth.

But in Permaculture they say that the problem is the solution. And every one of these issues is also a leverage point for transformation. And when you aggregate all of those problems into a comprehensive solution you are stacking a lot of functions, which means more leverage and potential energy for positive change.

And not only are there profound social, political and environmental reasons to become involved in the food system, but getting my hands in the dirt also made me happier,

healthier, more spiritually connected and more in touch with my community – not to mention improving the food security and nutrition of my household and neighborhood.

Urban farming is a commitment to mutualism, and to deepening our awareness of how we relate to the larger ecosystem. Food is our most direct connection to this larger web of life, and yet how would we know this when most of what we eat comes to us wrapped in glossy packaging from a bright-

What do we eat? Where does it come from? How is our diet affecting our lives? How are we developing our awareness about the complex web of people, machines, corporations, plants, pollinators, sun, rain, wind and earth that make up every meal that goes into our mouths? How do we engage with this awareness and make change?

ly lit sterile supermarket? Urban farming is a bridge towards alleviating our disconnection from the source of our most basic sustenance--and the antidote to our mass cultural amnesia resulting from this alienation.

All of this emerges from the garden, which is both the classroom and the laboratory for transformation. This cultural blind spot was too big to let go unnoticed. I needed to amplify the signal of the food movement by returning to my former craft and spreading the good word.

My first movie is called *Edible City*, a turn on the Edible Schoolyard meme created by Alice Waters and implemented in the Berkeley Unified School District as a way of bringing ecological literacy to children. As the name suggests, *Edible City* is a transmission of ecological, agricultural and social awareness that reveals the complex connections between our food system and a myriad of urban issues; from nutrition and wellness to economy and social justice.

Edible City offers a brief but comprehensive explanation of how our modern industrial food extraction mechanism has come into being. It offers multiple paths forward for engagement and transformation from within the (sub)urban context, which is where we must start, because it is where most of us live. From urban farming to community-supported kitchens, nutrition education, political initiatives, think tanks, entrepreneurs, community organizers, and guerrilla gardeners, the film offers a cross-section of the many layers of involvement that can take place before you even step foot on a piece of actual rural agricultural land.

The essence of thinking globally and acting locally is to take a bird's eye view of where we are right now. What do we eat? Where does it come from? How is our diet affecting our lives? How are we developing our awareness about the complex web of people, machines, corporations, plants, pollinators, sun, rain, wind and earth that make up every meal that goes into our mouths? How do we engage with this awareness and make change?

Got a lawn? Tear it up and plant a food forest.

No lawn? Grow strawberries and kale in containers on your porch.

No porch? Grow sprouts, spirulina and mushrooms in your kitchen or closet.

No space at home to grow anything? Volunteer at a community garden or plant a garden at a friend's house. Share food with your neighbors, build community, apply your growing ecological awareness into every aspect of your life.

Of all of the possible modes of engagement, urban farming is one of the most direct and effective means of transforming our relationship to food and community. Urban farming can account for as much as 30% of a city's food needs when brought to scale. In food deserts like West Oakland--where there are no grocery stores--urban farming can make the difference between health and malnourishment. And in places like Cuba, where they lost their oil in the early 1990's after their industrial partner the Soviet Union collapsed,

urban farming meant the difference between starvation and a national weight-loss program.

There are a multitude of ways to be an urban farmer. Got a lawn? Tear it up and plant a food forest. No lawn? Grow strawberries and kale in containers on your porch. No porch? Grow sprouts, spirulina and mushrooms in your kitchen or closet. No space at home to grow anything? Volunteer at a community garden or plant a garden at a friend's house. Share food with your neighbors, build community, apply your growing ecological awareness into every aspect of your life.

There are countless ways to begin creating change within our cities and start reinhabiting the urban village. Join a Community Supported Agriculture Farm (**CSA**) or start one of your own (one of the characters in *Edible City* feeds 20 families from one backyard farm!). Or experiment with innovative practices and technology for producing and distributing food within a city: rooftop gardens, vertical growing, aquaponics, hydroponics, rainwater catchment, greywater irrigation. Tear down your fences and develop cooperative gardens with your neighbors. Become a mutually beneficial urban symbiote.

Begin right where you live and spread out like mycelium. Cultivate mutualism with an eye for the larger web of life that sustains us all. Reinhabiting the farm means reclaiming the depleted soil of our mother earth and learning how to be good stewards in order to build a mutually-beneficial symbiotic civilization. This work starts in the microcosm of our own backyards, then spreads out to the macrocosm of our cities and beyond.

Carl Grether

Carl sprouted from the fertile soil of Berkeley, CA. He transplanted himself to Oberlin College where he received a BA in Psychology, and then to the University of Southern California for graduate studies in Film Production. He worked as an editor on the environmental documentary "A Snowmobile For George" and for MTV before leaving Hollywood to start a garden and refocus his attention on permaculture, yoga and integral wellness. He worked as a counselor, teacher, gardener and food activist before returning to filmmaking in 2008 to amplify the message of the sustainable food movement. He has since produced and edited two feature documentaries: **Edible City** and **Occupy The Farm** – as well as recently directed the video that launched The Polish Ambassador's Permaculture Action Tour. Carl also serves on the stewardship council for the Gill Tract Community Farm in Albany, CA and is on the Board of Directors for the Bay Area based nonprofit ERIE (Entheogenic Research Integration and Education). Carl still lives in Berkeley and continues to garden, make movies, practice yoga and provide consultation for individuals and organizations on a variety of subjects.

http://ediblecitythemovie.com
http://occupythefarmfilm.com

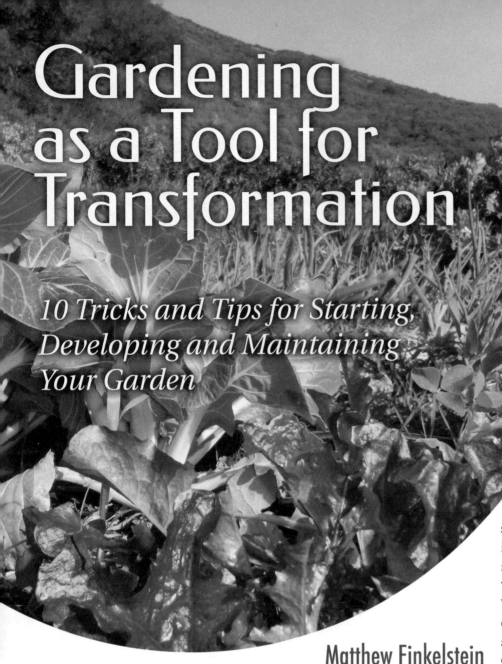

Gardening as a Tool for Transformation

10 Tricks and Tips for Starting, Developing and Maintaining Your Garden

Matthew Finkelstein

My transformation first happened in the garden. I was in second grade and my school had a garden that we occasionally visited. I remember being so intrigued by the compost piles and the blackberry brambles in the creek nearby. Surprisingly, this memory eluded me for many years. Reflecting back upon it, I was very fortunate. That same elementary school "forced" us to plant bulbs on a hillside in spring as part of a fifth grade tradition, ride horses, and go for regular hikes in the surrounding countryside. In high school, we were sent into the hills for two weeks during the spring to botanically catalogue wildflowers. I remember deeply enjoying these activities, however, I was still more turned on by sports and other things a young boy would typically enjoy.

Nature connection (or the lack thereof) is at the heart of many social and environmental issues we face today. In his book "Last Child in the Woods: Saving Our Children From Nature-Deficit Disorder," Richard Louv coined the term "*Nature Deficit Disorder*" to explore the individual and cultural impacts stemming from a lack of distinct experiences with nature. These impacts have affected us psychologically, emotionally, and spiritually, and this conundrum is both disturbing and motivating.

A strong movement exists to help facilitate deeper connections with nature. Organizations, community groups, and schools are working to help foster a greater understanding of the world and our place in it. The *nature connection movement* is rekindling both old and new ways through a variety of different ways and channels.

Gardening is among them and is a tremendous way to help foster deeper connection. Tangible and observational experiences help us to see and feel the patterns and cycles of nature. We come to understand our impacts on the earth and cultivate new ways of being and acting. The growing of food fosters this connection even deeper.

Developing a distinct relationship with our own sustenance cultivates a new feeling of care and responsibility for the well-being of the planet and the environment. We see the interconnectedness of all things and understand how our choices impact the world around us.

Whether for the first time as a child (or with children) or as a seasoned gardener, gardens provide a perfect place to engage with many different kinds of physical, emotional, and spiritual medicine. They provide a physical outlet and place to relax and meditate. They are a microcosm of the world at large. They are an integral part of ReInhabiting the Village.

Growing a garden will come naturally to many people. Whether intrinsic or not, we all have a green thumb. Simply put, the best way to start gardening or develop your garden further is just that: go garden!

There are a few tips and tricks that I have found in my ten years of farming and gardening that do make the process much easier. I hope to convey these here and also to provide you with resources or ideas for further discovery.

Simply put, the best way to start gardening or develop your garden further is just that: go garden!

1. Start with the soil

Your fundamental knowledge of the growing of plants and animals must begin with a fundamental knowledge of soil dynamics, development, and management. Healthy soil feeds healthy plants!

Before starting your garden, or if you'd like to understand your garden better to solve problems or engage in other projects, figure out what your soil type is (sand, silt, or clay) and where your inherent ecology and fertility is at. You can do an intuitive test with your hands (is it dark and rich? Compact? Does it fall apart?). Or you can send in a soil sample to your local garden/farm supply store (for $25-$50; this is generally worth it).

For more information, I recommend Grace Gershuny's book "Start With The Soil."

2. Create a fertility plan

Fertility is the essence of life. It is the dynamic cycle of growth and decomposition that feeds the plants and organisms in your garden.

Fertility is primarily focused on the transference of ionic minerals, mainly nitrogen (N), phosphorous (P), and potassium (K). Nitrogen primarily drives vigorous plant ground, phosphorous drives root growth and flower and seed production, and potassium is responsible for overall health and disease. There are many other trace minerals involved, each with their own purpose and impact, but for now, let's stick with these.

Green plants and generally nitrogen-rich and dead/desiccated plants are generally brown and carbon-rich (the nitrogen having escaped into the atmosphere). The carbon-nitrogen balance is integral to understanding and creating your fertility plan.

Humus is one of the most important parts of a soil's fertility. It is made of decomposed organic matter and provides the basis for a soil structure to hold on to water and minerals. An abundance of humus via compost, cover cropping, and normal plant growth and decay will provide the basis for your soil to maintain a healthy biology and fertility.

When thinking about your fertility plan, practices you may consider are:

- Composting
- Cover cropping
- Application of supplemental fertility (aka fertilizers)
- Active soil biology and ecology
- Vermicomposting (worms) & mycelium (mushrooms)
- Groworganic.com has a great section on fertilizers – I highly recommend checking out their offerings and informational guides

3. Understand your climate

Plants are very sensitive to your garden's climate and micro-climates. In choosing which plants to grow, it's integral to understand what their preferences are and what your garden offers. Are you tropical? Sub-tropical? Temperate – 4-seasons? Temperate – 2 seasons (i.e. warm/cold)? Within your garden, are there distinct microclimates due to south/north-facing exposure, wind or wind-blocks, elevation rises and cold sinks? These are all things to pay attention to over time.

When it comes to seasons, plants are also sensitive to day length. Concurrent climatic changes occur as we approach equinoxes and solstices. This is something to keep in mind when scheduling planting young seedlings or trees.

Both annuals and perennials are subject to short-term climate variations, but perennials are especially subject to long-term changes. This is especially true when it comes to deciduous (i.e. leaf-dropping) fruit trees. Consider your overall warmth and research your area's chill hours when planting deciduous fruit trees. This information can usually be found with your local Ag Extension or County Advisors, however, as the earth's climate changes, you may need to do your own tests observationally.

4. Different plant characteristics

All plants are different. There are, however, botanical trends that can make your understanding of growing and identifying plants much easier. I highly recommend developing a basic understanding of botany and the important plant families.

Check out Thomas J. Elpel's "Botany In A Day."

Before planting, it is important to do your best to understand the plant's inherent growing characteristics and needs. Is it a perennial or annual? Does it want full sun or partial shade? Will it tolerate frost or does it need high temperatures to produce?

There are many ways to develop this knowledge – I recommend receiving as many seed catalogues as you'd like and placing them around the house (bathroom and bedside are very nice). Please see the list of seed catalogues in the resources section below.

For gardeners on the West Coast, Sunset Magazine's "Western Garden Book" is a wonderful compendium of knowledge.

Pruning and training are often needed for some plant species – especially fruit trees (particularly stone and pome fruits). Check out Peaceful Valley's "Pruning 101" video for a quick intro (http://groworganic. com/organic-gardening/videos/pruning-101) and consider participating in a local workshop or class series through organizations like the Master Gardeners or Rare Fruit Growers in your area.

There are also optimal harvest techniques. These will vary widely based on the plant. Certain techniques, such as picking kale leaves from the bottom and down to the stem can not only prolong the life and overall quality of the harvest, but can also serve to minimize disease and pest issues. Observation from other farms and gardens along with your own experience and research will help tremendously!

5. Garden design

When it comes to garden design, one size does not fit all. Every garden is unique and must be approached as such. Variance in climate, topography, soil, and resources will all affect what your approach and design should be. Certain designs and techniques, while beautiful or seemingly more "eco-friendly," may not necessarily apply to your particular space, so come at it with a critical and creative eye.

The best gardens are designed through thoughtful observation and experience. You need not necessarily ascribe to the hottest trends in permaculture (like raised beds or hugelkultur) if those techniques don't serve your situation.

Be open and flexible and let your garden design over time! It's totally OK to take a season or two to just observe the land and let the design come to you, or design your garden in such a way that future changes will be easily done.

There are lots of books and resources out there to help you design your own garden, but to start, I recommend Toby Hemenway's "Gaia's Garden."

6. Pest management

Simply put, diversity in a garden is your best ally. Your role as a gardener is not to 'manage' or 'control' pests, but rather to create an environment that mimics a naturally thriving eco-system. The pest-predator relationship will naturally unfold.

Knowing your garden's inherent needs and characteristics, plant an abundance of variety. Include flowers and insectary habitats to provide food and shelter for predatory insects and birds. Encourage the world to come into your garden!

This is called integrated pest management – there is a great section of resources, books, articles, and videos on the subject at http://groworganic.com/organic-gardening/books/pest-management

7. Watering and irrigation

Water is our most precious resource – it's important to know how and when to water your garden.

There are many different techniques to watering, including: overhead, drip, micro-sprinklers, flood, etc. You should consider your budget, plant needs, and water availability as you explore your approach.

Different plants have different watering needs as well – some may need to be watered every day (i.e. leafy greens), some like to become established and then water-stressed (i.e. tomatoes, to develop a deep and vigorous root system), and some may need only periodic watering (i.e. perennial fruit trees or water-hardy plants).

If you're interested in automating your irrigation system on a timer, I recommend spending some time watering by hand first to better understand your watering needs.

Also, the best way to store water is in the ground! A healthy and vibrant soil ecology and poly-cropping of plants will hold and maintain much more water. Water catchment systems are only applicable if your particular situation calls and allows for it.

8. Seed starting, planting, and transplanting

Building upon your knowledge of the unique plant characteristics, it is also important to know a bit about how to plant seeds and transplant seedlings into your garden.

Some plants have preferences – as in, they would rather be directly seeded into the garden, whereas some may have more success being started in a greenhouse or in trays and transplanted into the garden.

Transplanting allows for greater organization of those crops – you can place them according to the spacing that they would prefer.

Direct-seeding is usually preferred by common crops such as carrots, beets, and corn. You can also direct-seed such things as lettuce/spring mixes, spinach and arugula to get a dense "carpet" and outcompete weeds.

Most seed catalogues will have good descriptions and explanations of how to plant – check out the list in the resources below.

Make sure to water freshly transplanted seeds and seedlings well! Don't overwater and don't underwater. The soil (especially below the surface) should be moist at all times – like a wrung-out sponge.

9. Observation – Plant health and upkeep

Observation and attendance to your garden is your best ally. I highly recommend visiting your garden daily, as often as you can.

Catching problem signs before they proliferate will help you immensely in the long run. (And also not stressing about little things – you may get a few holes in your leaves!)

As you grow in your gardening experience, you'll begin to learn the serious distress signals of plants – do you have a just a few aphids or are they taking over the plant? Are your leaves vigorous and green or yellowing? Is the plant continuously putting on new growth or does it look stunted?

Use your intuition – you'll know if it's a crisis or not. You may have to weigh out certain options, sacrificing some things in favor of others; that's OK!

And remember, your best starting point comes from all of the points listed above so far – healthy soil,

intelligent planting, right watering, diversity, etc. will give you an immense boost towards a thriving and vibrant garden.

10. Lastly, go experiment!

Curiosity, observation, and trial and error will help you develop your fundamental knowledge and skills as a gardener. No matter what, you can't go wrong!

It's always good to do a little bit of preliminary research before trying out new techniques or new plants, but don't get too caught up in it.

Go visit as many gardens and connect with other gardeners in your area – that will give you an abundance of ideas and knowledge.

Matthew Finkelstein

Matthew Finkelstein is an organic farmer, permaculturalist, holistic health advocate and educator. He brings nearly ten years of experience in the field, including expertise in production farming, urban farming, home gardening, herbalism & wildcrafting, holistic land management, the natural products industry, and holistic nutrition. He also has had extensive experience in organizational strategy and development from working with his family's management consulting firm for several years.

Much of Matthew's experience stems directly from the plants themselves. Growing up in the green hills of Marin County, he received his undergraduate degree in Psychology from UC San Diego and completed the Apprenticeship in Ecological Horticulture at the Center for Agroecology and Sustainable Food Systems at UC Santa Cruz. Matthew has chaired the board of non-profits, organized educational events, worked on and managed organic farms and land projects. Matthew continues to further his experience, knowledge, and understanding with both experiential training and education.

Currently, Matthew is directing his efforts towards what he calls garden & wellness empowerment – working directly with home gardeners and land-based organizations to develop their projects further. He will be teaching and travelling along the West Coast and abroad, including several festivals and gatherings. For more information or to contact him, please visit http://mateosol.com

Re-Building the Village

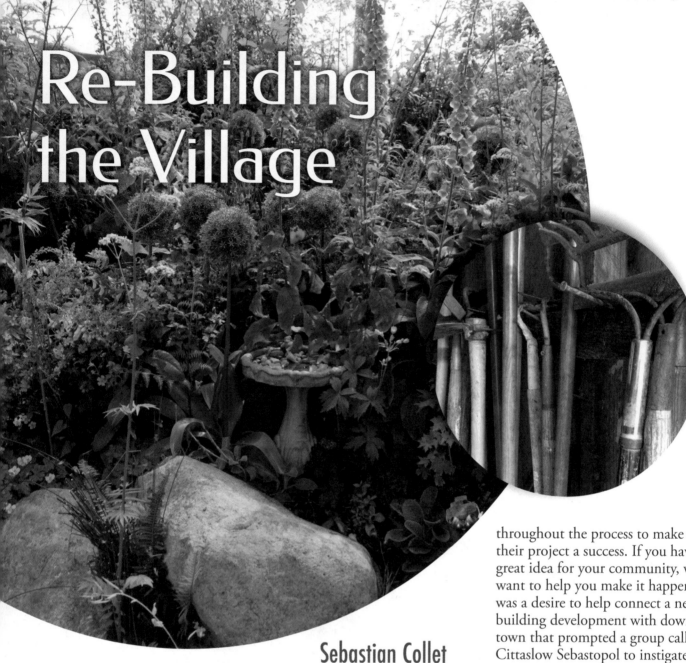

Sebastian Collet

The sun was just about to emerge over the misty morning horizon as we carefully drove up to our first day's target. With a trailer packed full of paint, brushes, barricades, brooms, mops, and chalk, we knew in our hearts that between this and the eleven other projects, we were planting seeds that would enliven this town for many years to come. It was the first day of our ten-day long Placemaking festival with the theme "Reclaim the Commons," and with a highly coordinated effort

we were about to literally paint the town. As I placed official barriers in the street with my friends, I felt for the first time like I was actually a contributing member of our town. It is our responsibility as citizens to be an active part of the village; it's up to each of us to create the future that we want to live in.

Placemaking is about empowering people to take ownership and responsibility for the place that they live in. We invite people to come up with ideas to improve their own neighborhood, and then help them

throughout the process to make their project a success. If you have a great idea for your community, we want to help you make it happen. It was a desire to help connect a new building development with downtown that prompted a group called Cittaslow Sebastopol to instigate the large murals we painted in over 300 downtown streets. Over 400 people showed up to help paint the murals, which included a 'Spirit Bird' followed by a dozen salmon swimming upstream to the 'Spirit Canine,' and ultimately the 'Spirit Tree' in an intersection. Not only was this the most diverse group of people I'd ever seen come together in Sebastopol, but I met five different neighbors who live on my street while we were painting! We had a diverse range of projects including: an earthen bus stop, with hand-sculpted salamanders and approved by Sonoma County Transit; community kale patches planted in various locations around town to give people access

to fresh, healthy food; a circular community bench and pizza oven made of adobe bricks and cob, where neighbors could come together to talk or host a pizza party for the whole neighborhood. Another project, led by Daily Acts and designed by the Permaculture Artisans, was a complete landscape transformation at City Hall, planting hundreds of plants including edibles, water-wise natives, and indigenous plants. Throughout the whole process we were meeting our neighbors and making friends by working together towards common goals.

It was at a Transition Sebastopol meeting that the Village Builders group first came together. If you're not familiar with the *Transition movement*, it is a "vibrant, grassroots movement that seeks to build community resilience in the face of such challenges as peak oil, climate change and the economic crisis."

The Transition movement... is a "vibrant, grassroots movement that seeks to build community resilience in the face of such challenges as peak oil, climate change and the economic crisis."

Our primary objective was to activate the community; to empower and inspire people to take action in improving their own community rather than waiting for anyone else to do it for them. At the heart of this work is *Placemaking*: the act of collaboratively transforming under-utilized 'spaces' into special and useful community 'places.' Based on the success of the City Repair Project's Village Building Convergence (VBC) in Portland, Oregon, we decided to host a 10-day long placemaking festival called the Sebastopol Village Building Convergence. I had the pleasure of working with the Portland VBC as the Placemaking coordinator, and several other members of our team had gone to take part in their event as well, so there was already a high level of familiarity with the event.

Village building is mostly about the social act of building community. Despite the rampant growth of social networking sites like Facebook, social isolation is at an all-time high. According to a Duke University Study, from

A Village Building Convergence (VBC) consists of four major components that help build community:

1. People Build Together.

At the heart of the VBC is the Placemaking projects where people plan, organize, design, build, garden and paint together, creating beautiful additions to their community while getting to know their neighbors.

2. People Eat Together.

Typically, every evening people will gather at a central venue and share food together and yummy desserts.

3. People Learn Together.

In addition to learning at the Placemaking sites by day, there are evening presentations and discussions led by local as well as some non-local leaders.

4. People Dance Together.

After a long day of building and learning, we celebrate with music and dance!

1984 to 2004 the number of people with no one to discuss important matters with has tripled to 25 percent.

Social isolation has been linked to high blood pressure, sleep disorders, increased risk of depression and suicide and even a 26% greater risk of death.

By rebuilding our village connections we increase civic engagement and improve the health of our community.

A Village Building Convergence is the greatest model I've seen for cultivating community and inspiring people to actively participate in the place that they live. If a small town like Sebastopol and a big City like Portland can host successful VBCs, I see no reason that it can't happen anywhere else. If you are interested in hosting your own VBC or something like it, feel free to contact us. Here are some tips that we've learned for any Village Building event:

No need to reinvent the wheel.

We are fortunate to have had so many great people come before us; it allowed us to be creative with how we wanted to craft this event for our town without having to reinvent the wheel. We used the Portland VBC as a template and then let the details fill in as we went. Our group began as a working group of our local Transition Initiative; this gave us a framework for meetings including ground rules and a decision-making process. For our evening venue we chose to utilize the local grange hall, which had almost everything we needed, including good indoor and outdoor space and a commercial kitchen. I mention these organizations because they were integral to our VBC here, and they exist in many locations and may be a good resource for your next event.

Reach out far, wide, and early for Partners.

There are so many great people and organizations out there working towards common goals. Make a point of reaching out to other organizations and let them know what you are planning before you have all the details figured out. Avoid potential conflict by giving appreciation to what other groups are doing and inviting them to help out if they want. When approaching the City and a couple other groups here in Sebastopol, we gave presentations of our vision to them and offered ways that they might help participate.

Be inclusive and allow others to shape the outcome.

I may be a competent designer on my own but the richness of a community-building project comes from designing with the community. Not only are there additional ideas to pull from when working together but there are also more people engaged with seeing the process through and taking ownership when they see their ideas represented. I've had the pleasure of working with quite a few schools, and I know for a fact that many of those kids are the ones looking after the project for years to come.

The sun was shining and despite spending the previous nine days working and dancing together, there were big smiles all around as we followed the marching band down the freshly-painted streets, taking long breaks to dance in celebration together at each mural and intersection along the way. It felt great to occupy the street for the fourth time that week, dancing with costumed friends old and new. The first Sebastopol Village Building Convergence was a great success! We are currently planning our next VBC and working on a Placemaking Ordinance with the City to make projects like these even easier to implement in the future.

Are you inspired to plant seeds of change in your town?

Get engaged!

You, who live in your own community, wherever it is are the one to make a change!

Sebastian Collet

I grew up in California and fell in love with the natural world at an early age.

During my teenage years in Sebastopol, CA, I awoke to the understanding that the way we are living on this planet is destructive and unsustainable.

So I moved to Oregon to study Architecture at the University of Oregon. I learned as much as I could about why we live the way we do and what other options we have.

After college I moved to Portland, OR, to work with Mark Lakeman and The City Repair Project. I served as Placemaking coordinator to help facilitate projects for the Village Building Convergence.

I am currently one of the main partners in a fabric architecture company called GuildWorks but am also working on other projects in Sebastopol, CA, such as the Sebastopol Village Building Convergence.

Please come join our next Sebastopol Village Building Convergence this September!

Find out more and sign up for our newsletter at http://SebastopolVBC.org.

Direct Action for a Regenerative World

Ryan Rising

The chain holding shut a gate to 20 acres of underutilized land gets cut with a pair of bolt cutters. Three thousand people pour beyond the fence. Farm rows begin to get tilled. A kitchen is established, music plays, and people begin to plant a community farm.

This is direct action–the process of directly creating the changes we want to see in the world around us. It is based on our ability to organize together to find common ground and mutually agreed upon projects to implement. Sometimes, direct action can be an impediment to the actions of another entity, such as standing in the way of a bulldozer. Sometimes it can be creative and constructive, such as planting an edible garden on land that is to be

developed. Sometimes it can be both, as embodied by the Gill Tract Farm Occupation, a direct action that reclaimed a piece of University of California owned land in the San Francisco Bay Area, before it was to be paved over in 2012.

The Great Turning

Joanna Macy, activist and author whose work examines the intersection of spirituality and psychology with the causes of peace, justice, and environment, offers the framework of "three spheres of action," through which we catalyze change in this time of transition. "The great turning," she calls it. What is this transition? It is a transition away from an economic model that extracts labor, topsoil, and minerals, uses

them as "resources," and then throws them "away." It is the transition toward a cyclical eco-nomy (literally, "management of the home") that distributes energy across pathways of interconnection and returns it to from where it originates. It is the transition away from highly centralized and hierarchical governance systems; and towards collaborative decision-making and more horizontal power structures. It is the transition away from an environmental framework that sees nature as the other; and toward an ecological framework that places humans within an interconnected web of relationships.

Buckminster Fuller is quoted as saying that, "You never change things by fighting the existing reality. To change something, build a new model that makes the existing model obsolete." Well Bucky only had it half right; or according to Joanna, a third. Yes, let us create the alternative systems and ways of relating that we want to step into. This is

the world we are creating every time we share our gifts with friends, when we communicate with compassion, when we grow food with our neighbors, craft medicines from wild plants to heal our children, and come to decisions through shared agreements. However, let us not leave essential tools out of our toolkit for social change. We also have the power to stop the damage that is being done, to stand in the way of ecological devastation and human harm, to resist. In our creation of the better world we are moving into, and our resistance to the harm being perpetuated by systems of domination, extraction, and exploitation, we change the story of the time we are living in—our "cultural narrative"—the third sphere of action according to Joanna.

Direct Action. It is not necessarily tree sits, blockades, and sabotage. This is only a small fraction of the tactics within this great compartment of our toolkit. Direct Action, rather, is those actions we take as individuals or collectively that take power into our own hands and craft the world we want to see. Embracing the tools of direct action frees us from being limited to the methods of social change that institutions of power have designed themselves to deal with—protests, petitions, cast votes, electoral politics—that at best advocate, and at worst beg, for change to be made from above by the same power structures that hold the barriers to the future we envision tightly shut. It is action to render these power structures obsolete through building consensus within our communities and making what we want realized—an empty city lot turned into a guerrilla garden, an abandoned bank transformed into a community center, the development of a Nestle water-bottling plant at the foot of Mt. Shasta halted and turned into an eco-village, our neighborhoods thriving with common space and a roof over everyone's head.

THIS IS WHAT WE'VE FORGOTTEN - THAT WE'RE NOT OBSERVERS AND CONSUMERS. WE ARE ACTORS, CREATORS, PARTICIPANTS, CO-ORGANIZERS, COLLABORATORS; AND WE ARE POWERFUL.

Resources to Relationships

The last fifty years have seen a great change in our worldview from that of capital, to that of resources. Where before we saw nothing but dollar signs; we now see resources: minerals, trees, water, soil, humans. The great challenge of our generation is to encourage this shift in perception one step further, and in doing so, bring us back full circle to a way of seeing the world practiced throughout many ancient traditions, and today in many indigenous cultures. We are the generation that will make the shift from "resources" to relationships. This is what permaculture is all about—relationships. More than permanent agriculture, permaculture is about the ecological design of permanent culture; culture being all the ways that we humans relate to one another, to the landbase and watersheds, to the systems we create and interact with for sustenance, and to our ecology as a whole.

People within a planetary ecology. That's what we are. We have somehow separated ourselves from this picture, or placed ourselves on top of the pyramid of all life. Now it is time to re-integrate, if not by choice, than by the forces of nature. Of all the great problems and inequities of our world—from war, police brutality, and other globalized methods of coercion, to economic exploitation and wealth inequality—climate change is the crisis that will remind us that we are of a common place, literally. We have largely left behind the long historical understanding of land as the commons—the places we all steward together. We have entered a time of dichotomization between what is private and what is public, with public lands having steep limitations on what kinds of truly subsistence-providing activities can take place on them; and private lands on which hydraulic fracking, mountaintop removal, agricultural contamination, and toxification of groundwater are now acceptable norms of an ever-growing and "efficient" global economy. Alterations we are making to our climate will not let us forget this one essential point–that the land, waters, and ecological systems compose a web of interconnection of which we are a part. We affect the whole; and the health of the whole affects each of us.

So what are we doing? We're writing the future through our action, and

those of us organizing, building, planting, and creating together invite you to join us. Be aware of the dominant narrative—the laws and limitations placed upon "what's possible" by the culture that pervades our perception of reality—but never let this limit our imaginations, nor our action. When Oakland Unified School District shut down four elementary schools in the City of Oakland, California in 2012, the teachers, students, and parents at Lakeview Elementary held a sit-in on the last day of school. While teachers and students across the state were packing up their desks for summer vacation, those at Lakeview Elementary opened up their tents to camp on the playground overnight, and through the weekend. When Monday came, the people opened up the school building and declared that if the city government couldn't afford to keep these four elementary schools open, they would do it themselves. The People's School for Public Education was born.

For the next three weeks, students of the People's School attended classes taught daily Monday through Friday on art, theatre, social justice, gardening, and physical education, all taught or staffed by unpaid volunteers. There were reading and writing assignments, group brainstorms, lunch and snack times. All meals were prepared by the community. There was no financial cost for food, materials, or classes to the participating students or their parents, and no government or corporate funding. The school was functioning through a voluntarily-constructed gift economy, and decisions were made by consensus of those who participated.

Parents of Lakeview Elementary were told by the City that come the following year their six to twelve year old students would have to attend schools more than a mile away and to which the city would not provide any transportation. And here, people were not only saying, but demonstrating, that they could implement the solutions themselves—and in many cases do it better, as the People's School showed through its integration of classes on social justice, gardening, and sustainability. And this is what we must learn—we are capable, we have everything we need to implement the alternative structures and ways of relating we imagine. In fact, the dominant institutions that claim to provide these services for us operate off of our energy. It's us who actually make them work, often to the advantage of the people at the top, but we can flip this pyramid, or flatten it if we wish. "Horizontalidad," as they say in the factory takeovers throughout South America where workers fire their bosses, make decisions together, and share equitably in the fruits of their labor. It's in our best interest to do so. People's movements and underground resistance throughout history demonstrates that this is commonly understood, and is the will of the people. But now, in this time where we stand on the edge of total ecological collapse, it is imperative that we do so.

And we are. In Jacksonville, Florida, during the Permaculture Action Tour with music producer The Polish Ambassador, a neighborhood of mostly black veterans turned an empty lot on their street into an edible forest garden—a food forest that would provide produce for the community. Denver's Action Day built a geodesic dome greenhouse in a community garden that had once been a landfill in the middle of the inner city, amongst four other project sites: an urban indoor farm that provided fresh greens; a collective closing the loop on food waste by turning it into compost and growing herbs and vegetables, a school campus implementing regenerative agriculture, and an edible garden surrounding the house of a man in a food desert neighborhood who now grows the healthier alternative to the food he wasn't able to afford at the grocery store.

In Our Own Hands

The March Against Monsanto in San Francisco comes to its destination on a sunny San Francisco afternoon in 2013. The lawn of the park in the financial district gets torn up. An affinity group turns the soil and plants kale, chard, and collards. Marchers who thought that at most they could use their voice, and maybe their protest signs, to demonstrate against the harm committed by the corporate agricultural giant, now take into their hands a seedling for the first time in their lives, and plant it in the ground at their feet. A moment of empowerment. A seed of the future.

Direct action bypasses the often ineffective and energy-draining avenues of change offered to us by political and economic institutions—protest, petitions, letter writing, political campaigns—and instead involves people self-organizing to directly implement the change they desire. We are casting off the narrative that's been forced upon us of what's possible in this world. We're re-imagining how we share space in common, and redesigning our relationship to the land. We're building autonomous communities and growing the bioregional, regenerative food systems that will foster biodiversity and total ecosystem health while meeting the needs of people. We're turning Bank of America's foreclosed homes into community houses and urban homesteads, reclaiming plots of land in the city from the back pockets of financial speculators, and turning them into medicine gardens and food forests. We're re-appropriating the infrastructure of a world that no longer serves our needs and we're creating a world based on care for people, care for the earth, and a socially just and equitable future for all. We're taking direct action, and we ask you to join us.

Ryan Rising

Ryan Rising is a community organizer and permaculture educator based out of the San Francisco Bay Area. Ryan most recently organized the Permaculture Action Tour with music producer The Polish Ambassador in the fall of 2014. The tour visited 32 cities across the country, holding an Action Day in each one and bringing together up to 400 people at a time to implement projects including public food forests, edible gardens, natural buildings, and greenhouses.

Ryan focuses on creating community access to land for local food growing and regenerative living – connecting people to take direct action and transition to a resilient way of life; and social permaculture – the ways we make decisions, build resilience, resolve conflicts, and organize in community.

Following the Permaculture Action Tour, Ryan continues to organize Permaculture Action Days for festivals and other events as well as permaculture courses and educational spaces. A certified permaculture designer with a degree in Peace and Social Justice Studies, he also co-founded a regenerative, productive urban farm in the East Bay and now organizes with the Omni Commons – a community space and education center in Oakland, CA.

Check out his work at

http://PermacultureAction.org

City Repair Project

and the Village Building Convergence

The Portland, OR based City Repair Project fosters thriving, inclusive and sustainable communities through the creative reclamation of public space. City Repair facilitates artistic and ecologically-oriented placemaking through projects that honor the interconnection of human communities and the natural world. We are an organized group action that educates and inspires communities and individuals to creatively transform the places where they live.

The many projects of City Repair have been accomplished by a mostly volunteer staff and thousands of volunteer citizen activists. One of our featured projects is the yearly Village Building Convergence which is a 10 day citywide Placemaking and Urban permaculture community festival.

We provide support, resources, and opportunities to help diverse communities reclaim the culture, power, and joy that we all deserve. We commonly work with Community Based Art, Natural Building, Permaculture.

The City Repair Project has over 18 years of experience, numerous awards and international recognition in the areas of placemaking, civic engagement, community-initiated ecological development, urban permaculture and natural building, urban planning policy and advocacy, and tactical urbanism. Their team provides consultancy services to help other community leaders with:

- Creating a Village Building Convergence (VBC) an annual multi-day placemaking festival in your city, town or community

- Navigating/developing policies and legalities for citizen-led public space initiatives

- Leading design charrettes

- Building a sense of place

- Design Services in: public gathering places, community art, and ecological landscaping, building, and architecture.

http://cityrepair.org
http://villagebuildingconvergence.com

Community Land Projects

THE NATURAL WORLD IS THE LARGER SACRED COMMUNITY TO WHICH WE BELONG. TO BE ALIENATED FROM THIS COMMUNITY IS TO BECOME DESTITUTE IN ALL THAT MAKES US HUMAN. TO DAMAGE THIS COMMUNITY IS TO DIMINISH OUR OWN EXISTENCE. Thomas Berry

Imagine this. In the morning you harvest ginger root to turn into an herbal tincture for supporting the immune system. After stopping by the children's center to drop off supplies for a science project, you walk across the field to your neighbor's house to share the peaches just picked from the orchards and food forests. She thanks you by offering you some honey she gathered from the hives, and you share inspiration for a new art studio that was sparked from the latest cob building workshop attended by several participants last Saturday. After tending to your weekly service project by turning the compost pile and mulching the flower beds, you join the council meeting scheduled to discuss plans for a future aquaponics site next to the greenhouse. Later that night, after the community meal, the young and the old gather by the fire under a blanket of stars celebrating with storytelling, music, and laughter. While this is a snapshot of community life in an ecovillage, versions of this exist all over the world, everyday, as people choose to share life intimately connected to each other's well-being in stewardship of a place they all call home.

The single greatest lesson the garden teaches is that our relationship to the planet need not be zero-sum, and that as long as the sun still shines and people still can plan and plant, think and do, we can, if we bother to try, find ways to provide for ourselves without diminishing the world. Michael Pollan

Ecovillages are comprised of those who see themselves as an intentional community collectively working to tend the land, share their resources and skills, agreeing to make decisions together based on an alignment of core values and ethics. Some ecovillages are connected through family bloodlines, were seeded from religious or spiritual affiliation, or are comprised from individuals, families, and peers who act as a "chosen family," engaging social and ecologically regenerative practices that minimize environmental impacts, provide for the needs of the collective, honor the purpose and value of each member, and strive for balance in the whole.

This phenomenon is a Global movement, with ecovillages in many countries around the world. Through organizations such as the Global Ecovillage Network (GEN) these communities are now connected to other ecovillages forming a network of shared intelligence, best practices,

and resources. Some examples of internationally renowned communities include Findhorn in the UK, Damanhur in Italy, Auroville in India, Christiania in Copenhagen, or Sarvodaya in Sri Lanka, which is made up of 2,000 villages working in partnership with Sarvodaya.

Usually housing around 30-500 people, ecovillages have become a small-scale version of traditional villages. By choosing small-scale micro systems that attend to needs in a holistic framework, these micro-villages often incorporate cottage industry that fuels the living economy providing nourishment to the stewards of the land, to the surrounding communities, and feeds the land itself. The aim is towards self-sufficiency and the regeneration of resources to continue ensuring the success of the community and bioregion.

These small scale communities also spark the integration of whole systems design incorporating practices from permaculture, biodynamic, or sustainable farming, natural building

and food preservation, small engine mechanics, innovative approaches to water, waste, energy systems, animal husbandry, transportation, ecological restoration and preservation, natural medicine, immersive skills based education, and an array of governance models to help members find cohesion in shared leadership and decision making.

The term ecovillage is not only rural based, but has also taken root in the urban landscape as adaptations of rural templates have been developed for a cooperative style of living practices by those inside city limits. There is so much potential to designing and redesigning our urban environments to model the successes and approaches of rural ecovillages and many examples exist of thriving community interdependence.

Many in mainstream culture attribute ecovillages with the "Back to the Land" movement of 1970's America, which compelled many

to leave the devastating impact of pollution and consumerism typical of urban sprawl to reconnect with a simpler way to live. Yet the idea of humans settling together is certainly not new and the thriving ecovillages of today hold key strategies for the challenges facing our current living systems. The scalable relevance of the systems practiced in these eco-villages, focused on local coherence yet connected to a global interface of intelligent design and implementa-tion, provide a blueprint for regional resilience and a pathway forward from our currently collapsing global infrastructure. Whether urban or rural, the lessons gathered by people learning to work together and live together in a symbiotic relationship to the ecosystem around them are invaluable and necessary for the jour-ney ahead.

Jamaica Stevens

TYPES OF COMMUNITIES

Intentional Community

An intentional community is a planned res-idential community designed from the start to have a high degree of social cohesion and teamwork. The members of an intentional community typically hold a common social, political, religious, or spiritual vision and often follow an alternative lifestyle.

Ecovillage

A community whose inhabitants seek to live according to ecological principles, causing as little impact on the environment as possible. Advocates seek infrastructural independence, a cooperatively sustainable lifestyle, and inte-grated living systems.

Urban Ecovillage

An Urban community whose inhabitants seek to live according to ecological principles, often sharing resources, garden space, and cluster housing.

Co-Housing

Intentional collaborative housing, with indi-vidual homes and shared outdoor and gather-ing spaces creating a community that makes decisions together about design and operation of their neighborhood.

Living and Learning Centers of the Future Now

A. Keala Young

SHIFTING PEOPLE BACK TO ACCESS FOR PRODUCTIVE AND CREATIVE LIVELIHOOD, BEYOND "ACCESS AND MANIPULATION," IS PART IF NOT PARCEL OF AN EVOLUTIONARY MOVEMENT. Paolo Soleri

Innovative Models for Living

The new self awareness and context for evolving our perspectives and our engagement happens through new systems of design and new models of ownership and endeavor that are also deeply grounded in natural systems. The dynamics of cooperation at all levels can be examined and discussed to consider how to leverage for the human potential movement to thrive in a way that is deeply integrated into a planetary context. How do these efforts grow and how do we continue to gather together the companions and resources to continue to weave a future of potential that sets the context for co-creating our future?

Creating a Life Together

Diana Leafe Christian's book, *Creating a Life Together,* was an early influential tool that I picked up during the Findhorn Ecovillage Design Training which we used as one of our maps for guiding the creation of Atlan. Reading this book combined with a thirst for a deeper study of permaculture led me to an internship at Lost Valley Educational Center in central Oregon where I had the opportunity to study with both Diana Leafe Christian and David Holmgren, co-originator of permaculture. Several key relationships at that time further led me toward grounding into the bioregion of Cascadia in the Pacific Northwest and the City of Portland including the inspiration of the City Repair Project and their annual Village Building Convergence which I had also learned about at Findhorn. Our Atlan founding group formed through these synergistic relationships and included our collaborations in the production of the Beloved Festival.

Learning and Leading from Examples

During the early formative phase of Altan, to compliment the existing foundation of ecovillage design study, we sent two of our founding members to Occidental Arts and Ecology Center (OAEC), to participate in one of their ongoing empowering workshops for creating community; Starting and Sustaining Intentional Communities. OAEC has a rich history and is an inspiring example of the commitment to weaving the cultural values of the arts and ecology for creating a thriving model of sustainability. Now at Atlan we are joining the communities that offer developing programs as well to support other groups and individuals through experiential learning programs.

Agreement and Resonance as a basis for Governance

Across many communities the study and adoption of practices for building resonance and agreement through mutual understanding have proven essential components for success. Driven by the need for clear processes of collaborative decision-making and communicating differing viewpoints which were often lacking in our cultural education, we are led to further study and develop cultural modes of communication, conflict resolution and resonance building.

At Altan we have adapted a variation on Sociocracy and Dynamic Governance which is a consent based model for decision-making. The model of Dynamic Governance with it's process of double-linking between decision-making groups fits well

with designing for multiple centers of initiative and honors the spirit of consensus while also empowering individuals and groups.

We have formed a community council for building resonance and continue to hone our skills through practices of compassionate communication, including recent studies and application of restorative circles, a group discovery process for addressing conflict.

Emergent Networks

The quest for community and land-connected living is alive in the hearts and minds of many of us. There is indeed a legacy from many who have followed this quest to reclaim that connection. Many resources abound to support these endeavors including the Global Ecovillage Network and the Fellowship of Intentional Communities as examples of networks that have formed to create a support system for emergent communities. Indeed for those interested there are many communities existing today that are seeking more members and inspired energy. A great part of our work moving forward continues through developing better tools for matching up individuals and families with opportunities in existing ecovillages and communities.

The work that we continue through Atlan and through this network of emergent community Living and Learning Centers is a further flowering of the human potential move-

Atlan

A brief story of Atlan: A Living and Learning Center is offered as a case example in the context of existing and emergent networks of communities. We have taken great inspiration and direct study of other projects as a fuel toward our success which we further promote as a model for other groups and communities.

Mission: Atlan is a living and learning ecovillage dedicated to the artful co-creation of healthy living systems, celebrating the connectedness and diversity of all Life.

Vision: Atlan provides sanctuary for the creation of sustainable culture through the holistic integration of healing, art, and design. Our ecovillage demonstrates permaculture and regenerative principles while engaging a network of resonant communities.

http://atlancenter.org/

ment. Sharing in the inspiration of VillabeLab's concept for a "distributed innovation laboratory and incubator based in existing ecovillages and other sustainable and regenerative communities and networks," we are actively engaging with projects and partnerships that exemplify this synergistic network pattern of growth and resilience.

Resources

Inspiring resources abound. Throughtout this book and through the Resources section on http://reinhabitingthevillage.com/ we encourage you to explore. Further exploration through online networks, Communities Magazine, and books such as ***Sustainable [R]Evolution: Permaculture in Ecovillages, Urban Farms and Communities Worldwide*** also feature inspiring examples from around the world. Whether you travel to learn from others by example, host others to learn from or weave through online and media resources, we invite you to find the great places to continue following your dreams and creative potential for building community *on the ground* everyday.

What is a Village?

People, in Places, Doing Projects... Together

It isn't the cliché, standing out there alone, saying "It takes a Village to ...". Or this one, that "Many Hands Make Light Work." It's the actions behind the words, the wordless energetic exchange that occurs when people show up and work together to accomplish a particular endeavor that supports a whole system functioning well.

A village is many people, collaborating in many places, each doing their work in harmony with others toward a common purpose. It is working together for the regeneration of the village, for the well-being of the people which is dependent on the and for the well-being of the places in which we live, work, raise our children, celebrate life, share growing old, die and go back to the soil.

It isn't a physically isolated location on the planet, with fences or walls, borders drawn on maps. It's a vibrantly alive, constantly shifting network of supportive relationships built on an underlying foundation of trust.

Trust is not a commodity, something that can be traded, bought or sold. To inhabit the village, to engage in the network you must show up, participate, and acquire your standing on the foundation of trust through the shared experience of collaboration.

Experience can only be shared when we show up, in the same place, at the same time. And that is where we build that trust–it begins with simply showing up, and it is cultivated with active participation, in good faith (and humor, preferably!), again and again, whatever the outcomes may be through your individual perception.

The village experience may be many things, and history is the telling of tales made up about experiences as seen from one or several sides. We can call it a good or a bad experience, a healthy or unhealthy exchange; there will always be differences of opinion and perspective on a shared experience, as in The Fiddler on the Roof, "It was a horse! It was a mule!" Language often creates divides, polarizing a story into opposing parts played out by each participant; however that is not what our story is about! The story of reinhabiting the village is about learning and growing together as a whole. It is our collective story, the sum of all our parts, adding up to… together.

This particular story is of these past seven years, and of the people that are now connected through shared experiences of home at "Our Sacred Acres" or "OSA." Here on this land, we have built a foundation of trust in Cascadia, a cultural bioregion of the Pacific Northwest, with a mission to cooperatively self-organize our village. OSA is in the heart of Cascadia, and through love for our home and land, the people are uniting on organically inspired, collaborative community projects to restore and protect our places… and ourselves.

Our Sacred Acres, located in the Coast Salish/Snohomish area of Northwest Washington State, is more than these three acres for which my legal entity has assumed responsibility. OSA is more than just a place! Its a hub in our village, with many spokes and persons and councils that spiral out geographically and energetically. It is also all our neighbors and the neighboring lands, it is the waters that move through her and bubble out of the mountain-fed springs to fill our well, and as much as all that, Our Sacred Acres is the people who've labored (for love) here. It has become what you brought, what you got, and it goes with each of those people to the next place and project, rippling out that frequency of home.

This place, and each person that carries that village experience outward and right back in again, are but one tiny little fractal of what's possible and emergent in stewardship of the astonishingly beautiful, natural gifts here on this planet of which we are ALL caretakers. One place in time and space of many in this global village, (re)inhabited by people who consciously choose to show up here and participate now in community. People just like me and you!

There are many amazing and awesome alchemists, architects and artisans at work here at OSA, on neighboring farms, across the bioregion, and around the globe. We are drafting the new/old ways to harmoniously tend the common grounds and waters such as these here at OSA that I am so fortunate as to steward. Some are consciously crafting the tools of their trades or cultivating the fields of their farms; others are apprenticing and advocating on behalf of our village, and still more are ready and willing to teach and learn how to practice the arts of living well together, in respectful ways.

We are truly blessed to be one of the places for these projects to be seeded, cultivated, and harvested! Tomorrow, we will be the stories of these transformational yesterdays. We are, right now, writing the history of what happened, as it happens at these places where our energies converge with aligned intention.

We are weaving an interconnected web of networks, organizations, communities, clans, and individuals, into what can truly be described as a global culture for local actions. All in shared purpose, with alignment of vision and goals toward building relationships that are mutually supportive of our collective intention:

Caring for one another as one family, sharing our planet earth as one village.

Dannielle Gennety

Danielle Gennety, the proprietress and home/land steward of a community impact and outreach center in Snohomish, Washington, founded a non-profit permaculture education project in 2014 named for her beloved forest sanctuary, Our Sacred Acres. She is a land stewardship educator, project therapist, team facilitator, community organizer, and nodal network ambassador in service to the earth, all beings, and love. She is also a mother, grandmother, daughter, sister and cousin to a vast family forest of ances-trees, with deep roots in many waters.

Her corporate history spans nearly 3 decades of customer care as an information technology (IT) systems analyst, software deployment liaison, process development consultant, and network infrastructure project manager on data stream innovations for the travel & hospitality, automotive, telecommunications, entertainment, and child welfare industries. She received her Project Management Professional Certification in 2007. Danielle also serves as a community builder and activist organizer, with well-established networks spiraling from her home town of Hartland, MI in the Great Lakes basin, through the unceded Coast Salish territories of Cascadia, Pacifica, and across North and Central America.

http://oursacredacres.com
http://cascadianw.com
http://theofficemystic.blogspot.com

Community Architecture: Designing for Success

Geofrey Collins

This is much more difficult for a large group of committed residents, than it is for one family. Imagine an apartment building you've lived in and the workings of it. Did it serve the needs of the community or did it isolate individuals, for example? The design of spaces, through uses of sacred geometry, permaculture and energy flow can coalesce into solutions that set the stage for harmonious and intentional living.

There is so much to the process of design, from the position of the stars (as is the case with biodynamic design and the writings of Vitruvius), to the acquisition of permits, all of which have their importance in successful community design.

The most unique thing in the toolbox of an architect though is the training of how to frame the vision. After seven years of education, I came away with an understanding of how to design from what worked historically, how culture and society create the context of design and how to illustrate and diagram the buildings for construction. Buildings and spaces are containers for the socio-political, cultural and spiritual needs of the community. The architectural process is a crucible for the creation of these aspects.

The dream of developing community takes a village. What happens when you put your heads together toward an intentional living experience? When the social and physical group mind come together, that is, when a group of people together in one place have the same experience, the intention can be formed into a reality that changes the context of the way people live together. That is the dream we can actualize!

My Backstory

I was indoctrinated into intelligent design for groups of people, since the age of 18. Among my many community architecture experiences over the years, I have had experience managing a multi-use warehouse and event space,hosting gallery exhibits / salons, running an architecture design firm, designing, building, and managing over 8 major theme camps/villages at Burning Man, and more. From Burning Man, I co-founded a permanent project, called "Tribal Oasis" in Topanga Canyon, outside of L.A. in 2005, and it continues somewhat today. Therein I learned the hardships and triumphs of leadership and building community.

Considering Design

When an architect starts to design a project, one first must get to understand the client very well. You watch how they live, and the specifics of their needs before you start sketching. This is the beginning of an alchemical process, whereby these ideas become the built reality.

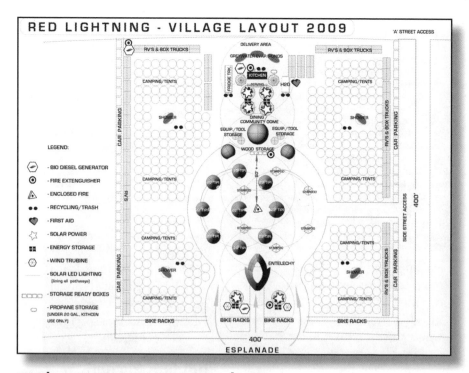

RED LIGHTNING - VILLAGE LAYOUT 2009

Exploring Community Typologies

I want to explore three types of intentional community designs that we can pursue:

A. The most intimate scale is simply living in a house with housemates which many of us

have done. This is inhabiting the existing version of a village, and integrating conscious practices to make your community successful. This intimate scale is what we can all start doing now, finding systems for our homes in the village to work together.

At this scale, it may be 6 people living in a house with community aspirations, or it may be a collection of buildings that pre exist and can be stitched together with 25 souls. This is the model of the Emerald Village Organization in San Diego, CA as well as the original idea of the Tribal Oasis community in Topanga that I co-founded. It is a very accessible possibility for us now, and is best when purchased, but can work when rented.

Although there may be challenges to community purchased property, with clear agreements and shared leadership, combining financial resources can lead to an easier entry into property ownership. Yona Fy, of Emerald Village, has a real estate practice called Activated Villages, which has inspiring models for successful community land purchases.

B. The next scale would be creating a space for 40 to 140 people all in one big gesture.
As permaculture suggests, we have to work with what we have, and what many of us have is urban infrastructure to inhabit. Similar these previously mentioned communities, you can get into property with multiple investors on a larger scale. An example of this approach would be to buy a couple warehouses or apartment buildings and redesigning them to create communal aspects. I think it'd work well to buy an old hotel with a pool and have like 50 friends living in it, for example.

This was the original concept for Tribal Oasis. A group of about a dozen folks had multiple meetings back in 2002 to try to bring it about. We explored purchasing a huge warehouse and outfitting it to have multiple meeting spaces and workshops on the first floor, and apartments for about 100 members on the second floor with outdoor roof decks and gardens above that. When you have so many people you have to break down the spaces into smaller 'families' that can dine and work together. So I designed into it a series of pods that each housed about 13 people where each pod was a different 'guild' in the community.

Based on a system used by Barbara Marx Hubbard and others from her generation, in the guild model we divide up a group into smaller pods focused on different professions or pas-

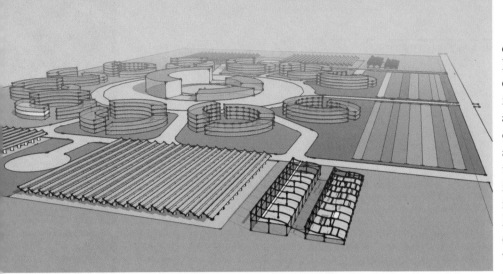

sions. Artists over here, educators here, gardeners here, etc; then you start many conversations and organize on smaller levels. This breaks up the hierarchical structure and distributes it through the larger group which is crucial to keeping everyone empowered and the happiness of the society sustainable.

With the diligence of Dlight Sky, Ben O'dell, Gregg Egg, Angya Additi and others, we pushed this concept into a business proposal to attract investment. A sophisticated 25 page report was created that proved how it could make money like a business while empowering its residents. We discussed a business model to keep the wealth distributed by creating an 'S' corporation, where investors would receive stock in the company as well as people working sweat equity and long term renters. With this model, if you helped build it and lived there for a year you would receive stock for those efforts alone. Making it a co-op model like that is also key to keeping the happiness quotient sustainable for the long term.

When the buildings we were considering buying sold to someone else, we then looked into a smaller model, and worked together to buy the land for "Tribal Oasis" in Topanga, CA on 13 acres and live in many small cabins together.

Whether it's many houses in an existing neighborhood who co-create their needs, or a larger multi-unit dwelling of intentional community, it is very possible to make a whole town that is more sustainable and livable. The Transition Town movement is a resource that has a blueprint for how neighbors can link up and help one another. Also, with internet groups, and apps, we must create 'needs and resources' lists and trade professional services and goods with other locals rather than driving across town and wasting money. A great example of this kind of resource is Next Door.com. A recommendation for multiple methods of building community together harmoniously is "Creating a Life Together," by Diana Leafe Christian.

These methods are just a taste of what you can do now, to reinhabit your village and build community with what exists.

C. The last and largest scale village model is a concept that is still being dreamed about or visioned because we are developing these ideas for our future. These are the utopian communities built on land somewhere from scratch. These high tech, master plans could house hundreds or thousands, into a harmonious new village, using all the tools we have to bring about this idealistic vision.

But these are the most difficult and rarely existent communities to date. Designers like Jacque Fresco, Paolo Soleri or Buckminster Fuller envisioned ambitious plans, and never realized them. However, they are possible! One cool project I designed was a concept of creating a business and building housing for the workers around it. In this case, investors would build a little factory for making solar panels, and then housing surrounding that for residents and workers. Outside of that would lie energy collection systems and gardens to grow food. This is a socialist concept, that doesn't necessarily fit the financial models in this country, so it wasn't realized, but the potential exists if carried through by the right people.

Building Resilience

However you may choose to live, it is vital to create self sustainability, which is key to the growth of any community. The application of permaculture principles, green design and perhaps most importantly the social architecture that allows harmony between the people living together must all be consciously designed and woven together to create a system of abundance that endures.

Consider applying these principles as a deposit in the bank. Once you get the functional systems in place, they should continue to pay for themselves and reap returns in the future.

Create systems for your communities' Energy, Water and Food needs. It's always good to "diversify the supply", so if one fails you have a back up. Maybe part of your solar is grid tied, but make sure you have battery back up as well. Include other sources of power such as wind or hydro to round out your portfolio. The freedom one senses of not paying electric bills, is profound and totally attainable.

Water is perhaps the most important commodity. I think all land projects should have this goal in mind: access to a lot of healthy water. I have built several earthships in arid areas, we have found ways to harvest a lot of water from the sky and live well without other water sources. Ideally you have access to well water and back up with rain water collection for irrigation. There are some great books you can reference in the science of water collection. In one rain event you can capture enough water to last for months, if you have a big enough cistern storage. Just like free money in the bank!

Finally, and sometimes most difficult, is to be self-sustainable with your own food. This really takes a lot more physical effort than most realize. Good gardens require excellent soil. I experimented with this a lot in Topanga at the Tribal Oasis land and it can be back breaking work.

This type of work is key to the regeneration of the garden, your health and the function of a homestead or community. The permaculture principles walk you through this process, in a much deeper way. For example the first thing to look at, on a new property, after observing the patterns of the land for a year first, is to find where the water runs across the landscape, build swales on contour, put the good soil there and plant fruit and nut trees to start a "food forest." This one activity may shape the layout of your whole place, and can ensure long term sustenance. There is so much to discuss in this area, such as the success of hydroponic systems for harvesting fish, and best practices for keeping animals, at least for dairy.

While I can only touch on the basics here, I urge anyone to familiarize yourselves, take classes and research all of these and many more methods. Investing one's energy and resources into getting off the grid, supplying your own healthy food together, and finding solutions to peaceful co-creation, is the highest work.

I hope this inspires you as a roadmap to successful forays into the work of community design and building. Inevitably, there will be great challenges yet engaging this is the best work you can do for your own growth and the growth of your community. Good luck!

Geofrey Collins

Geofrey was born in Cincinnati, Ohio and raised on a farm there, as well as 6 years in Germany and Belgium. He experienced the ancient architecture of the old world, and began a career of hand-building structures.

In the 1980's Geofrey attended the UC architecture program. During this time, he worked for a variety of architectural firms, including Mike Reynolds, building earthship solar housing in Taos, New Mexico. It was these self-sustainable houses that molded. Geofrey's green thinking.

In 1990 Geofrey left the firm SOM in LA to attend SCI-Arc. After receiving his second degree in Architecture, he worked for Frank Gehry. He started SCO Studio with a partner and began a career of multi-disciplinary design, including planning, furniture and architecture.

In 2000, he worked with a group to found the intentional community, Tribal Oasis. The project was realized in 2005 on land in Topanga, CA where he lived until 2013. He continued to design and permit dozens of houses and offices in LA. Known as 'Geo' on the playa, he cofounded Red Lightning, a transformational village that has had 7 years of different physical and spiritual forms. 'Geofrey, now living in Taos, NM, is currently designing and planning multiple conscious buildings. In 2015, GC Living Design is working on communal land planning projects in CA and NM. He is building houses off-grid in Taos, and continues to work on house designs in LA.

Body and Soil

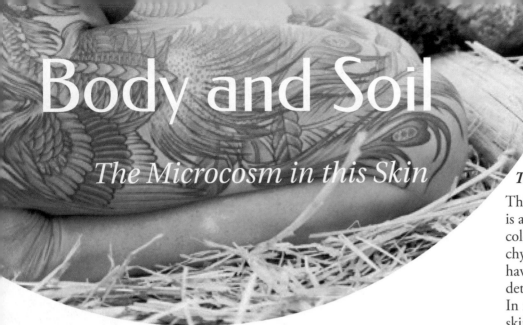

The Microcosm in this Skin

This is not the age for hiding from our skin.
Light is a spectrum that knows no division.
From a single perfect seed all life unfurled,
I am this vision, this secret, this black pearl.

If you have ever stared at a sunflower, you have seen one of the secrets of life: that everything is a fractal. In the perfectly ordered double spiral at the flower's center lies one of the primary codes for the movement of energy and consciousness in our Universe. Artists know this, and attempt to convey this mystery through paintings that perfectly reflect the proportions of the golden mean. The gardener's dirt-stained hand explores this ratio every time it caresses a growing tree or harvests a head of cauliflower. Scientists have long studied this natural geometric phenomenon, observing the similarity between the rhythms of nature's unfolding and the structure of the stars. And this same pattern lives in the proportions of the work of art known as the human body.

We live in an age when we can see the tiniest atoms or faraway galaxies through advanced instruments. We can separate atoms into their infinitesimal parts. The incredible gift of technology allows us to understand the detailed workings of life's systems in miniature. But the cost of this microscopic analysis is that we separate ourselves, picking apart life from life.

But we are not meant to live divided. Though our penchant for analysis gives us many gifts, our addiction to separation is also harming our living home planet and our own bodies. The more we pull things apart, the more separated we feel and speak and act. If we see things only in terms of their shards, we miss the wisdom of the whole. It is like seeing only one angle of a tetrahedron and believing that we are looking at the letter "v." And an organism cannot survive if its individual organs and systems do not work in harmony.

This Skin

This tendency towards divisiveness is apparent in our perception of color with the addition of hierarchy. Though we are one species, we have let the color of a person's skin determine his or her relative worth. In most instances, the darker the skin, the less valuable the person. Through centuries of propaganda, from religious doctrine to modern movies, black has been associated with evil or considered less valuable than white. This argument has been used to justify slavery, racism, oppression, and all manner of phobias. This fear of the dark is so ingrained, so infused in the collective subconscious, that it permeates all of our conceptions.

Soil is the skin of the earth. Farmers pay attention to color, as well. But when growing food, the darker the soil, the more fertile it is, the more nutrients it will provide to the crops, the healthier the people will be who eat that food. If you have ever held rich black soil in your hand, you probably remember the crumbly texture, the fecund smell, that particular quality of comforting moistness. The original name of Egypt was Kemet, "the black land," so-called because of the darkness of one of its most precious resources.

We are losing our fecund soil, and our pervasive fear of the darkness is robbing us of the ability to understand its true gifts. With modern agricultural and food distribution practices that focus on providing cheap low-quality goods to uninformed consumers, many people are unaware of the origins of their food and the intricate workings of soil fertility.

When the thing that gives life is perceived as unworthy of our love and respect, lost in the realm of unwanted dark things, preserving it is no longer a priority. When we fear the darkness within ourselves, we lose access to our wholeness. If we allow judgment and ignorance to separate us from our life-giving skin soil too thoroughly, we will no longer be able to survive because we will have no food. If we refuse to welcome home the dark things inside of ourselves, we will perpetually live a half life, a life devoid of creative fertility.

Coming Home

To bring ourselves back into balance we would be served by exploring the protocol that organic gardeners use to grow food.

* Composting and Preparing the Soil

In nature, everything that rots or dies is integrated back into the soil to become fertilizer. Shifting to a new way of being requires that we understand what is unhealthy, what is decaying, what is not serving life in our current paradigm. Explore your own relationship to darkness, especially that which is hidden or dark about yourself. How do you feel about things that are black? Are you willing to release your current perceptions?

Planting the Seeds

Understanding that our own bodies are one of the fractals of life, and that how we treat ourselves is how we treat our planet, what is your vision of wholeness and fertility in your own life and in the world?

Tending the Garden

Transitioning to a paradigm free of color hierarchy will require tenacious vigilance. A willingness to pull the weeds of hatred, racism, self-judgment, spiritual bypassing, and all the ways that we place white above black. Can you remain committed to your vision of wholeness no matter what challenges you face?

Sharing the Harvest

In nature, all life feeds and supports other life, and is fed and supported in turn. What can you give to others, and to life itself, that would contribute to wholeness?

Look into a flower long enough, and you will see the stars. Look at your body with equal patience, and you will see the entire Universe. Look into your mind and acknowledge your tendency to judge based on surface and not essence. And look into your heart and remember that there is no division. That your skin, the earth's skin, are fractals of one interconnected whole, endlessly reflecting and spiraling around each other until they can no longer be told apart.

Niema Lightseed

Niema Lightseed plays in the places where words, body, and spirit meet. As a trained theatre artist, transformational yoga teacher, medicinal poetess, mythologist, priestess, and facilitator dedicated to authentic embodied living she has explored the edges of human evolution.

Niema has published two collections of poetry in the genre of Cosmonautica, the sensual quest for divine connection and transformation. She also teaches yoga, offers performance art, facilitates group ritual, writes, edits, and ghostwrites various projects, and helps poets and writers birth their offerings into the world.

http://niemalightseed.com

The Vale Village

Born of a lifelong dream to create genuine spiritual sanctuary amidst the beauty of nature, The Vale Village is a land-project with distinct vision: saving the world (or at least part of it) through a mixture of permaculture, transformative art, inspiring theatre, community gatherings, and direct social action.

This vision would find one of its best expressions in the annual Vale Village festival; a cash-free community campout which uses a theme of "village life" as a basis for exploring community skill--sharing, gift-giving and celebration. Over four days, campers and other attendants are welcomed, oriented, and given the opportunity to create the festival content themselves. There are no paid performers, no paid staff, and no activities save what the villagers themselves offer. While that might

seem like a dull weekend, it has proven to be quite the opposite. The format and environment are built around participation, and few people show up without some mind-blowing offering to share (who knew uncle Ted could play the fiddle?). Some people set up small enterprises like healing tents, free tattoo stations or mini barber-shops. Others bring live music, yoga and tai-chi instruction, meditation practices, group art projects, games, and workshops of many kinds, all offered as gifts to the village.

campers are encouraged to share their goods, and to offer their skills and talents in whatever way they can. To help facilitate these offerings, the Vale continues to develop areas for workshops, performances, meditation, movement, and outdoor games. A schedule displaying all the free events, games, and workshops hosted by the villagers is posted in a central area, and can be added to at any time. Participation on every level is greatly encouraged and everyone has the clear opportunity to be of direct benefit to the whole Village, whether by offering a gift or service, hosting an event, or simply taking a turn at washing dishes.

This spirit of (and opportunity for) participation seems to strike a deep chord, not merely for its social ramifications, but also on a more personal level. More and more dissatisfied with being merely spectators and consumers, people of all ages are expressing the desire for opportunities to really contribute to their festival and celebratory experiences. Being and feeling like an active, important part of such events can greatly reduce that sense of alienation so prevalent in modern consumerist society and in consumer-oriented festivals as well. More and more people are seeking genuine involvement and collaboration, not simply passive spectating.

Fun and games are serious business in the Village, but it's in addressing

MORE AND MORE PEOPLE ARE SEEKING GENUINE INVOLVEMENT AND COLLABORATION, NOT SIMPLY PASSIVE SPECTATING.

Because the Vale Village, by its very concept, is intended to explore non-monetary community models, the event is completely free to attend, and no money can be used or traded on the grounds. Instead,

the real needs of the villagers where the experiment really gets interesting. Primal issues like housing, access to food and water, and community governance are all touched on; some explored in the form of workshops, others tackled more practically during the course of the festival. Food, for example, is addressed with daily brunches and dinners, which are potluck-style collaborative efforts made entirely by volunteers in the communal kitchen and saloon. Though campers are not obligated to participate in all meals, most do. The meals are consistently amazing, and so far, no-one has gone hungry.

Another example of the Gift Economy at work in the Village is the installation of a "free-store" where villagers can offer their old or unwanted clothes, books, or other items as gifts to whoever might want them. As villagers come and go over the course of the festival, the contents of this Gift-Market grow and change, and a surprising array of still-useful items avoid the landfill and make their way to new homes.

Ideas like these, and many others explored in the Village, are not new ones, but seldom have they been placed at the centre of such gatherings. Producing any quality festival or modern community event usually requires some considerable investment, and ticket prices inevitably must reflect those costs. Also, food and most other consumer needs within the festivals are addressed largely through commercial vendors, which means that even in the festival environment, options are fundamentally limited by how much money you have to spend. Of course, volunteering has long presented a way around these limitations for some. Those who do pursue this option often discover that working in a team of volunteers, rather than detracting from their leisure time, can instead greatly enhance the festival experience as a whole. It would seem there is something in each of us that prefers to be actively involved, if given the chance. This desire for involvement is what fuels the Village concept; a festival built and inhabited entirely by volunteers.

The Vale Village is, of course, only a small, temporary model of what might be a true moneyless community, but its intentions are set toward exploring the concept in a fun and inspiring way. Whether it can aid in the eventual transformation of the greater society into a healthier economic system remains to be seen, but the Vale is optimistic that events like these can at least respond to our need for more inclusive community gatherings, can test the possibilities of moneyless living, can harken us back to simpler times and just possibly, show the way forward.

The Vale is an organic farm, sanctuary and event centre set on fifteen acres near Powell River on the upper Sunshine Coast of BC. Though still under construction, the Vale began hosting spirit-centred community gatherings in 2011, and continues to develop its utopian vision in service to humanity.

Blake Drezet

Blake Drezet is a painter, author and theatre artist living on the wild west coast of British Columbia. Like much of his generation, he grew up with a gnawing certainty that something was seriously wrong with the world. Thankfully, he re-discovered that through love, self-acceptance, and creativity it might all, just possibly, be fixed.

Considering himself a well-rounded visionary, Blake dedicated the last two decades to a free exploration of the artistic imperative. His way would wind through the venerable paths of painting, creative writing, theatre arts, and event production. In 2002 he co-founded Spectral Theatre Society in Vancouver, and in 2012 began the Vale; a utopian land-project north of Powell River with collaborators Inger-lise Burns and Brian Drezet. The Vale portends the culmination of many of his greatest aspirations; as it strives toward being a genuine example of human-driven development in the direction of Peace, Beauty, Freedom, and Respect for all life.

Despite dubious academic records, Blake remains an eager full-time student of Life.

Retrospective on the Rainbow Gathering Invitation

TO BELONG, TO BE ACCEPTED, TO PARTAKE IN, TO ACCOMPLISH, TO EXIST AS A FAMILY TRIBE; SHARING, SWEATING AND LOVING TOGETHER, A POTENT REAL LIFE EXPERIENCE WITH NO SPACE FOR LONELINESS.

The Original Invitation from Rainbow Gathering

On July 1, 1972, near Aspen, Colorado, hopefully on 3000 acres of land set up for the purpose – there is going to be a gathering for all people – worldwide, and the invitation reads:

We, who are brothers and sisters, children of God, families of life on earth, friends of nature and of all people, children of humankind calling ourselves Rainbow Family Tribe, humbly invite:

All races, peoples, tribes, communes, men, women, children, individuals – out of Love.

All nations and national leaders – out of respect.

All religions and religious leaders – out of faith.

All politicians – out of charity.

To join with us in gathering together for the purpose of expressing our sincere desire that there shall be peace on earth, harmony among all people. This gathering to take place beginning July 1,1972, near Aspen, Colorado – or between Aspen and the Hopi and Navaho lands – on 3000 acres of land that we hope to purchase or acquire for this gathering and to hold open worship, prayer, chanting or whatever is the want or desire of the people, for three days, but upon the fourth day of July at noon to ask that there be a meditative, contemplative silence wherein we, the invited people of the world may consider and give honor and respect to anyone or anything that has aided in the positive evolution of humankind and nature upon this our most beloved and beautiful world – asking blessing upon we people of this world and hope that we people can effectively proceed to evolve, expand and live in harmony and peace.

Amen

Excerpt from the Upcoming Novel "Road to Peace Mountain"

With the swell of buds and bulbs pushing from a sunny temperate window in late February, came the sprouting gardening ideas of the Family. They had waited long enough, and were brimming over from the cooped up anxiety of waiting… waiting. Mac wanted to plant a strawberry patch. They needed to get an early start to bear in June. He envisioned a large heart-shaped patch, red, full of berries, down below the house. They had made lists of crops to plant and listed again what seeds to plant early and which to get started in the hot box, made from old recycled windows. A run was made to a local nursery, the strawberry starts obtained, along with seeds and tools for making and tending their endeavors.

A circle was formed in the garden, initiating a blessing ceremony, and the sod busting began. The river valley meadow soil was perfectly loamy, very different from the heavy clay found up on the hillside. Hardly a rock or root to be found; it cut like butter. After only a couple of days of steady laboring, a massive plot had been turned under. Praises were sung to the loamy texture and apparent fertility of the virgin plot that had most likely been pasture for decades in the past.

With the blooms of the newly-pruned orchard, the early spring crops sprang to life. The rising sap of nature flowed in the people's blood, invigorating and motivating the fulfillment of their plans. The Farm came alive in a hum of building and cultivating activity. Each finding their place to contribute to the whole: making goat and chicken pens, strawberry patch, additions of young fruit trees and berries to the orchard along with the tending of spring crops and starts under glass. A bee man had showed up, his hives opening to pollinate and gather from the array of wild and cultivated blooms.

A profound sense of group accomplishment and manifestation of a dream abided in their hearts. Joy and happiness led to laughter and bright smiles accompanied the work projects. To belong, to be accepted, to partake in, to accomplish, to exist as a family tribe; sharing, sweating and loving together, a potent real life experience with no space for loneliness. Their gratitude and thanks was outwardly felt and expressed in the depth of their handheld prayer each evening. Courses of energy passed around their circle as they recognized the oneness of the Universe, and it in them. No small thing. One could say, utopian in nature.

However, as always, the task remaining at hand was: *how to make that love stay.*

Rob Roy Rowley

Arriving into the Pacific Northwest in 1969, Rob Roy was involved in the initial surge of 'Back To The Land' communal activism in the then, very 'Red' territory.

As one of the original core Rainbow Family members he assisted in the fulfillment of the first Rainbow Gathering in 1972, which united peace loving, Mother Earth loving and community building peoples from across the nation.

After years of living in a variety of American states, London, UK and tropical farming in the West Indies he is now residing back in his beloved Northwest where he is meshing with community, working with the land and pursuing his writing.

Holistic Event Production

THIS IS THE POWER OF GATHERING: IT INSPIRES US,
DELIGHTFULLY, TO BE MORE HOPEFUL, MORE JOYFUL, MORE THOUGHTFUL:
IN A WORD, MORE ALIVE. ALICE WATERS

From the dawning of human culture, we have partaken in rituals to align us with the rhythms of the earth, to create a relationship to the Spirit of a place, to mark the passage of time, to initiate our youth, to honor the dead, to welcome new life. These rituals have long served to acknowledge our living connection to the plants, the animals, the seasons, the stars, and to each other. Celebrations reflect our cultural values and creative expression as we navigate the constant process of transformation, dancing around the wheel of time, flowing from one cycle to the next in connection and continuity.

Cultures are built on a shared mythos, keeping stories alive through song, dance, feasting, and a sense of tradition which ensures that wisdom is passed on to the next generations.

Cultural celebration is not a new phenomenon, it is fundamental to human expression. So what is unique to celebrations that are emerging now? What is a Transformational Festival and how is it transforming the landscape of global culture?

From the revolutionary spirit of Woodstock and Grateful Dead shows, to the international open source Rainbow Gatherings, to the rise of underground electronic dance music in urban warehouses, to the radical expressions of Burning Man, to the West Coast outdoor Visionary Art and Music festivals, and everything in between, we see millions of people traveling around the world to converge yearly for these gatherings. They come for all variety of reasons, to share a collective peak experience, to live outside "the box," to express personal creativity, to learn and explore new ways to live, to exchange arts and goods, to get inspired, to be a part of a scene, to meet new people, and to "party." The diversity of these festivals is reflective of how many different people and cultures there are on the planet. There is no "one" festival culture to speak of, it is a movement of many cultures each with their own values, focal points, audience of participants, music genre, and ingredients that make up their version of "festival."

What makes Transformational Festivals a new breed of celebration is the focus on things like health and wellness, movement or flow arts, Visionary and interactive art installations, intentional community ritual, health conscious food vendors, alcohol free, artisan crafts and marketplace, intergenerational and multi-cultural inclusion, and the temporary creation of a "Village" type setting with infrastructure set up to sustain 500- 60,000 people meeting their basic needs for shelter, water, food, energy, not to mention PLAY! Through these festivals, production teams and volunteers are learning collaboration, healthy community communication, reskilling, design and building, effective project management, successful business/legal/financial practices, safety and conflict resolution, marketing/media, land management, waste management, and so much more!

Yet in order to be "Transformational" the onus has come back to festival producers to align the actions of their productions with the ethos of their branding. There is a distinction to be made between "party and play with purpose." One of the latest evolutions of the Transformational Festival movement is a focus for some producers on environmentally conscious "green principals" from composting, recycling, upcycling re-useable materials, leave no waste, end single use, local sourcing for food and suppliers, imposing impact fees and encourging carpooling to reduce cars, creating systems for gray water and drought considerate water practices, etc.

Also emergent is the concept of grassroots leaders taking the inspiration and momentum that Transformational Festivals stir up in participants and galvanizing volunteer forces to take part in direct action and social changemaking through examples of urban permaculture, placemaking,

disaster relief, and service to the local community near a festival site.

Among some Transformational Festival producers, there is a desire to take things a step further, addressing the amount of time, energy, resources, consumption and waste that go into creating a temporal experience only to be torn down after a week and built again the next year. It's time to own the land that our events are held on, and steward that land in consideration of the environmental impacts and the long term viability.

There are groups partnering to pursue the purchase of shared venues, aligning with the intention of using permaculture practices and whole systems design to collaboratively create the regenerative systems for meeting food, water, shelter and power needs. Building permanent art installations, using natural building techniques, solar and appropriate technologies, installing compostable toilets, building outdoor kitchens and growing

the food used to feed participants, working with restoration practices to renew the land after the impact of many people, and also designing and building infrastructure to work with the landscape instead of imprinting on top of the natural systems in place. In between festivals, the idea is to use the land as year-round immersive education centers hosting permaculture design, wellness, and hands-on education courses, while modeling permanent Village culture.

Also, a few Transformational Festivals are beginning to tend the delicate consideration of building relationships with the original "People of the Place" on sites where festivals are held. Consulting with Indigenous representatives of an area and working to honor the Spirit and Land in respect for those who have long stewarded the earth is starting to change the approach to HOW and WHY we are gathering. While this Transformational Festival Culture is still young, full of inspiration and with so much yet to learn, there is a power and undeniable force that can be harnessed for the sake of bridging our past and our future; older culture with younger culture, mainstream with alternative, with indigenous. We all have much to teach each other.

Festivals are a celebration of life and the act of celebration is about wholeness, renewal, gratitude, lifting us from creatures focused merely on survival, to beings capable of making an offering of reverence for the world around us and for ourselves as a part of that world. The call is to reforge a relationship to the Spirit of Life and that as we gather we remember our roots, remember our place in the great dance, and through our actions that we honor all we hold as important, as valuable, as Sacred.

Jamaica Stevens

We Are One

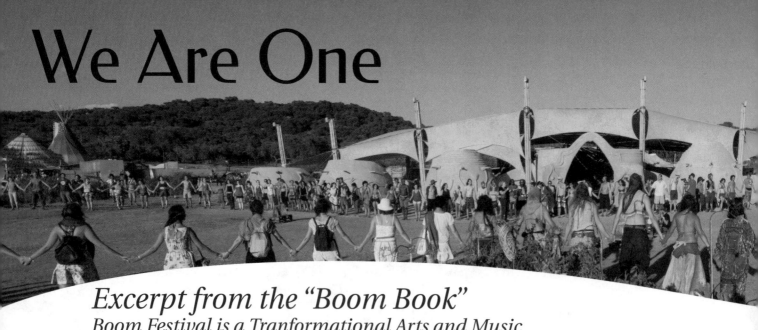

Excerpt from the "Boom Book"
Boom Festival is a Tranformational Arts and Music Festival held every two years in Portugal.

"We are one," "Nos somos um" and "Somos todo uno" is a mantra heard in many languages in religious celebrations and dance festivals around the globe. But what does this actually mean in an ecological sense? In theory, many of us know that "we are one" and that the world is more than people, resources, and an environment to be vaguely protected. However, in practice, many of us demonstrate otherwise. We continue to lose precious species, cut down forests and cause pollution that spills into the skies. But the sacred energies are rising, and many of us feel that this is the time, the time to deal critically and creatively with reality and discover how to participate in the transformation of the world where there is a fusion between new ethics and positive action.

In today's society many of our values spring from egocentric ethics. An egocentric ethic or grounded in the self, is historically associated with the rise of capitalistic and the mechanistic worldview and is the ethic of mainstream industrial capitalism today. Environmentally, an egocentric ethic permits individuals (or

corporations) to extract and use natural resources to enhance their own lives and those of other members of society, limited only are the effects on their neighbours. Eco-feminist, Carol Merchant states that ecocentric ethics often reflect on the Protestant ethic where an individual is responsible for his or her own salvation through good actions. It focuses on "I am one" as the center of the universe. Egocentrism is a danger to the planet and to use because of its excessive narrowness.

This varies from a homocentric (also known as Anthropocentrism) ethic or grounded in the social good. From these feelings has come an ethic of humanism that urges sympathy for others regardless of race, color, sex, or creed. The homocentric ethic served humanity well when numbers were small and techology limited as it taught benevolence, justice, and mercy toward one another. This ethic also underlines the social interest model of politics and the approach of environmental regulatory agencies that protect human health. It focuses on "We are one," "we" as a symbol of humans, yet we do not exist separate from "them."

The root meaning of the "eco" is "home," and the revealed Ecosphere is the home-sphere from which all life came and in which all life exists. The "we are one" egocentric ethic recognizes that the ecosphere, rather than any individual organism, is the source and support of all life. An ecocentric philosophy changes the role of homo sapiens from conqueror of the land to plain member and citizen of the Earth by providing a new basis from which to examine the questions of how we should value the natural Earth and its systems and of how people should live. This outside-the-human focus brings with it new standards for thought, conduct, and action on such seemingly intractable problems as the mountains of rubbish we continue to hide, urban sprawl, impoverishment of ecosystems by globalization, loss of cultural diversity, and the ethical duties to the varied natural ecosystems and their wild species.

It transfers the reality spotlight from humans to the Ecosphere which includes inanimate elements, rocks, and minerals along with animate plants and animals.

This outward vision, reminds humanity again that Nature is the source of the creativity called "life." It shatters the illusion of separation and encourages us to take our eyes off our species and ourselves, presenting a new philosophy, based on the first ethic of care for the earth, followed by care for people, challenging the foundation principles and current ethics of society. It implies that we, as Earthlings, can learn something from our source and support: the Earth.

Stan Rowe, geo-ecologist once said that perhaps the strange human hankering to fly away from the planet, to "slip the surly bonds of Earth," is a necessary impulse for discovering who we are. Perhaps we can never be satisfied on Earth until we have travelled away in space and come back home, back to our roots, to where we belong to understand "we are one". The realities of the world are not people and separate "other things." Nor is the Earth a machine whose secrets lie in its fragmented parts. It is beyond all understanding. It is an integrated Ecosphere of marvelous creativity, "we are one."

"We are one" the Earth ethic, resonates the new paradigm. By replacing the destructive ethics of homocentric and egocentric, we shift the focus of importance from the one species to the living Ecosphere. The enhanced physical and spiritual qualities of our relations to the earth will ensure steady improvement in the quality of life for this and future generations that respects our common heritage, a ball of living star-dust, a four and a half billion year old miracle; the planet on which we live.

The challenge lies in a willingness to do things differently than we have in the past.

Lucy Legan

Lucy Legan was born in Adelaide, Australia. Being an educator, permaculture designer and gardening expert, Lucy has always worked in the educational field and with permaculture. Lucy Legan has traveled the world, working and developing projects for education and sustainable development. In Brazil was co-founder of the Ecocentre IPEC. Author of the book **Sustainable School**, Lucy showed a practical and objective way to involve educators and children to a new environmental awareness, where sustainable development in schools goes from just a dream to a sustainable reality. She also wrote books **Creating Sustainable Habitats in Your School, Sustainable Solutions and The Little Book of Seeds.**

http://boomfestival.org/

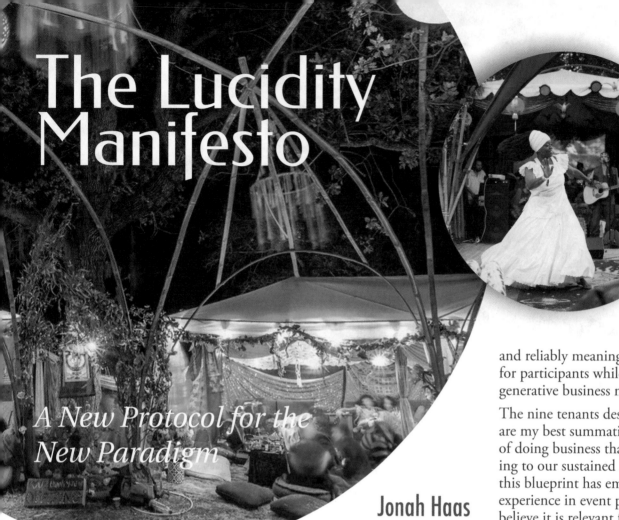

The Lucidity Manifesto

*A New Protocol for the
New Paradigm*

Jonah Haas

YOU NEVER CHANGE THINGS BY FIGHTING THE EXISTING REALITY. TO CHANGE SOMETHING, BUILD A NEW MODEL THAT MAKES THE EXISTING MODEL OBSOLETE. R. Buckminster Fuller

Transforming the world that currently IS by modeling the world we wish to see is a consistent thread throughout this book. I recognize that calling what I've outlined below a "New Protocol for the New Paradigm" is certainly a lofty way to describe what is really just a list of things that have worked for my organization in creating the new world that we wish to live in. But hey, let's think in lofty ways, yes?

Lucidity Festival LLC is a transformational events production company that I co-founded and co-developed with some beloved members of my community over the last four years. We are finding success in a growing industry that sees many mega festivals dominating the market, one-hit wonders that create fun experiences but lose their shirts, production crews that experience rapid burn-out, inexperienced newcomers flopping out of the gates, and very few organizations creating consistent and reliably meaningful experiences for participants while maintaining generative business models.

The nine tenants described below are my best summation of our way of doing business that are contributing to our sustained success. While this blueprint has emerged from our experience in event production, I believe it is relevant for any organization dedicated to supporting full-spectrum and significant change at the personal, collective, and global levels of scale within the socio-cultural, spiritual, economic, and environmental realms.

1. **Lead with Values First**
2. **Collaboration is Key**
3. **Embrace a collective ownership model, empower your community**
4. **Transparency, Trust, and Accountability**
5. **Sacred Commerce and Spiritual Economics**
6. **Cultural Literacy, Forging Allies**
7. **Sustainable and Regenerative Best Practices and Procedures**
8. **Embrace and Develop a Deep Mythos**
9. **People aligned with Purpose**

Lead with Values First

At Lucidity we've found that focusing on core values is a great way to ground into why we're playing and growing together. Before we ever dreamed of making a festival, a group of artists and leaders from our community got together and brainstormed the values at the foundation of our urge to come together in co-creation. This brainstorm session yielded seven core values that remain our touchstone and compass for why we do what we do in the world.

Lucidity's values are:

- Participation and Immersion in the Artistic Process
- Personal Growth and Global Healing
- Awake and Aware Consciousness
- Environmental and Social Responsibility
- Family Fun and Creative Play
- Communal Reciprocity
- Transparency

Starting with the WHY and focusing on values allows us to deepen and enrich the WHAT. By getting clear on the things that bind us together, we realized that our organizational identity is so much richer than just a festival production company. Instead we're an interconnected family of artists, healers, entrepreneurs and dreamers committed to making space for the transformation of the relationships between individuals, communities, the planet, and the universe. Producing festivals just happens to be one of the things we get to do together.

Collaboration is Key

At Lucidity 2014, I facilitated a panel discussion called, "Organizing in the new Paradigm." I introduced the panel with the bold statement that the single most defining feature of what people are referring to as "the new paradigm," at least in regards to how we are relating to each other on an organizational level, is an enthusiastic embrace of collaboration as a Modus Operandi over the "old paradigm" proclivity toward competition.

Indeed it was Aristotle who told us that, "The whole is greater than the sum of its parts." Through collaboration, we arrive at outcomes that are greater than could be accomplished in isolation. Depending on the type, mission, and objectives of the organizations, collaboration may mean greater social connectivity, increased access to resources, dialogic creative processes, and systems of feedback and reflection that act as quality control.

Collaboration certainly comes with its fair share of challenges, especially when there are a multitude of perspectives and voices. We've all heard the cliché of "Too many cooks in the kitchen," which is a very real concern when operating in a collaborative environment. This is why it's imperative that organizations dial in their systems for facilitating a collaborative approach to working. For example, large roundtable meetings benefit greatly from having clearly established agendas in place, trained facilitators guiding the flow, dedicated time-keepers, note-takers, and task-managers. Project management software platforms and other web-based cloud technologies (see Asana, Podio, Google Drive, Dropbox, etc.) are showing up to help streamline communications within and across large groups of humans in complex organizations.

Collective and Cooperative Ownership Models

Saying an organization embraces collaboration is one thing. Actually codifying the principle of collaboration into an organizational model, an approach to equity distribution, decision-making processes, and systems of accountability, is quite another.

Having studied social and anthropological theories of ownership and resource management in graduate school, I learned that there are plenty of alternative ways to organizing that don't necessarily assume that isolated entities operating in competition within a free market economy is the best and only way to do business. For organizations that are intending to build community and operate from a foundation of collaboration, a collective ownership model is desirable for a variety of reasons:

Why Collective Ownership?

- Building a business or organization from the ground up without a cent of investment capital requires a ton of energy and labor. When there are not financial resources to compensate people for their time, pathways to equity provide a form of non-monetary energetic exchange that works for people, especially when they feel passionate about the mission and vision of the business.

- By becoming a partial owner, our co-creators become highly invested in the success of the business or organization. Equity becomes fuel!

- If we hope to transform the world into a more equitable place and diminish the wide gap between the "haves" and "have-nots" then our organizations must reflect that intent from the inside-out. Top-heavy hierarchies will not create anything other than more top-heavy hierarchies. Grassroots, bottom-up endeavors consistently experience resource scarcity and uphill battles for their causes. We encourage meet-in-the-middle approaches.

- When decisions are made by the collective, even though the process for making decisions is slower, the support and energy behind those decisions once they are made is considerable and unwavering.

Lucidity is member-owned and operated and organizes within a heterarchical structure.

A **heterarchy** (as opposed to a hierarchy) is defined as "a system of organization replete with overlap, multiplicity, mixed ascendancy, and/or divergent-but-coexistent patterns of relation." While there are hierarchical relationships within a heterarchy, there is frequently two-way communication in decision-making and policy. Heterarchies are collaborative at their core and instead of "command-and-control" they tend to "sense-and-respond." As a visual, imagine concentric circles within which there are meshworks of interconnected individuals as opposed to a pyramidal structure with one or a few at the top, wielding all the power and reaping all the benefits.

Our most dedicated workers are our owners. There are no large stake investors who own the assets, projects, and work that we all do. Each of our members are offered equity in the company and new shares on an annual basis. The "membership points" that represent ownership shares in the company also function as voting power in a variety of core business functions.

Prospective members are put through a long and rigorous vetting period where we "get to know" them. This period lasts no less than one year. Pending positive performance review, prospective members are invited to start the process of accumulating units. At year's end, our collective engages in a performance review where everyone reviews the work done by everyone else, the results of which impact the number of ownership points vested by each member. Thus, we have implemented a system that holds all of our members accountable to all the rest of our members.

Accountability, Trust and Transparency

Perhaps the most interesting dimension of our Collective Ownership model is the accountability system for determining how many ownership units are distributed to each member. We've based our distribution system on the principle of **Do-ocracy**, meaning "the more you Do, the more you vest, and the more voting weight you receive." This is elsewhere referred to as a **meritocracy**. The challenge that arises, as you can imagine, is how do we track and accurately assess how much someone does? Our answer is threefold:

1. Develop systems that track and acknowledge the vast amount of contributions made in terms of hours, output, relationships, and quality of character.

2. Put our trust into the collective **groupmind**.

3. Keep the groupmind in check by allowing final review and oversight from a board of directors.

We TRUST. Trust is an important component to a truly transformational organization. We trust one another to do our best. We trust that when an obstacle comes up it is coming up to push us to grow. We trust that interpersonal challenges with collaborators is not a reason to lose trust in them. It's important to note that this trust is not blind trust.

Sacred Commerce

Exchanging information is just one of many forms of value-generating exchanges that occurs in our day-to-day interactions and business. Transformational organizations are gleaning the understanding that it's important

LARGE ROUNDTABLE MEETINGS BENEFIT GREATLY FROM HAVING CLEARLY ESTABLISHED AGENDAS IN PLACE, TRAINED FACILITATORS GUIDING THE FLOW, DEDICATED TIME-KEEPERS, NOTE-TAKERS, AND TASK-MANAGERS.

This trust comes hand in hand with accountability, which is made easier by *transparency*.

Transparency is a non-negotiable part of a truly transformational organization. To foster transparency it takes some time and positive habit forming. At Lucidity, all of our meetings are open. Our finances are available to anyone who asks. Internally, all of our meeting notes are made available to the entire collective at every rung of the decision-making process. Open communication channels make transparency easy and with the free flow of information comes a greater sense of trust. Confidentiality and discernment are also valued as we recognize that some information is shared between individuals and organizations in confidence. Part of developing a good trust framework is having strong discernment around what information is appropriate to share and when.

to acknowledge these non-monetary forms of exchange and take them into greater consideration throughout all aspects of doing business. At the foundation of the concept of Sacred Commerce is the notion that all exchanges represent the transfer of energy between individuals or groups. Within Lucidity we refer to the concept of Equal Energetic Exchange or E cubed. In all of our interactions, collaborations, and transactions we are seeking to reach the equilibrium of E cubed.

In our collective human past, there was a time within a variety of geographical places and cultural contexts, wherein the transfer of energy between individuals was considered a sacred act of communion. We seek to bring about a re-awakening of the recognition of this teaching. In the very first sentence of Charles Eisenstein's seminal work on the subject, entitled "Sacred Economics," he states his intention "...to make money and human economy as sacred as everything else in the universe." Indeed there are others who share this mission. Barbara Wilder, in her book "Money is Love," observes that money has developed an energetic imprint from our relationship with it. Most people's relationship with money is a dirty one, and is associated with fear, lack, and greed. However, we can, she tells us, with intention, transform our relationship to money, which is after all only a symbolic proxy for our exchangeable energy, into a vibration of love, abundance, and joy.

Cultural Literacy, Forging Allies

In the face of great inequity, we stand for more than just perspective shift. We are calling for a profound healing of our collective past, transforming all our relations through the cultivation of widespread cultural literacy. The dark shadow of traumas past and the echos of genocide, holocaust, colonialism, apartheid, racial and ethnic segregation, and cultural appropriation weigh heavy on our souls. We see a great need for full acknowledgement, truthful education, compassionate listening, and deep forgiveness to take place between members of "dominant" and "subordinate" cultural groups.

As *rainbow people*, we celebrate each of our unique *ancestories*. We honor all color, appreciate and celebrate diversity, and hold space for every being to come into their highest potential as sovereign members of our One Human Family, with strong roots in their distinct lineage and bloodline. Empowered with where we come from, we may more fully know who we are.

There are a variety of practices that support the fostering of *cultural literacy*. The first and most important, especially when approaching embedded wounds, is deep listening--simple, open, non-reactionary LISTENING. We also embrace a practice of *Gender Alchemy,* which is a powerful set of processes designed to heal gender wounding and bring balance of masculine and feminine essences within each of us. It is from this place that we may find sacred union in our life relationships. *Ho'oponopono*, a Huna ceremony that facilitates the radical forgiveness of self and other, is inte-

gral to the process. Letting go of the stories of hurt and forgiving our own judgements and defense mechanisms and allowing ourselves to be seen by our community in our vulnerability is a profound wellspring to draw strength from.

Sustainable and Regenerative Practices and Policies

Arguably one of the most important realms where widespread and rather dramatic transformation is necessary is within our socio-ecological relationships and interactions. A majority of humans' lives, and our dominant global systems, are unsustainable in terms of their resource use, their waste management, and their overall carbon footprint. It's time for us to take a firm look at these challenges and arrive with solutions for new ways for humans to interact with the natural world that are sustainable or regenerative, leaving a more synergized and functioning environmental system in our wake.

Lucidity approaches sustainability realistically. We understand that we do not yet operate a fully sustainable festival and our aims are directed toward making it MORE sustainable and closer to our ideals. Our efforts are visible in a variety of ways. We implement a dishwashing program which eliminates single-use dishware and utensils. We empower a green team to handle composting and sort recyclables. We recently signed a multiyear contract to power our Alive stage completely off solar power, reducing our reliance on petroleum-powered generators and proving that large-scale production CAN and WILL be powered by the sun. Instead of printing thousands of throw-away paper maps and schedules, we offer our famous

Mapdana (our festival map printed on a bandana) and our schedule can be accessed on a mobile app. These efforts, and a number of others, demonstrate that it is not only possible to "green" large festivals, but it is something that our participants love and appreciate. What's good for the earth can be good for business too.

Deep Mythos

Similarly to the way dreams speak in story and symbol, emerging from our subconscious, we believe our waking lives are understood through a similar mythic language. Humans are storytellers. We develop stories about our lives and who we are, about our families and communities, our states and nations. We tell stories to understand where we've been and to trust in where we are going. Entrenched cultural stories teach us how to behave, while new stories offer us fresh possibilities.

At Lucidity we understand that people love stories and want to be a part of a story that is bigger than themselves. We created our six year story to act as a *meta-narrative*, within which each festival year is a chapter, telling a grand saga of personal and collective transformation. Inspired by Joseph Campbell's Hero's Journey and much of the work that describes the Heroine's Journey, we've created a storyline that informs each year's theme, aesthetic, content curation, and vibe. Even if people who show up to Lucidity aren't aware of our six year story, they can feel something special because the mythos is speaking to their subconscious.

We encourage organizations to contextualize what they are doing in the world within a larger narrative. By doing so audiences are able to connect to the work in deeper ways, oftentimes finding greater meaning in their lives through the process.

People Aligned with their Passions

When organizations hold space for their members to align themselves with their passions, their higher calling, and their purpose, magical things happen. We encourage our members, our staff, our volunteers, and our participants to engage our company and community in the ways that bring their brightest heartsong forward into being. We notice that when people fulfill a job that doesn't stoke their inner fire, a host of problems arise. When we see people not aligned with their passion, we see productivity and efficiency decline, interpersonal communication can become strained, and ultimately the final product suffers in quality.

In addition to making sure people in their official roles are aligned with their personal passions, we also make sure they are aligned with the core values, mission, and vision of the organization. All too often organizations hire people primarily because of their skill sets. Companies make partnerships with other organizations strictly for financial reasons while overlooking the question of alignment. When people align the work they do in the world with their dreams, and the dreams of the people they work with, this is when we see people become capable of previously unprecedented amazingness.

Bringing it Home

By embracing these nine tenants and striving toward them, we have found our way, and we are happy to share this way with the world. None of it is new wisdom and we don't profess to be pioneers in generating any of the suggestions offered above. We do, however, take pride in our whole-systems weaving of this new protocol for the new paradigm into every level of our operations. Certainly, it's not perfect. There are many challenges. Sometimes things get clunky. Yet, because there is a deep sense of love and family among us, because we share values and always strive to collaborate, because there is shared ownership and a foundation of trust and transparency, because our people are aligned with their purpose, we make it work. And we're just getting started!

Jonah Haas

Jonah Gabriel is a messenger, an anthropologist, and a lucid dreamer. He holds a Master's Degree from the University of California in Cultural Anthropology and is currently the Marketing Director and a co-founder of Lucidity Festival. Jonah is dedicated to calling in the people who create safe and magical spaces that facilitate transformational experiences for individuals, communities, and the planet. He's also experimenting in what he refers to as emergent lived mythology, storytelling the reality that he wishes to call into existence through archetypal narratives that enroll large audiences into living their purpose and fulfilling a mission greater than themselves.

http://2015.lucidityfestival.com

Building Positive Impact Festivals

Festivals stimulate the spirit, rejuvenate our sense of connection and open our hearts to new possibilities. Through managing educational platforms and villages at events for nearly a decade and a half, I have witnessed this transformation. Inspiration comes in many forms, a variation in the divide between apathy and action. The spark and glimmer in the eye of a participant after a thought-provoking exchange. Individuals returning to an event a year later to excitedly report the progress of their project that was motivated by a 20-minute banter had while at the festival. Beyond a space for education and dialogue, festivals represent an even broader occasion for systemic shifts within our culture. A collision of energy, resources, finances and social capital, events afford community the prospect to stimulate a legacy.

Whole systems theory thinking–it's in our DNA and we have forgotten it. Consumer culture has shifted a perception in our worldview that says we are apart from nature, but we are not. The collective energy and captive attention of large gatherings represent an unparalleled opportunity to expose this myth and map new means by which we can

produce social wealth and regenerative economies that give back locally and perpetually. From my perspective, the "sustainability" of an event starts after the event ends. The imperative of today is action; compassionate action, swift action and symbiotic action that alters our landscapes and builds new perspectives of our evolving world.

Meaningful changes occur when individuals arrive to their own conclusion and act upon their own accord and free will, contrary to when something is done for someone. In order to reach these self-made conclusions, the introduction of the realities that will make up our future must surface again and again. It all starts with a conversation and perpetuates with persistent affirmations. The more a concept is witnessed, the sooner that meaningful change will come.

Along with the personal implications, there are community implications as well. The forethought and desire of the event promoters to effect change is paramount to the success of the potential impact. Beyond the industry standards of donation drives, fundraising auctions, and staffing plans that benefit local nonprofits, the composition of opportunities is immense. Integrated waste management offers a variety of materials recovery initiatives that are capable of providing long-term benefit. Advanced learning and certification programs can facilitate

a workforce that creates infrastructure, serving as continuing education platforms and new growing mediums to support a local food movement. Interactive installations can fuel conversations on-site, and if left behind in communities these demonstrations can nurture ongoing discussions of innovation, intrigue and possibility.

In nature, there is no waste. "Waste"equals food, or an input to a parallel configuration. In nature there is an order and symmetry to all functions; systems structured to support corresponding systems. The human habit of throwing something "away" is one of the most pompous manufactured practices of our current system. Introducing **biomimicry** into the operations and design of events not only works to correct the undesirable patterns of waste, it also sends a clear message to festival patrons that there are alternatives to consumptive behaviors. Building materials used in the event can be repurposed and gifted to organizations building homes for low-income families. Banners can be manufactured into bags, creating revenue and a brand ambassador for years to come. Integrated compost programs can turn paper products and post-consumer food waste into nutrient-rich soil that can benefit local communities long after the festival is gone. Outside-of-the-box thinking around waste management begins with the procurement plan. Intimate knowledge of the inputs will define what is possible and the limits are only defined by old routines.

Permaculture is gaining headlines and exposure among the public discourse. Incorporating permaculture into events can create an additional revenue stream, and build added value into the positive impact an event can bring to the site location. Examples exist of pre-event workshops where participants pay for hands-on learning that certifies individuals in a specialized field. This win-win scenario is the optimized definition of a regenerative model. Everyone gains an advantage. If the promoter owns the property, land improvements can enhance the quality of the festival participant experience; including improved drainage to reduce impacts of mud, natural building installations that produce unique resting places, and colorful guild plantings that develop the brand identity of the event. If the promoter does not own the land, the installations created by the course can serve as food-growing mediums bequeathed to local civic organizations such as a veterans group, a food bank or community garden group. Not only do such installations serve long-term needs of individuals near the event location, they empower the course participants into future leadership roles and actively engage festivalgoers on real-world solutions to some of today's most pressing issues.

Art and interactive demonstrations are becoming a standard within the industry. The distinct differentiation of your brand is essential to the acclaim needed to compete and keep pace with the growing marketplace. Why not incorporate sustainability into this trend? In some circles, wood scraps could be a chance to support the local ecology and boost skill-building for local youth. A scouts group might jump at the opportunity to earn a woodworking badge and fill the local town with colorful birdhouses boasting the event brand. Abundance is real. Is your garbage can half empty or half full?

Festivals are a temporary village setting that can model and showcase systems and innovation that can also be applied to inspire transformation in our culture. They have the power to convert our landscapes into foodscapes, host conversations that matter and etch new ethos into our philosophies. We gather to connect. Uniting together to collaborate and celebrate our potential is perhaps the highest calling of service. Talk isn't cheap; it is costly. It comes at a high cost; that of the world we leave to generations that follow and the quality of life they will, or will not have. How can your event make an impact? What issues resonate with your fan base? The future is bright … let's build it together.

Nick Algee

Nick Algee is a community organizer, project manager and consultant with a focus on events and companies that encompass social and ecological awareness within their mission. A skilled event producer with a passion for people, Nick's background is multidisciplinary with experiences that range from incubating networks, non-profit organizations and businesses that still flourish today, to leading teams within large-scale music festivals and advocacy campaigns that support healthy communities. His creative problem solving, holistic systems theory thinking and courageous spirit have afforded him numerous leadership opportunities. A proponent for initiatives that act with future generations in mind, Nick is an avid collaborator, an outdoor enthusiast, and engages in mindfulness daily. Hailing from generations of engineers and builders, he enjoys carpentry and has a deep appreciation for local food, often helping others start gardens and compost systems in his frequent travels.

Transformational Festival 2.0

Transformational festivals. There's quite a buzz about them. And with good reason. They're reshaping the way a lot of people gather, think and collaborate in an artistic environment, oftentimes transforming a party into a celebration, ripe with tools for empowerment.

When I look around at transformational festivals though, I see one thing missing, or perhaps it's not that there is something missing but rather, there is an opportunity.

How about long-lasting physical transformation of the festival sites and nearby communities? Patron creation of something that adds value or instigates life after the festival is complete.

Human beings are going to these festivals by the tens of thousands. That is a ton of hands, many of which would love to get their hands dirty and create something. How about planting trees? An orchard? Reforesting land ravaged by fire? Cleaning up an island riddled with plastic? How about creating ***earth bag*** homes for a community in need? A massive 5,000 person work party DJ'd by your favorite artists.

I personally know and am friends with a lot of festival organizers, and I know how hard they work. None of this is an expectation but rather a discussion point for future festival organization.

I'm excited to see the festivals that start to weave this mentality into their design and culture. I'm sure there will be challenges, but it seems to me this idea can be experimented on a small-scale and expanded upon.

Perhaps this is a model to think about for Transformational Festival 2.0?

Imagine, what could we create or heal with the power of 10,000 people in one location at one time, with one focused intention? And… that's just one festival.

I'm dreaming of festivals where instead of people saying, "That festival was so awesome, I got so wasted!", they say "That festival was so awesome, we built all those amazing things."

I'm dreaming of festivals that are smaller 300 - 500 person gatherings where energy is focused and directed. By the end of the festival you might actually have said hello to every single participant.

I'm dreaming of festivals with one stage, with a well curated and spacious lineup. Music is secondary and impact is primary. Imagine festivals that are permaculture design courses, orchard planting, build parties, cleanup parties, with a few key musical performances to help fuel creation and celebration.

I'm dreaming of festivals where all participants form a giant line spanning the length of the grounds to clean up the entire site with a single sweep on the final day of the event.

I'm dreaming of festivals that minimize vending and instead create a "gift mall" to encourage people to offer their intangible gifts whatever they may be.

I'm dreaming of festivals that end amplified sound at midnight or earlier to allow for proper rest, ensuring peeps are charged up for a sunrise swim at the river.

I'm dreaming of festival grounds that are owned and operated by people that are 100% committed to reverence for Mother Earth. True stewardship of land held down by community that lives on the grounds year-round.

I imagine this resonates with many of you, and maybe some of it does and other pieces do not. I'm totally OK with that. Let's inspire each other to take festivals in more conscious directions.

I suppose I'm of the mindset that we need to share our dreams whatever they may be, even if it feels a bit vulnerable to find out who our true community is.

It's through the community that dreams are actualized with the quickness and I want to see some epic change in this lifetime.

Are you part of this community? Are you part of this Transformational Festival movement?

David Sugalski
The Polish Ambassador

For the Polish Ambassador project, born on the richly-colored streets of Chicago and San Francisco, infectious melody is paramount. However, this never overshadows the depth and harmonic complexity that have made the Ambassador a favorite amongst festival curators and beat aficionados worldwide. Over the course of 6 years, his sound has dipped and swirled through a staggering range of styles, with each album exploring uncharted sonic territory. Warm, analogue dreamwave; mind-altering glitch; world-infused groove; bass-fueled breaks; sexified down-tempo; electric lullabies; and psy-fi funk are just a few of the genres that have poured from the Ambassador's soul into earbuds and ghetto-blasters across the galaxy.

Despite the political connotations of his name, the Ambassador is not a political partisan. Rather, he is a diplomat for a new paradigm rooted in creative joy, radical self-expression, and ecological principles. As part of this mission, the Ambassador has committed to carbon-neutral touring, instigating exploratory dance, and igniting fan participation.

Learn more about our Permaculture & Music tour at:

http://thepolishambassador.com

Urban Festivals

Since 2009, the Evolver Network has been forming an international network of local hubs for collaboration, called Evolver Spores. The Spores connect together the initiatives in a region that help people practice sustainable and just lifestyles. Currently in 32 cities, they support resource sharing, outreach and decolonization efforts.

As a local "Sporeganizer," I see grassroots communities focused on resilience who are struggling to make ends meet within the dominant capitalist infrastructure. At the same time, I meet healers and artists who have visions for community businesses that would restore healthy relationships with land and our bodies, but who lack the business skills to bring these ideas to fruition.

As a solution, I developed an urban festival that could reach across cultural divides and restore a true public commons space where people could skillshare with each other and match resources. Initially called Hive Mind, it is an open-source brand: any organization can adopt the business model intact or in part, freely copy or modify the graphics and sponsorship packet. It can happen under any brand umbrella, as serves the needs and interests of a community. My goal is to offer this festival yearly in a different city each year, and to share the model with all who would benefit from it.

I offer these examples of what's been shared at Hive Minds so far, as inspiration for festival producers and local community organizers.

Healing

One can reach out to invite the Indigenous people of the area and ask what would be of support to them at this time. Classes on growing and preparing herbal medicines can promote access to preventative healthcare. A company can sponsor a mini healing village, so participants can access it for free.

While illegal in the U.S., entheogenic plant ceremonies, such as with San Pedro, mushrooms and Ayahuasca, have long been a foundational practice for healthy community. Information about the benefits of legalization, habitat preservation and integration tools can support harm reduction and public awareness. A group at the forefront of this effort is ERIE, the Entheogenic Research, Integration and Education Group.

Alternative Economic Models

For many people, money is a barrier to healthy choices. Alternatives exist to meet our needs and nurture the plants, animals and waterways that support us. Time banks empower people to trade services based on hours. Crypto-currencies like Bitcoin bypass the banking system and ensure digital privacy. The Open Money app makes it possible to instantly create an alternative currency

for a community. Worker-owned cooperatives are a fantastic way to pool resources and share profits. The U.S. Federation of Worker Cooperatives offers resources for how to start one.

Local Governance

Festivals can teach practices like dynamic facilitation, where people from opposing viewpoints to come to a mutual understanding. Restorative justice can be introduced, where instead of sending someone to prison, a community talks through what they would need to rectify the harm done. Frequently this leads to healing and lower financial impact for the entire society. Social justice workshops can introduce frameworks of liberation to address systemic racism, sexism, poverty and more. People can role play to "call out" and "call in" when one experiences forms of social violence.

For further resources, The Evolver Editions book, ***Empowering Public Wisdom*** by Tom Atlee gives an introduction to community-guided policy tools. Bay Localize has a set of replicable city policy templates for urban farming and local clean energy.

The festival itself becomes a community think tank. By keeping the presentations short and the dialogue long, the whole event shifts from being a dichotomy between presenter and audience, to a community sharing their wisdom and creativity together! A hackathon can follow to build technical infrastructure for specific projects.

Urban Farming

Many cities now have permaculture guilds and Transition Towns where people can plug in to learn skills and build gardens. Long Beach, California has their entire local food business community networked to-

gether via a program called Long Beach Fresh! My favorite permaculture practice is permaculture finance: making regenerative financial systems that benefit people, planet and community. While hundreds of people gather at the festival, they could plant a food forest together in a park.

Science & Technology

Inviting science and technology practitioners into a holistic commons space bridges political and cultural divides. Scientists can teach about local watersheds and innovations in materials science. Engineers are great problem solvers, from water issues to civic software. Indigenous sciences present a deeply holistic perspective of land management. Demos might include small-scale sustainable energy production such as rocket stoves and biogasification. RE-Volv presents a revolving solar fund to install solar in cooperatives and nonprofits around the country.

The Arts

It's helpful if the music and art at the event represents the diversity of the people living in the region, to make the event accessible. Traditional world music instruments often provide a connection to ancient rhythms and songs from times when people lived more in balance with plants and animals. Poetry about social justice and empowerment lifts people's spirits up and builds community. A scheduled festival-wide moment of silence can create a space to sense and feel each other outside of our egoic identities, spoken language, and the loud noise of electronics. Shutting off the electricity grid at the festival to the extent possible can support this tuning in.

Local-grown food and drink vendors create a high vibe for the event with legal medicinal plants that nourish and foster connection.

Finally, people can record audio of the presentations and package them for sale or gift. I'm happy to host recordings via the Evolver Network website.

In three years of hosting this festival in the Bay Area, I've seen people's lives transform upon being immersed in a culture of empowerment, access, and DIY community building. We also designed a collaborative traveling version of this offering that can plug in to festivals and conferences.

When you watch a beehive in action, there is a synergy to their movements, a harmony to the way the individual bodies work together for the sake of the whole. We have all the resources we need to restore healthy relationships and inspire each other to greater creativity, so let's do it together!

Magenta Ceiba

Magenta is the executive director of the Evolver Network. She co-designed their international decentralized governance system, and spearheads global media and action campaigns. She has been a community organizer and event producer for 13 years. Magenta founded HiveMind, an urban festival that supports coalition building among people and initiatives devoted to restoring ecology, right use of technology and holistic healing. She has served as editor and curator for Aorta Magazine, a magazine for female and trans-identified radical political artists. As a practicing artist, healer and community leader, she describes herself as an "Imagination Healer," reminding people of our collective capacity to create the world we truly want to live in.

http://imaginationhealer.com
http://evolvernetwork.org

Catching and Storing Festival Energy

Clayton Gaar

Designing Festivals with Permaculture

There are a growing number of people who recognize the full scope of opportunity inherent in festivals to inspire, educate, and connect people on a large scale and in a lasting way. These people want more than a blowout party--they want community, and they see festivals as a vehicle to bringing these communities into form. There is a growing consensus among this contingent of the festival community that it is time to initiate the next step in festival evolution by redirecting the vast amount of resources generated by these gatherings into more permanent, regenerative structures and systems. In order to accomplish this, we must revision our festivals' goals, reexamine the fundamental patterns of our outdated festival model, and redesign our gatherings from a holistic, integrative perspective.

This essay applies permaculture principles to the idea of festival redesign, and presents some of the emerging themes related to using festivals as a technology to develop land-based community projects.

Permaculture: A Whole-Systems Design Science

Permaculture is a design science that applies whole systems thinking to produce highly efficient, integrated, and regenerative systems to meet human needs, while simultaneously enhancing ecological functioning of the natural environment. Put another way, "Permaculture is a philosophy of working with, rather than against nature; of protracted and thoughtful observation, rather than protracted and thoughtless labor; and of looking at plants and animals in all their functions, rather than treating any area as a single product system."[1] While the sheer complexity and all-encompassing nature of such an integrated system often takes some time to explain, and nevertheless to grasp, permaculture can be described simply as a holistic framework for solving problems. What makes permaculture unique from other design sciences is the fact that it is ethically based, with its three core tenets being:

1. ***Care for the Earth,***
2. ***Care for the People, and***
3. ***Return of Surplus.[2]***

In addition to considering the core ethics, permaculture applies a list of design principles that are based on observations of the natural world, such as "catch and store energy," "use and value renewable resources," "produce no waste," etc. Some of these key principles will be covered in more detail later as they relate to festival redesign, which will begin to illustrate the adaptability of these principles to a wide variety of problem-solving scenarios far beyond the realm of agriculture.

Permaculture design is often summarized by the age-old adage, "Work smarter, not harder," and the most successful festival producers are the ones who employ this type of holistic thinking in their production work. For a paradigm shift in festival design, we need the festival community to shift from short-term to long-term planning, from high to low input, from synthetic to biological product sourcing, from industrial to ecological mentality, and compartmentalized to holistic thinking.

Revisioning Goals and Core Values

The preliminary step in any design process is to get clear on what the primary goals are, and it is the same with festival design. It is time festival producers and patrons alike start asking the question more openly, "What is the purpose of this festival?" Far too often the answer is money or fame, and a majority of the time the primary goal seems to be throwing a massive party. While there may not be anything inherently wrong with celebrating for the sake of having a good time, one must consider the ecological effects of an event recklessly consuming large quantities of resources and the detrimental health effects of encouraging, or even tolerating, widespread substance abuse among its patrons. Imagine how different festivals could be if their goals became things such as provision of alternative education through hands-on learning, building community resilience by planting gardens, ecological restoration through planting trees, honoring our ancestors by sharing indigenous wisdom or building lasting infrastructure for future eco-villages. Some of the more progressive festivals have begun publicizing vague intention statements about transformation, expression, and celebration, but imagine taking it a step further to the point where each festival's website lists both the explicit goals and the core values of the gathering.

Having explicit goals outlined from the outset will not only change the entire design process of the event and attract participants aligned with the goals, but it will also give patrons the ability to hold the event accountable to its stated mission. While many festivals share common goals like providing peak experiences and encouraging creativity, the tangible lasting benefits of such events are not realized because they were never intended in the first place. This is the ultimate design flaw at the core of our current festival paradigm, and it is changing slowly. To help aid in this paradigm shift, festivalgoers must also start demanding more than a party from the organizers, or they can always create the event they want to see.

Needs and Yields Analysis: Identifying Leverage Points

A useful design tool from the permaculturist's toolbox is the Needs and Yields Analysis. This method identifies the characteristics of various components in a system both in terms of what resources each component needs in order to function as well as what useful resources each component yields produces. This analysis is useful because it can help identify potential functional overlap and interconnections between components, thus weaving the integrated web of resource flow within a system and maximizing overall efficiency. A comparative analysis of the needs and yields of a current paradigm festival (a party) with those of a new paradigm festival (a party with a purpose) reveals a powerful insight for festival redesign.

All festivals, assuming a similar attendance size, require the same basic needs to be met in order to function on a basic level. These standard needs are things like potable water, food catering, parking and camping space, electricity, toilets, staff, volunteers, artists, and likely at least one stage and sound system. While there are plenty of additional features and details that certainly add ease and comfort to a festival experience, this is a simplified list of the bare necessities for event production for illustrative purposes. When these basic needs alone are met, more than likely the yield is going to be a basic party.

Now, if the desired yield is a party with a purpose, perhaps one that plants a garden or builds an outdoor straw-bale kitchen, what are the additional needs that must be met? Some experienced instructors will be needed to guide participants through the process along with the provision of tools, materials, soil amendments, and plants. In addition to these tangible needs are the more invisible needs like focused attention, participants' desire to complete the project (often related to perceived "cool" factor), and of course the strong-willed intention to complete the project from the start. Therefore, we see that a slight increase in inputs from event producers and an even slighter increase from participants result in a much higher yield from the gathering. In other words, it does not take much more effort beyond the norm to accomplish a lasting project that

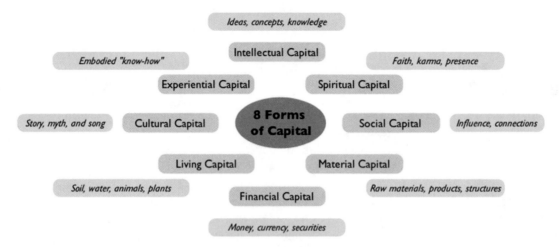

Ideas, concepts, knowledge — Intellectual Capital
Embodied "know-how" — Experiential Capital
Faith, karma, presence — Spiritual Capital
Story, myth, and song — Cultural Capital
8 Forms of Capital
Social Capital — Influence, connections
Living Capital — Soil, water, animals, plants
Material Capital — Raw materials, products, structures
Financial Capital
Money, currency, securities

yields a multitude of benefits and transform a mere party into a party with a true purpose. Festivals are already handling 95% of the work to gather people into one place and care for them; all that is left is to put the shovel in their hand and teach them how to plant a tree. This hypothetical situation of how a festival can take small steps to make big changes illustrates the idea of leverage points. Leverage points are the hidden gold in designs where relatively low inputs yield high outputs, and these leverage points are the key to efficient festival redesign.

The Grand Vision: Seeding Land-based Community Projects

How far can we leverage festivals with the proper intelligent design? This question is currently discussed so frequently in the forward-thinking festival community that David Casey, co-founder of Project Nuevo Mundo and co-founder of Cosmic Convergence Festival, refers to it simply as "The Thing." It is the thing that the festival community seems to be simultaneously dreaming into reality as if the idea just tipped the scales of the 100th monkey effect when an idea is birthed instantly into the consciousness of an entire population.[3] While there are many different versions of The

Thing, the common thread woven among them is the call for a redesign of our gatherings, to catch and store their myriad forms of energy in the form of lasting regenerative structures and systems. The result over time is a land-based project that is built using the experiential, physical, and financial capital generated by festivals, or at least the development and maintenance of which is greatly expedited and enhanced by the resources of a festival community. Different iterations resemble various aspects of permaculture demonstration farms, holistic retreat centers, event venues, intentional communities, and ecovillages, while some ambitious models incorporate elements of all of the above. Admittedly, this vision represents the extreme end of festival redesign layered with complexities and challenges. However, aspiring to these lofty goals may be the best way to transcend the limits of what we think is possible to achieve as a community.

Now that the design framework has been introduced, the goals have been reassessed, and the ultimate vision has been identified, let us explore some of the strategies and details of festival redesign through the application of permaculture principles, as articulated by David Holmgren, co-originator of permaculture:[4]

1. Observe and interact

The act of thoughtful observation is what gives us insight into the state of system functioning and is often referred to as the primary design principle upon which all others expand. The most obvious observation from anyone's point of view, especially from that of the builders and festival staff, is that it is highly inefficient to build the entire infrastructure for these temporary villages only to tear it down again at the end of the weekend. Other commonly cited observations include the patterns of festivals losing money year after year, festival grounds getting trashed, participants leaving unhealthy and exhausted, and the fact that thousands of people leave the grounds at the end of the weekend with nothing tangible to show for it. These are but a few of the major observable patterns informing us that a redesign needs to address these areas of concern.

2. Creatively use and respond to change

The ability to creatively respond to change is crucial for successful event production. Consider two dynamic market forces that festival producers face today. First, every year more new festivals pop up on the map as festival culture spreads and people want to attempt to create their own event. There are only so many weekends in a year, and the calendar is getting so full of events that competition for participants is increasing each year. While the festival market is simultaneously expanding, this competition is still putting pressure on festivals to find innovative ways to distinguish their brands and find their unique niche in the festival world. The second force is the increasing demand for festivals to be multi-dimensional in their offerings (i.e. provide more educational workshops), reduce their ecological footprint, and show efforts towards actively improving the greater community outside of festival culture. Visionary event producers will recognize that the second force offers a viable solution to the perceived challenge of competition created by the first change. Pioneering green businesses have repeatedly proven that being early innovators can result in strong competitive advantages and emerging as leaders in their field, and the field of progressive festivals is no different.[5]

3. Catch and store energy

Permaculture systems seek to catch and store energy in all forms (i.e. water, nutrients) as it flows through a system. Festivals attract abundant flows of energy in a variety of forms, but currently the opportunity to harness and utilize these resources in a regenerative fashion is mostly squandered. A useful way to track these energy flows is by categorizing them according to Ethan Roland's Eight Forms of Capital.[6] As classified by Roland, festivals involve large flows of financial capital (money), social capital (connections), material capital (natural resources), intellectual capital (ideas, knowledge), experiential capital (action), and a variable amount of spiritual capital (prayer, intention) and cultural capital (story, ritual) depending on the festival. While some strategies are mentioned in this essay, there are enough strategies and techniques for harnessing this enormous amount of festival capital to warrant writing a book or even an open source forum on the subject. Some examples include land purchases through crowd-funding, creation of land trusts, natural building intensive workshops, food forest planting, land cultivation, permaculture design certifications, any permanent infrastructure construction and land improvements (i.e. composting toilets, water systems, stages, kitchens, housing), and timeshares and access to venue property year-round. Above all, it is imperative that more gatherings experiment with these strategies and openly share their successes and failures with the larger community to accelerate the collective learning curve.

4. Design from patterns to details

Thoughtful design on a pattern level first helps ensure that the details are in alignment with the bigger picture of what it is our festivals seek to create. Event scheduling is a good example of this. If a gathering has the goal of encouraging holistic health and education, the gathering needs to design its schedule to reflect that. Having music play 24 hours a day for multiple days on end, by design, encourages participants to consume stimulants, stay up for extended periods of time, and sleep when their bodies reach their limits. On the other hand, scheduling music to end earlier each night and start in the afternoon helps guide the overall behavior of the population towards getting a good night's rest and attending educational workshops during the day. This intentional design comes with additional benefits of staffing less late night staff, decreasing the incidents for security and medical staff, and ultimately lowering the overall liability of the event. Beloved Festival and Kinnection Campout are two examples of events that use similar scheduling patterns.

5. Integrate rather than segregate

Integration is a fundamental theme to the new festival paradigm. Rather than festivals being just about "play," our gatherings are shifting towards being microcosms of real village life where a community of people play, work, learn, and live together in an integrated, multidimensional experience, even if only for a short period of time. A ubiquitous pattern in modern society is the tendency to compartmentalize, separating our "work" from our "education" and our "vacation," which one may argue is contributing to our increasing symptoms of collective schizophrenia. The benefits of reintegrating these aspects of our existence back into one continual experience that is "life" has myriad benefits. A work project can be fun and educational at the same time, and when it is all said and done, there will be progress to show for it, as well as new skills and new friendships. Moreover, productivity, mental health, and community resilience will likely see significant increases. This is the way many villages have fundamentally operated for thousands of years, and there is a growing contingent of people who see the merit in it.

Weaving it All Together: Multiple Functions and Succession

One of the easiest ways to leverage systems is by taking actions that have multiple functions, and this is the key to designing new land-based festival models that are financially regenerative. Hands-on classes and workshop intensives can generate revenue, provide cheap labor in exchange for educational experiences, and ultimately produce permanent land improvements. The festival itself is a highly effective existing platform to crowdsource funding and market both the overall project and the individual workshops to a large, targeted demographic that is interested in progressive living, has access to resources, and is generally excited to invest in new community projects. For the more elaborately designed land projects, on-site infrastructure can be used for a variety of year-round businesses after they are built by the festival intensive workshops.

For example, micro-cabins, yurts, and bunkhouses can be used for staff and artist housing during events, and throughout the rest of the year these buildings can be used to house year-round community staff, to rent to members of the community, to operate as an eco-hostel, rent to other event producers as part of a retreat center package, or time-shared to members who pay annual dues. Large fields can double as parking lots for events and grazing paddocks

for farm livestock. A community center building can be used year-round as the primary council hall for the community residents and as indoor event space for concerts, weddings, classes, and retreats. Residents can be core staff that handle administrative duties of the various businesses, operate the farm, maintain the grounds, teach classes, lead tours, and build new structures. Some of these residents can be employed while others participate in long-term work-trade internships. The farm can produce organic produce to feed residents, to create a Community Support Agriculture (CSA) for the surrounding community, and to cater retreats and festivals that use the venue. Farm trucks and golf carts can also be rented by events at the land, composting toilets will be used year-round to produce compost for the orchards, and the list goes on and on.

This highly integrated business model produces diverse year-round revenue streams and uses existing material, financial, and human capital for multiple functions throughout the process. The result is a substantially faster return on investment, optimized resource utilization, and a highly resilient web of profitable business ventures that can be simultaneously developed and operated on the same piece of land. The development timeline of these ventures can and should be planned in such a way that establishes an intentionally designed economic succession, or phasing in, of Return on Investment (ROI). For example, ventures with more immediate returns depending on less infrastructure--like festivals and weddings--should be launched first, while ventures with long-term returns depending on infrastructure development--like polyculture tree crops and rental cabins--will phase

in later in the project's timeline. This economic succession follows nature's pattern of ecological succession, whereby an ecosystem gradually passes through various characteristic phases of maturity with corresponding complexity and plant communities until it eventually reaches its climax state as a stable, thriving old growth forest.

Final Notes of Consideration

While there is an increasing number of festivals that employ various components of the models discussed in this essay, there are few, if any, large-scale examples designed to catch and store their tremendous available capital using the integrated polyculture business models proposed herein. This is likely because the models described in this essay are complex, dynamic systems with many moving parts that have far-reaching legal, financial, and social implications. Major challenges to consider include: being unrecognizable in the eyes of participants used to the existing music festival model, zoning limitations, county regulations, building codes, securing land with sympathetic neighbors, noise complaints, quality control of student and volunteer work, managing large groups of people, creating a sense of personal investment in the project for individuals involved, creating equitable yet efficient decision-making structures, and fostering ongoing community cohesion and cooperation.

According to Diana Leaf Christian's extensive global research on intentional communities and ecovillages, nine out of ten communities fail due to either finances or social conflict. [7] Therefore, choosing a solid core team to begin with whose members have aligned core values and the ability to communicate and resolve conflict is vital for creating these types of festivals and projects. As suggested in this essay, festivals aid greatly in generating the capital flow for such projects, so finding the individuals with the financial capital and technical know-how will likely be the easy part of the process. The true challenge will lie in finding the group of people who will work well together and keep the project functioning in a healthy way.

Conclusion

Many non-corporate festival producers find it challenging to establish a financially successful music festival within the existing festival paradigm. Moreover, progressive event producers may even feel pigeon-holed and unable to break out of the current mold that is recognizable as a "music festival." Adopting the strategies and techniques of a truly progressive holistic festival paradigm not only increases the chance of financial success, but also gives event producers a replicable way to create thriving community land-based projects that are ecologically and socially regenerative. Our current festivals are underutilizing the vast capital flows available to them, and permaculture design is one useful framework for redesigning our festivals to adapt, utilize, and benefit from changes like evolving consumer preferences, more competition in the festival market, and an increasingly unsure economic and ecological future.

Music festivals and their culture of celebration are here to stay, so we might as well use them to achieve great things. Our festivals are still viewed as consumptive "parties" in the eyes of the mainstream, and to a large extent that belief is true, but there is a way to legitimize our "fringe" culture in the mainstream. It is time our progressive festival community up-levels the way we do things so that we can model a way to create communities of the future and have a great time doing it. It will take teams of individuals that are detail-oriented, business-savvy, and truly visionary thinkers. Festival producers must become land stewards and community planners at heart, and patrons must become invested participants rather than mere observers showing up to consume the festival experience like a product handed to them. Most importantly, we need the entire festival community working cooperatively towards this greater vision with open dialogue about what is working and what can be improved. Open source templates, business models, and methodologies will greatly accelerate our realization towards these goals, and the ReInhabiting the Village Project gives us a way to do this.

Clayton Gaar

Clayton is the founder and director of Tribal Council Collective, an Asheville-based event production collective and an ever-evolving community rooted in the southeast U.S. He is also founder and co-producer of Kinnection Campout and co-founder of Tribal Alliance Gatherings. Clayton is a certified permaculture designer specializing in small-scale homestead design, and he shares his passion for ecological living through festival workshops and permaculture intensive courses. Clayton is a U.S. Regional Ambassador for Project Nuevo Mundo and was a member of the inaugural 2013 Earth Odyssey caravan through Mexico and Guatemala. Between travels, Clayton manages his family's budding permaculture homestead in the Three Forks Watershed of North Georgia. In alignment with his mission to unite festival culture and permaculture, Clayton is working on a diverse gene pool of projects that catch and store the energy of human gatherings in the form of lasting holistic communities and regenerative land projects. In addition to applying permaculture principles to festival and community design, Clayton spends his time exploring ways to integrate ancient ways of living into modern visionary culture in a way that serves all life.

http://kinnectioncampout.org
http://tribalcouncilcollective.org

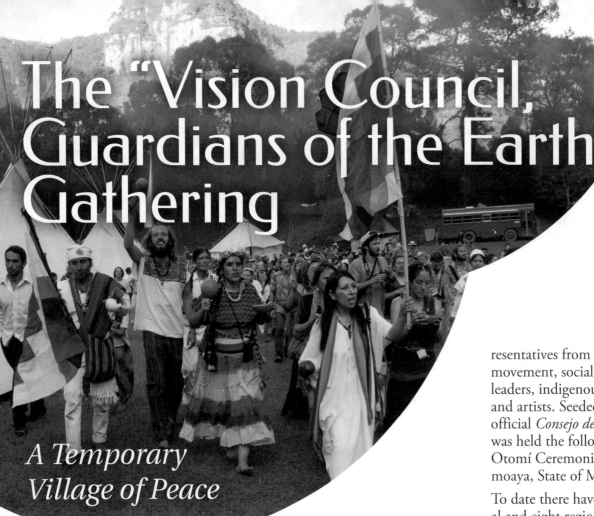

The "Vision Council, Guardians of the Earth" Gathering

A Temporary Village of Peace

Ivan Sawyer Garcia

The Consejo de Visiones, *Guardianes de la Tierra* or "Vision Council, Guardians of the Earth" is a unique gathering of sustainable communities, artists, healers, activists and alternative culture in general, developed over the past two decades in Mexico and different parts of Latin America.

Since the beginning of human history we have been gathering together in tribes to strengthen our relationship with each other, to council about important community matters, to exchange seeds, medicines, gifts and to simply celebrate life; to sit by the fire and hear the stories of our elders; to heal and honor the spirit and the body. Following this very ancient tradition the "Vision Council, Guardians of the Earth" has been gathering representatives from the Bioregional, Ecovillage and Permaculture movements, Indigenous leaders, artists and healers for over 20 years. These gatherings also have the purpose of supporting the host community by improving infrastructure, boosting the local economy with influx of visitors and also supporting specific environmental protection and restoration efforts.

It was in the spring of 1990 that Huehuecoyotl Ecovillage, located in the State of Morelos in Mexico, hosted the *Encuentro sobre la Naturaleza de los Guardianes de las Tradiciones Sagradas y Científicas* or "Guardians of Sacred and Scientific Traditions Nature Gathering." This event gathered a wide range of representatives from the environmental movement, social activists, spiritual leaders, indigenous elders, scientists and artists. Seeded there, the first official *Consejo de Visiones* gathering was held the following year in the Otomí Ceremonial Center in Temoaya, State of México, México.

To date there have been 13 national and eight regional gatherings in Mexico, as well as three international gatherings; the First Continental Bioregional Gathering in Meztitla, Tepoztlan, Mexico in 1996, the "Call of the Condor" in Cuzco, Peru in 2003 and the "Call of the Hummingbird" in Brazil (2005). Since 2012, there have been annual events in Colombia known as the *Llamado de la Montaña* or "Call of the Mountain," and small versions of the event have also been held in Puerto Rico, Costa Rica and Ecuador. More recent events have been held for four consecutive years in *Ecoaldea Nierika* in Chalmita, Mexico from 2010 to 2014. All of these events have been of great relevance and significance for the Bioregional and Ecovillage Movements as they have served to link and cross-pollinate different movements and organizations that include everything from social justice and sustainable communities, to indigenous wisdom, alternative economy, music and art.

Event Organizational Structure

The Consejos' constantly evolving organizational structure based on different Consejos, or Councils, has been able in many ways to successfully implement an Ecotopian model of organization inspired by the Deep Ecology philosophy, The Bioregional Movement of North America and the Rainbow Gatherings, in which many of the original organizers participated in the 70's and 80's.

The organizational model also includes a consensus and council-based form of organization in which everyone participates in the creation and maintaining of all areas of our temporary ecovillage, which has been deeply influenced by a recovering of ancient governance models and spiritual practices of Native American and tribal cultures worldwide. In this event there are no spectators; everyone is an active member and has an important role in the community. Everybody, even the event organizers, artists, performers and healers are expected to participate in the different Clans and Work Committees that are in charge of maintaining different areas such as the kitchen, waste management systems, security and other areas of the event. From maintaining the composting toilets, to preparing the food, to participating in councils and talking circles to dream up the next gathering, festival or theatre presentation, we are all responsible for the building and maintaining of the community and all threads that keep it together.

The "Consejos" or Councils

The "Consejos" or Councils are the specific theme groups or areas of interest that create their own schedule of activities during the entire event. Each Consejo also has certain responsibilities within the community depending on their area of expertise. For example the Health Council is in charge of the clinic and the Kids Council is in charge of the kids' play area, etc. All Consejos need a coordinator, a physical space (tent, tepee, tree, etc.), workshop materials and facilitator or group of facilitators. All Councils prepare a presentation using performance, ceremony or music to present in the final plenary, on the last day of the gathering. The number and focus of the Councils have evolved, but here is a list of the most common focus groups.

Ecology Council

The Ecology Council has always held a very central part of the Consejos de Visiones, as it has been a meeting place for many leading environmental activist and leaders throughout the years. Some examples of activities organized by the Ecology Council include local environmental protection initiatives, Ecovillage Network meetings and hands-on permaculture/natural building workshop. Some of the preferred topics of the Ecology Council include: *bioregionalism*, *permaculture*, natural construction, renewable energy, ecological sanitation, watershed protection, children's workshops and more. As such, the Ecology Council brings together important representatives of the Ecovillage and Permaculture movement in Mexico and all of Latin America.

Traditions Council

The Tradition Council holds the spiritual center of the event - the Sacred Fire. In Mexico we have a rich heritage of oral tradition and Indigenous traditions and the Traditions Council is a space to listen to the voice of our elders and ceremonial guests, to sing, pray and talk late into the night around Grandfather Fire. The central fire is the first fire to be lit at the beginning and is kept going for the duration of the entire event. Some of the activities held in the Traditions Council include the ceremonial inauguration of the events, Elder Council, sweat lodges in the mornings and in the afternoon, sacred

pipe ceremony and many others. Over the years the Traditions Council has hosted representatives of Indigenous traditions from all over North, Central and South America.

Arts and Culture Council

The Arts and Culture Council employs self-expression as a means of honoring the earth and the world's cultural traditions. The musicians and the artists in the community ignite the fire in our souls through music, dance and the arts both on the stage and around the fire. The Arts Council is in charge of the nightly program of artistic performances that include music, dance, poetry and theatre and any form of art. During the day the Arts Council organizes different workshops that include things like mask-making, stilts, drum circles, silks, capoeira among other activities. In addition to performance the artists are invited to teach workshops and host dialogues such as how to use art and culture to leverage social change and environmental consciousness.

Health Council

The Health Council is in charge of providing alternative and holistic health services as well as medical assistance when needed for the entire duration of the event. Also different workshops are offered by the Health Council that include local traditional medicine forum, plant walks and workshops of all sorts including herbal medicine, reiki, women's health, Ayurveda and many more.

Youth Council

This Council is a place for youth to gather and express themselves. Activities include daily meetings, creativity and performance arts workshops and any activities organized spontaneously by the youth themselves. Never doubt the power of spontaneous youth organization!

Kids Council

The Kids council is not a place to leave the kids while the parents go play; instead it serves as a place where children can become more involved in the community and also have their voice heard. It's also a place for children to make art, learn about nature, learn songs, play games, and make fun of the adults, among other things.

The Peace Village

The Consejos and its vision of creating a temporary "Peace Village" has been the source of inspiration for a great number of regional events, gatherings, ceremonies, festivals, exhibitions and caravans in Mexico and worldwide. In the same manner the Consejos has been greatly inspired by the Bioregional Movement North America, which has held 10 Continental Bioregional Congresses since 1984.

The Rainbow Peace Caravan was partly responsible for bringing this idea of traveling Peace Villages to South America during the past decade; organizing a Peace Village of Indigenous Women in Ecuador in 2002, el *Llamado del Condor* or "Call of the Condor" in Peru in 2003, Chile 2004 and later in Brazil for 3rd World Social Forum in Porto Alegre in 2003, as a space exhibiting and practicing sustainable and regenerative solutions--a living space that promotes a Culture of Sustainability with a foundation on a Culture of Peace, Community living, Permaculture and Alternative Economy.

In this event there are no spectators; everyone is an active member and has an important role in the community.

Areas of Convergence

1 – Central Camp

The Central Camp is the village's main community space and is used for cultural presentations, workshops, circle dances in the evenings, films at night and many other activities. Daily community plenary is held here every morning and there is always a large board informing participants of the program and different activities.

2 – Eco-kitchen and Café

Catering all-natural and organic food, the Eco-kitchen will be managed by a team of professionals in the field of sustainable food and serves two or three healthy vegetarian meals. The Café sells different snack and food items as well as teas and natural drinks that are the basis for the financial maintenance of the kitchen, or to support local community projects such as the Ecomundi School in Chalmita.

3 – Youth Camp

The Youth Camp held by the Youth Council aims to be a space for youth to gather, organize workshops and

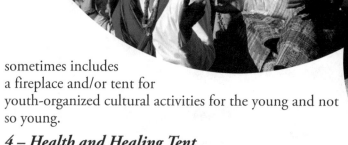

sometimes includes
a fireplace and/or tent for
youth-organized cultural activities for the young and not
so young.

4 – Health and Healing Tent

Area used by the Health Council for their activities. This
place also includes a first aid and medical response team
as well as holistic health practitioners that can offer their
services for event participants.

5 – Moon Tent

The moon tent is a space dedicated to the honouring of
the sacred feminine energy. In this space different activities take place such as women's circles, ceremonies,
discussions and workshops about women, *eco-feminism*,
women's spirituality, etc.

6 – Flower Beehive

Children are the new consciousness and will be the future
guardians of the planet. In the Village, reserve a room
for mothers, fathers and children, which will develop
Ecopedagogy activities, workshops, circus, theatre, plays,
children's games and presentations.

7 – Bioregional Village

Area for internal management of the temporary Ecovillage. The Bioregional Village will focalize such tasks as
management of composting toilets, solid waste, composting systems, security, water, communication and internal
programming of the village.

8 – Fair Trade and Exchange Market

Convergence point for the commercialization of goods
such as crafts, clothes and natural foods mostly consisting
of local and natural products and incentivising the use of
local currencies and barter.

9 – Sacred Fire

Our central fire is always lit on the first night of the event
and is watched over by a team of fire keepers. During the
day and in the evenings this fire is used to light the sweat
lodge and receive the community's prayers as well as help
purify our minds and bodies.

10 – Permacultural Technology

Scattered around the village, numerous environmental
low-cost technologies are available for any visitors to use
such as: composting toilets, compost, greywater filtering
systems, green roofs, recycling station, solar oven, earth
oven, bicycle generators, bamboo construction and other
natural building techniques, among others.

Ivan Sawyer Garcia

Ivan Sawyer García is Logistics Coordinator and Impact Centre Liaison for Project
Nuevo Mundo. He is also a regional mo-
bilizer for C.A.S.A Continental, the Latin
American branch of the Global Ecovillage
Network. Ivan considers himself an event
producer, sustainability educator and
social entrepreneur. Among other things
Ivan has been participating in the development of different cultural exchange and
indigenous knowledge preservation initiatives in different parts of North and South
America for almost a decade. In the realm
of festival culture Ivan is dedicated to
bridging festival culture with sustainability and Indigenous wisdom. He currently
collaborates with the annual Consejo de
Visiones gathering in Mexico, Cosmic
Convergence New Year's celebration in
Guatemala, Festival Ometeotl in Mexico,
and Tribal Alliance Retreat, among others.
He loves Cacao.

http://ancient-futures.com

http://consejodevisiones.org

Living Economy

EVEN THE MOST THOROUGH CHANGE HAPPENS ONE CHOICE AT A TIME.
Charles Eisenstein

What do you value? What is sacred to you? What does abundance equate to in your life?

Every day we make a decision to agree, collectively and individually, to participate in a story. One of the most widely accepted stories ever told. It's the story of a thing called Money, a piece of paper, printed with a different symbol or cultural archetype, assigned a number value, and then circulated to fuel our consumption, exchanging "this for that" as a means of meeting not only our survival needs, but our wildest dreams. For many a sense of freedom and security is deeply embedded in the amassment of these pieces of paper. And yet, the only difference between a $100 US monetary note and say, a text book of permaculture principles, is that WE have agreed to a system in which those particular pieces of paper called "money" have a VALUE that gains one access to goods, services, experiences, and even power. Each day, each one of us agrees to a system that allows the commodification of everything that was once an inherent part of nature, given freely by the Earth, infinitely accessible to all who would tend the conditions necessary for the regeneration of the resources of our ecosystem.

IN THE WORDS OF THIS BEAUTIFUL CANTICLE, SAINT FRANCIS OF ASSISI REMINDS US THAT OUR COMMON HOME IS LIKE A SISTER WITH WHOM WE SHARE OUR LIFE AND A BEAUTIFUL MOTHER WHO OPENS HER ARMS TO EMBRACE US. "PRAISE BE TO YOU, MY LORD, THROUGH OUR SISTER, MOTHER EARTH, WHO SUSTAINS AND GOVERNS US, AND WHO PRODUCES VARIOUS FRUIT WITH COLORED FLOWERS AND HERBS." Pope Francis

Our current economic system has escalated our species to an epidemic of disconnection from the very Nature of our Being. Money itself is a neutral force, another tree turned into stack of green notes, yet how we USE money and RELATE to money has given it the power to either be a force of creation or destruction.

Changing the way we use, share, and relate to "money" will change every fundemental system of life on this planet. For at this point, all infrastructure, all government, all cities, all food, all human rights, all of the resources of the planet, in fact our entire ecology is bound to this collective story of money, power, and value.

Although at times it seems that those in government or corporate hierarchies have the dominance of power, it is important to recognize that as consumers, WE have the power to guide the marketplace based on our choices of what business to support and where to place our "value." We have a choice to buy from companies that commit to Fair Trade practices, ecological consciousness, and ethical responsibilty. Hold-

ing companies accountable for their resource extraction, for their treatment of workers, for their environmental impact is one way to insight a shift in the approach these companies take. There are so many cases where consumer choices influenced the practices of businesses.

One step further would be to empower and fortify local economy, choosing to support local organic farms, artisan goods, co-ops, and shop local businesses. Yet another step would be to consume less, to trade, barter or gift goods and services more, to share resources with neighbors, to build or fix things for ourselves, to upcycle materials, and to simplify our lives by choice.

Since money is so woven into the fabric of our existence, there is not one simple solution to resolve the current scenario. It will take a systems over-haul, one sector at a time, one company at a time, one government at a time, one consumer at a time to dismantle the current construct and collectively decide to tell a new story with our everyday actions and choices.

The translation of the word "Economy" originates from Eco- "Home" and Nomy- "Stewardship or Management." literally the Stewardship of our Home. It is vital to come into ethical stewardship of our planet's resources and abundance. With intelligence, compassion, and service we can redesign what we value and how to attend to the needs of all. To value life and to choose a Life-affirming economy is the highest expression of human potential.

Jamaica Stevens

Excerpt from
A Circle of Gifts

Wherever I go and ask people what is missing from their lives, the most common answer (if they are not impoverished or seriously ill) is "community." What happened to community, and why don't we have it any more? There are many reasons – the layout of suburbia, the disappearance of public space, the automobile and the television, the high mobility of people and jobs – and, if you trace the "why's" a few levels down, they all implicate the money system.

More directly posed: community is nearly impossible in a highly monetized society like our own. That is because community is woven from gifts, which is ultimately why poor people often have stronger communities than rich people. If you are financially independent, then you really don't depend on your neighbors – or indeed on any specific person – for anything. You can just pay someone to do it, or pay someone else to do it.

In former times, people depended for all of life's necessities and pleasures on people they knew personally. If you alienated the local blacksmith, brewer, or doctor, there was no replacement. Your quality of life would be much lower. If you alienated your neighbors then you might not have help if you sprained your ankle during harvest season, or if your barn burnt down. Community was not an add-on to life, it was a way of life. Today, with only slight exaggeration, we could say we don't need anyone. I don't need the farmer who grew my food – I can pay someone else to do it. I don't need the mechanic who fixed my car. I don't need the trucker who brought my shoes to the store.

I don't need any of the people who produced any of the things I use. I need someone to do their jobs, but not the unique individual people. They are replaceable and, by the same token, so am I.

That is one reason for the universally recognized superficiality of most social gatherings. How authentic can it be, when the unconscious knowledge, "I don't need you," lurks under the surface? When we get together to consume – food, drink, or entertainment – do we really draw on the gifts of anyone present? Anyone can consume. Intimacy comes from co-creation, not co-consumption, as anyone in a band can tell you, and it is different from liking or disliking someone. But in a monetized society, our creativity happens in specialized domains, for money.

To forge community then, we must do more than simply get people together. While that is a start, soon we get tired of just talking, and we want to do something, to create something. It is a very tepid community indeed, when the only need being met is the need to air opinions and feel that we are right, that we get it, and isn't it too bad that other people don't … hey, I know! Let's collect each others' email addresses and start a listserv!

Community is woven from gifts. Unlike today's market system, whose built-in scarcity compels competition in which more for me is less for you, in a gift economy the opposite holds. Because people in gift culture pass on their surplus rather than accumulating it, your good fortune is my good fortune: more for you is more for me. Wealth circulates, gravitating toward the greatest need. In a gift community, people know that their gifts will eventually come back to them, albeit often in a new form. Such a community might be called a "circle of the gift."

Charles Eisenstein

Charles Eisenstein is a progressive author and public speaker. He is the author of several books including **The Ascent of Humanity (2007), Sacred Economics (2011), and The More Beautiful World Our Hearts Know Is Possible (2013).**

http://charleseisenstein.net/

http://facebook.com/groups/TheMoreBeautifulWorld/

Holonomic Perspective

A rising tide lifts all boats.
John F. Kennedy

Steven Michael Ehlinger II

We are living in evolutionary times where the conventional mechanisms of economics, including how co-creation is funded, and more intimately, our own relationship to money and each other, are being deeply examined, redefined, and reconfigured.

There is a return and movement toward something that feels more aligned, more resonant, more honoring, more fulfilling, more liberating, and more sustaining for ALL of the participants in the newly-forming arc of value creation and its underlying economic premise.

I like to refer to this premise as the **Holonomic perspective.**

What is this Holonomic perspective? Let's break it down…

HOLO /ˈhälō/ = a combining form meaning "whole" or "entire."

We intuitively know that the approaches that we have been taking relative to living systems, business practices, and resource allocation are missing something, and that they are not attending to important aspects of life as a whole. So, HOLOing anything means we are inviting a much larger and more comprehensive approach to be embraced. We see the use of HOLO integrating into many of the progressive organizing models that we are drawn to today, whether that be The HO-LOgraphic Living Model™ (http://ultimatedestinyuniversity.org) or **HOLOcracy** (http://holacracy.org) (adopted by Zappos last year). Therefore, HOLOing out an economic perspective is long overdue, as it seems the very DNA encoded in the seeds of current economic perspectives have left something (or many things) out of the equation.

ECONOMICS /ˌēkəˈnämiks/ = a social science that studies how individuals, governments, organizations and nations make choices on allocating scarce resources to satisfy their unlimited wants. It can generally be broken down into: macroeconomics, which concentrates on the behavior of the aggregate economy; and microeconomics, which focuses on the individual.

When there is a large enough body of perspectives at work, we tend to classify it as a science. Taking a closer look at the subject of economics and its history, we see that it emerged

Holonomic

first from political science and the larger body of social sciences. These political and so-cial science roots of economics are what have, for the largest part, governed humanity's economic approach, and in particular, for the ruling class and politically empowered, to create a huge divide between the have(s) and the have not(s).

So, the Holonomic perspective invites us to both widen the lens and deepen our focus with which we see the world, and to incorporate that vision in the design, implementa-tion, operation, and e(valu)ation of the ongoing moment of creation, of which we are all a living and vibrant part. At the heart of the Holonomic perspective is a connected-ness to where we are making choices from and where those choices will lead us.

In delving deeper into my own journey of discovering what Holonomics meant to me, I explored some of the related terminology that exists in our current lexicon. I found the juxtaposition of the various terminologies we use to describe these aspects of our lives to be quite revealing as to the nature of many of the challenges humanity faces, individual-ly, collectively, and globally.

> ***ECONOMY*** /əˈkänəmē/ = *the arrangement or mode of operation of something; a system of interaction and exchange.*

So, ECONOMICS is the science of choice-making relative to resource allocation, yet an ECONOMY is a system of interaction and exchange. Most systems are based in what I like to refer to as an "Agreement Field," meaning, in order for that system to have come into being, a long lineage of agreements were required to have been made in order for that system to even exist. Many of the systems we were born into consist of agreements that were made by our ancestors well before we were even born. To that end, there exists an innate tension in us relative to operating under an agreement field in which we didn't directly participate in the myriad choices that were made during the adoption of a particular system, especially when those systems have not incorporated newer insights that have emerged since the system became the de facto protocol.

> ***ECONOMIC(al)*** /ˌekəˈnämik(ə)l/ = *avoiding waste or extravagance.*

So, ECONOMIC(al), as juxtaposed against ECONOMICS, would suggest our choices relative to resource allocation not be wasteful or extravagant. For me, this suggests an honoring of the deep relationship we have to resource; a coming into right relationship, so to speak. Sadly, whole civilizations could(have) exist(ed) and thrive(d) on the waste and extravagance we've agreed to give certain sectors within society.

Holonomics suggests there are other perspectives for choice points we might begin to embrace that are rooted in our core values of life, lasting fulfillment, and connected wholeness. This deep desire for wholeness (holo-ness) has provided the fertile soil for ever-progressive and restorative models of economics that are guiding us into what might be possible:

- Multiple-bottom lines
- Money as love[2]
- Top-line economics[3]
- Yin-Yang currency models[4]
- Gift & compassion economies[5]
- Radical sufficiency[6]
- Sharing economies
- Sacred commerce[7]
- Resource-based economics[8]
- Sacred economics[9]
- Co-operative & generative models
- A return to barter & trade
- Sacred marketplace[2]
- Use of alternative currency models
- The activation of the real wealth of nations[10]

All are serving their part to open us up to new possibilities, return us to our deepest roots where the alchemy of our collective soul's yearning can begin to take form. From the Holonomic perspective, all are welcome and serve as bridges back and into an economic perspective that works for ALL.

The Holonomic perspective suggests an opportunity exists for us to seed and root our economic models and operating systems in a set of core values that will deeply resonate with all of life, to provide a new basis for ascribing value. It also suggests us to come into a profound honoring and embodied respect with each and every participant and resource being utilized in the creation of the value contained within the respective products and services that we co-create, offer, recommend, support, and utilize.

HOLONOMICS /ˈhälōˈnämiks/ = an economic(al) whole systems perspective that is seeded, rooted, and deeply connected to the vibrancy of universal core values and promotes the emergence of new and adaptive systems of governance, operation, interaction, and exchange that provide harmony, true balance, and an innate and generative prosperity for all.

An interactive technology platform incorporating the core tenants of the Holonomic perspective is currently in development to assist organizations, communities, and families to ground, market, operate, and transform our collective experience into one that is impeccably aligned, congruently life-affirming, and deeply fulfilling.

As with any endeavor, there are a set of guiding principles or underlying premises on which it is based. The following core values form the basic or primary GPS for "how" the Holonomic platform will create and interact, both inwardly, and as part of birthing a more sacred marketplace for all of life to enjoy and mutually benefit. With a deep attenuation to "how" we do what we do, the Holonomic perspective and its core values strive for it to be done:

Authentically

Clearly with Clarity

Collaboratively

Creatively

Generatively

Inspiringly

Integrally

Intuitively

Lovingly

Playfully

Prosperously

Responsibly

Synergistically

Unifyingly

From a functional perspective, the Holonomic platform will serve as an integrative solutions platform for visionary and socially conscious enterprises that will:

- Feature a unique blend of values alignment technology

- Serve multi-dimensional collaborative environments

- Offer shareable wealth interoperability

- Provide an adaptive Holonomic dashboarding system, and

- Allow for multivariant success factoring

There exists an invitation for the Holonomic perspective to come alive in our collective experience. Therefore, you are invited to envision your deepest core values being integratively and integrally woven into the visioning, formation, start up, and operating of organizations whose products and services carry that deeply-rooted and retained value into the lives of everyone participating in the entire value creation chain and beyond. The time has come for a transformational approach to how resources are allocated and value is co-created. This calls for new perspectives and new toolsets to come online to support our journey. May the Holonomic platform offer to serve such a noble quest.

Stephen Michael Ehlinger II

Steven founded Integrated Management Consulting in 2007 to fulfill a deepening curiosity about how an intersecting of spiritual principles and business mastery principles could benefit key executives in embracing new approaches to organizational design.

He brings over 25 years of experience in the areas of finance and operations; serving in CFO positions in the environmental, media, and technology sectors. He was founding partner and CFO of EcoMedia, an environmental media company (acquired by CBS Corporation in 2010), and as Managing Director for the asset management firms of Adams O'Connell, Inc. and Morgan Adams, Inc. honed expertise in entity structuring, policy formulation, capital structuring, and financial reporting.

He is passionate about the sacred commerce/economics movements in both form and function and is passionately developing a tool, based on what he is calling Holonomic (Holo=whole or entire + Economic) to assist organizations to embody the core values of the sacred marketplace.

He has presented and co-presented and facilitated workshops with his beloved Samavesha-Gayatri Devi on the topics of: Sacred Erotic Commerce, Corporate Shamanism, The Gift Economy, Yin/Yang Complimentary Currencies, The Soul Tantra of Money, Consciously Conceiving the Soul of Money, The Yoga of Money, and Gender Alchemy.

http://facebook.com/IntegratedManagementConsulting

http://linkedin.com/pub/steven-ehlinger-ii/14/195/934

The Sharing Economy Is Healing Our World

Brandi M. Veil

When we were kids, "sharing is caring" was taught in pre-school. But as we grew up, we eventually grew into a competitive "adult world" with other competing adults. Sooner than later the concept of "sharing is caring" seemed to vanish–along with the tooth fairy–overnight.

Whatever happened to this basic principle, that sharing is in fact, caring? Why did we stop believing in and practicing this mantra?

I believe we became consumed with new ideas, ideas of ownership and egoistic advancement. It wasn't long before we realized we wanted to protect these new ideas, so naturally, we sought out ways to do so. We discovered the awesome power of secrecy and fear, and we used these powers to tear a path toward becoming card-carrying members of an erroneous "survival of the fittest" society.

Then 2008 happened: The Great Recession.

As one of the "lucky ones," I lost only my business in the crash, and I found myself in severe debt. By chance, I'd already been doing a version of home sharing with friends since a number of them had lost their homes, jobs, or both. To be frank, my couchsurfing friends were not always welcome for long periods of time nor was I benefiting from them financially. While I did have assets to share, I still needed a way to survive financially.

Then I discovered home sharing online–and my home sharing efforts began to take on a business model of their own. I became a true host, a caring concierge, a bed and breakfast, a healer, and a referral service to my local restaurants and community stores.

Soon a greater sense of community began to take hold in my life. I was beginning to see that something

more was happening in this new economy (aside from all the opposition). Optimism was lurking. The sharing economy was giving us hope. Allowing others to stay in my home, sleep in my rooms and experience my personal belongings as if they were family was, and continues to be, rewarding. I get a sense of gratitude each time a guest leaves a note that says, "Thank you for all that you have done, please come and visit me at my home sometime."

I have experienced a more connected sense of self when I host a guest. It's spiritual. It's financially rewarding. It's a new eco-system that reuses, restores, revitalizes community. And yes, as expected, some of it has to do with economics, human economics.

"Economics is much broader than managing personal finances; it's more than that, it's about how people make choices under scarcity." Professor Alex Tabarrok of Principles of Economics online series makes a true and valid notation.

Although the new tech movement for sharing may seem like another convincing way to manipulate a new market for corporations to get rich off "the people," it has its truths. In 2014, Airbnb--leader in home sharing with a $10 billion valuation corporation--faced laws that potentially limit the home sharing movement, and even take away the people's right to share. Like many freedoms in our society, there are laws that limit our happiness, our freedom of expression, our creativity, and our rights, but is it a necessary evil?

The importance of this statement is to present that without the voice of community leaders, advocates, ambassadors, and local city and neighborhood councils, our rights have the potential to be wiped out from underneath us. In order to retain the right to share, "the people" had to speak up, and they did! In April 2014 in San Francisco, thousands of Airbnb community members came out to speak about their rights to share.

"The short-term rental market is exploding and cries out for some regulation," Welch told the Chronicle. "People are stunned to find out that a house on their block is now a hotel."

It's important that we evaluate the effects of change as they occur, involving our personal rights, truths and beliefs starting with our local community. To have a voice for a better world, start in our own backyards and present our beliefs, individually and collectively, as a community. It's not enough to "hope" things fall into order for the greater good; WE must do the work!

Economics is much broader than managing personal finances. It's more than that; it's about how people make choices under scarcity.

Get involved.

If you feel called, become knowledgeable about your neighborhood council, understand who is running for the city official, know what is on the ballots when it is time to vote, BE A CITIZEN. If we desire change, we must work with the change makers.

What does all this have to do with the sharing economy healing our world?

As a mother, educator, community leader and a practitioner of the healing arts, I witness breakthroughs all the time with my clients. The same goes when a stranger becomes a guest in my home; I witness a similar breakthrough. I get a sense of healing that is occurring through sharing, a sense of reconnection, vulnerability, and openness. This experience is powerful for change. Real human experiences that open our hearts and our minds to a healthier more connected world "is" powerful enough to change the world--one heart at a time!

In today's new age of crowdsourcing and open sourced technology, collaborative, peer-to-peer and startup communities are reminding us of the lessons we learned as children. The rise of the sharing economy is happening, and just like we were taught to be kind and share, we are getting the opportunity once again to revisit this valuable lesson in leadership. There is more to this new economy than what you are hearing.

Sharing is, in fact, healing our world—and it is uniting us once again.

Brandi M. Veil

Brandi M. Veil is a Fun-TastiK social entrepreneur focused on creating a better world through social good and event technology. Brandi is an event producer, public speaker, and holistic practitioner. Inspired by the growing number of festivals, Brandi has chosen to pioneer the next generation of stewards through an event marketing platform using technology to grow social good community online and offline as a CSR driven mission for clients. Brandi's powerful drive to move the audience to a greater purpose awarded her the opportunity to speak at the United Nations (WIT) World Information Transfer Conference, Winter Music Conference (WMC), Lucidity Festival, and AEG Smarter Shows. She believes the shared economy is healing our economic system one person at a time.

http://lht.la
http://brandiveil.com

Cultivating a Culture of Contribution

La Laurrien

At difficult times, I have experienced my inheritance as a weight, separating me from others, and putting me in a particular box. I felt that no matter how big my personal creations, they would always be small when measured against what was given. This had the effect of belittling my abilities and sometimes threatened to cast a shadow of meaninglessness over my best efforts. I eventually solved the psychic and social schism in a very simple, easeful way.

As I turned toward receiving the inheritance, I noticed that it was made out of hard work and heartfelt giving that passed through several generations of my family. I had been seeking a meaningful response. I began to recognize myself as recipient of a powerful gift of legacy with the ***freesponsibility*** to send a powerful message back into the world. But first I needed more experience in up-leveling my capacity.

I have accepted the benefit of this gift, using energy and time to grow my consciousness, utilizing the leverage it has offered me. I have taken risks to live more fully, daring to become a more potent being… my ultimate goal, to Awaken within Community… while modeling a new dream for ***Gaia*** and Humanity.

I have realized that my "life is art"

One area of life in great need of cultural translation is our economy. This includes the perceptions we hold about our social environment, new ways to frame and encourage energetic exchange… and the hard work of moving through transition… as one world dies, another is being born. As members of a First World economy, many of us have experienced enormous privilege as a result of our birthplace and circumstance. For this reason, we hold freesponsibility to leverage our power to make change possible for the benefit of all. I feel great hope that a new economic reality can be forged starting here and now, steadily moving toward a vision that includes recognizing social capital, engaging in gifting exchange and sharing our super natural resources, within a ***Culture of Contribution***.

In my lifetime, I have been gifted by circumstance. I was born into a family of wealth, which put me in an awkward position. I was never comfortable in spite of the comforts, instead always trying to comprehend the inequities of the world… why some people have so much while others have so little? This question haunted my childhood and adolescence, encouraging me toward living a life of radical inquiry, within exploration of new cultural norms and forms.

and that my emergent being is my most precious art piece, the prima materia, crafted by Source and co-created with my own self-love and ongoing joyful expression. I have accepted the family gift that was given to me by becoming worthy of receiving it. The gift continues giving through me, as I become a conduit for a *Gifting Economy*, cultivating a Culture of Contribution.

In my life, this has looked many different ways. First, I have been reminded to be grateful for everything. I seek giving to others in a balanced way, without drama or karma attached. Although charities are beautiful, I have been more interested in giving directly to those who come into my sphere, offering support during moments of need. Whether it is the person on the street selling papers or a friend in a Third World country making a fresh life start, or a person in my community launching a visionary new enterprise, the intimacy of sharing directly has been important to me.

My current *creationship* has led me to participation in the lifelong dream of intentional land-based community. The gift that passes through me has allowed our group to purchase land and the means to begin seminal building projects there. I feel excited to share, noticing how those in the receiving roles must, each in their own way, grapple with the gift just as I did. What does it mean? Am I worthy? Am I doing enough? Too much? How can we keep this energy most alive as a river of giving that flows without end…?

Regarding Gifting Economy

In our current worldview, *dana,* or charity, exists as a purely given offering with no intention for reciprocity. Our alternative offspring culture has departed from the mainstream culture of commodity or fiat exchange. Another feature of Gifting Economy is that it is based on and encourages abundance, rather than competition and scarcity.

By imagining and enacting such things as free stores, free Craig's listings, open source permissiveness and copyleft rather than copyright, people can engage in free use of knowledge and promotion of "the commons." Burning Man has demonstrated the power of a Gifting Reality, where the field is activated by giving and not by sales. Collaboration is a brilliant means of expressing synergy while creating new forms of ritual, art, healing, culture and Integral Thinking. Our *synergenius* qualities are making new realities.

In transition, we can explore various spheres of exchange, allowing us to move between gifting and commodity realities with relative fluidity. In an ethical framework, humanity is moving from the paradigm of "unequal" to a visionary realm where *uniqual* is the perfect meme! A Culture of Contribution reflects the truth that everyone has a unique contribution for Community. Through self-reflection and sharing what is learned, we can easily hone our skills, offer our inner and outer assets, our time, energy and love to creating a more vibrant life and culture. Michael Tellinger, South African author and visionary, says: "Let each citizen contribute their natural talents or acquired skills to the greater benefit of all in the community."

The Win Mill is a *solutionary* concept I have introduced casually into our community. The Win Mill gives small gifts or loans similar to the microfinance model. These gifts/loans have been offered to community members with businesses in need of

By imagining and enacting such things as free stores, free Craig's listings, open source permissiveness and copyleft rather than copyright, people can engage in free use of knowledge and promotion of "the commons." Burning Man has demonstrated the power of a Gifting Reality...

support at key moments of growth. They have also been granted to people with visionary ideas, seeking coaching as well as financial support to get started. Eventually The Win Mill Foundation will serve the purpose of establishing a local economy of triple bottom lines, indicating success without sacrificing values. I can see a future where people who have plenty, will realize the value of gifting to the next generation carrying the intuition and intelligence to be our new ecology and economy leaders, steering the course toward a thrivable future. Egocentric human-

ity is failing; ecocentric humanity will thrive! The key is "eco" as we move from egonomy to economy! Nothing could be more timely than the generosity needed to finance a more conscious world.

There are many in our existing culture who have far more resources than they can ever personally utilize. The old paradigm system is dynastic in nature and most wealth is passed down through blood in honor of tribal allegiance as the primary reality. Having lived this lifetime without procreating, I have instead co-created extended soul family with many on my path. Loving and including others as kin, my heart opens wide to recognize all beings as my family. It is time to expand our current realities to include the Human Family, all Gaian creatures and the myriad of life forms on our planet… to include our galaxy, our galactic kin and beyond what we have known. Step by step, we are moving toward a full honoring of Creation, stretching to include all of life, even beyond our current comprehension.

As we reinvigorate our self-knowing, we become energy-rich participants in generating an economy based on giving. We are redefining success to include a full diversity of contribution that support the ***thrivlihood*** of all people regardless of class, race or historical roles. In these and other ways we will find our path through the current state of despair into the light of a new day.

I can imagine a time when our measure of wealth will be described by what we give, as in many native traditions. And even more importantly, that our vast human family will each discover what "enough" feels like and to honor the ideal of "enough for everyone." Let's turn up the volume on our creativity, breathe deeply into our self-making creations and co-inspire together. Let's do the necessary work to recognize our highest offerings and gift them to the world.

In the end, this conversation is not about money or finances or economy at all. We are each given life and the life force that pours through us is our energetic bank account. How we live enhances or decreases our supply. Whether we call it Energy, or Love, or the inexhaustible Mystery Herself, the truest Culture of Contribution is the offering of our enlightened selves to one another and to our shifting world.

May we Reinhabit the Village, and thrive in the sweet nectar of Community Life!

La Laurrien

I spend my life creating in the world: Art, Business, Community. It is important to honor the many relationships and creationships that occur along the way. My spiritual practices include a variety of modalities such as Meditation, Yoga, Ayahuasca Shamanism, Hindu Tantra, Mayan Tzolkin. I live in celebration of the meme: Life Is Art.

As an artist I have spent many years making ceramics, showing and selling, making batiks and paintings, as well as having a henna body art business. As a writer I have focused on transformational writing practice, participating in writing groups and leading workshops at festivals. My writing creativity features the formation of new words. Word Magic aka Meta-poetics is a powerful tool for solidifying a new world view based on harmony, freedom and creativity… My website called MirrOracle will include art, writing and new language and will be up and running in a few months.

I am part of the dynamic co-creation of an intentional land-based community in the Columbia Gorge. This is where the experiences of my entire life are called forth to help birth such an enormous project. Atlan is becoming an Eco-Village with a Learning Center and while our learning curve is steep, our progress is steady.

At times, my role is inspirational~helping people move past blocks to access transformation. I have experience with some simple ritual containers for the purpose of providing personal counsel, and I engage in one-on-one mentor relationships as needed.

What matters most to me is 'learning how to live' and sharing the process with others. My ongoing inquiry into the nature of Life and how it works, fuels my spiritual path of Art and Visionary Community. And these pursuits fuel my Spirit unfolding.

I believe that this quest is something we all share however varied our approach.

A Life-Affirming Economy

Full Spectrum Flourishing at the Intersection of Self, Culture and Science

**Ferananda Ibara &
Crystal Arnold**

THE FUTURE IS NOT SOME PLACE WE ARE GOING TO, BUT ONE WE ARE CREATING. THE PATHS ARE NOT TO BE FOUND, BUT MADE, AND THE ACTIVITY OF MAKING THEM CHANGES BOTH THE MAKER AND THE DESTINATION. John Schaar

This is an invitation to stand at the intersection of Self, Culture, and Science and imagine what flourishing in a life-affirming economy would be like. Interrelatedness, diversity, and creativity are patterns of nature; and this intelligent design is reflected in a life-affirming economy. Through wise use of technology people are moving from fragmentation to interrelatedness. Decentralized virtual platforms have great potential to democratize value creation and exchange. While using complementary and reputation currencies, communities are able to create transparent mutual exchange systems that do not operate under the spell of compounding interest; thus developing a more cooperative culture. This intersection, the focal point of a life-affirming economy, is a place of wealth creation, where individuals choose to live and act in a way that enhances the well-being of all life forms on the planet. Here we yearn to make a valuable contribution, with freedom to create wealth and access the collective intelligence. Together we address not only our present challenges, we create a world an octave higher in all which is Good, True, and Beautiful. As humans stand at the intersection of Self, Culture, and Science we each participate in meaningful and valuable ways. This is a story of how it's happening.

We need each other. This plain and simple reality of the human experience has been denied through marketing, as more and more of what was not monetized has become for sale at a price at the expense of our relationships. Our global economy, with its mandate to grow built into the very design of the currency, is quickly reaching the boundaries of a finite planet. **Yet, simultaneously creative pathways are being forged by the pioneers who are defining a more holistic sense of wealth that orients people towards a full-spectrum flourishing.**

Steve McIntosh says, "Evolution creates value by progressing towards ever-widening realizations of what is True, Good, and Beautiful." The evolutionary impulse to improve that we feel in our hearts and minds is the very same impulse that has been driving the unfolding of the universe from the beginning.

THE PRIMARY VALUES OF GOODNESS, TRUTH AND BEAUTY ARE LIKE THREE SIDES OF A PRISM THAT REFLECT TO US THE ENTIRE SPECTRUM OF HUMAN EXPERIENCE: ART, MORALS, AND SCIENCE; SELF, CULTURE, AND NATURE; I, WE, AND IT.
Ken Wilber

Truth, Goodness, and Beauty present themselves in every culture and wisdom tradition with different words. They appear in the triads of science, morals, and art; nature, culture, and self; it, we, and I. We must understand that these primary values guide our actions, and are not an abstract construction but a compass, a map, a code, and a set of lenses.

Each dimension must improve for us to experience a full-spectrum flourishing. Here every Self can be a conscious creator, Nature is encouraged to regenerate through wise stewardship and the support of Science, and each Culture thrives in its uniqueness and glory; fully supported by our values in action, our technologies, social systems, and all our relationships. They are all intimately woven; a shift in one dimension impacts the rest. As an example, as 'I' value healthier and more sustainable lifestyle, my values impact the market which in response 'It' tries to satisfy the needs of conscious consumers by creating more sustainable choices. 'We' all give feedback with our transactions and conversations. As social technologies evolve, flows will become more visible, we will be able to see where every small part of your cellphone came from, and observe corporate behavior and its impact in nature and our communities. With this information 'I' can make more conscious choices.

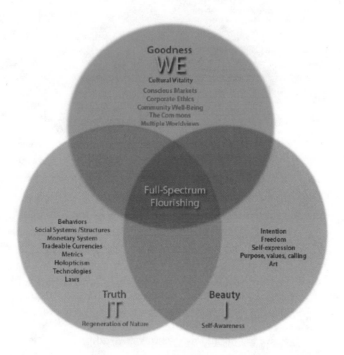

Here is what we call the life-affirming economy, that which invites a full-spectrum flourishing at the intersection of Goodness, Truth, and Beauty. There is no need to react negatively and fight against the old system. Instead, we are choosing to create an economic system with rules that generate infinite games that serve the next stage of our material, spiritual, and cultural evolution.

There are many themes, principles, intentions, and players emerging and addressing economic issues to create greater resilience and lasting value. Together we can move from a materialist-centered sustainability to a full-spectrum flourishing. This is a largely decentralized movement, which has been created by advances in communications technology,

our collective consciousness, and people's intentionality. Individuals are participating in information sharing and value creation through smart devices using the internet. Together we embody a unique cultural vitality resulting from transparency and intelligently designed interdependence that encourages mutually-beneficial cooperation. Step after step we must not lose sight of what is True, Good, and Beautiful, and acknowledge the actions large and small that are building a Life-Affirming Economy.

Example of themes and some organizations in the New Economy scene:

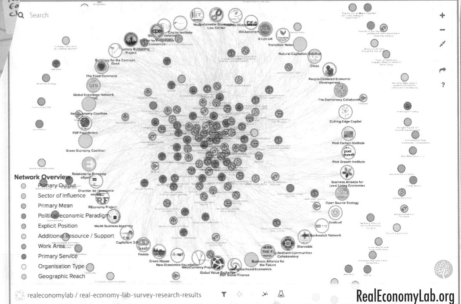

It is an experimental time, mutable and messy. Asking questions is now more important than repeating answers. We invite you to imagine the life-affirming economy to be the highest expression of human potential. We are building technologies to manage our complexity and augment our intelligence through collective stewardship, decentralization, and open source values. For example, many people have effectively organized using the internet to care for the Earth and address the environmental, social and economic challenges we are facing. A new field of consciousness is being born from the pioneers in economics today. And, as Willis Harman said, "Because of the interconnectedness of all minds, affirming a positive vision may be about the most sophisticated action any one of us can take."

As we evolve our values and what we value evolves, consequently the ways we measure and trade value change too. Today there are many cash-deprived Americans, including an emerging generation crushed by student loan debt and the baby-boomers, many of whom watched their life savings evaporate during the 2008 crisis, who are learning the true meaning of wealth and prosperity. These challenging situations have invited many to question the status quo and ask, what is enough? And what is true wealth beyond dollars? Many are inquiring into the prosperity that can be experienced as a balanced flow of energy, where a sense of enoughness is experienced. **We are learning that human satisfaction doesn't arise from having more but by needing less and learning how to access what you value without having to possess it.**

We are redefining and improving our notions of value and wealth. When money changes a lot more changes. Almost everything can become possible. With such a fundamental shift will come the opportunity for innovation far beyond what previous generations could even imagine. Bernard Lietaer

Complementary currency systems and other ways of exchanging and measuring value are an essential component of the life-affirming economy. According to Mark Pesce, "I consider the block chain to be the innovation of Bitcoin… I see that distributed validation is the essence of this invention, and will become a pervasive part of the connected economy…this distributed authorization of a contractual element." These cryptocurrencies are but one aspect of how decentralizing forces open new pathways for humanity. In the next years we will see millions of currencies, not as tokens but as symbols that will allow us to shape, trade, measure, and make flows visible. What flows? Any flow. Current-sees to see communication patterns, social dynamics, production processes, and environmental levels. It's happening now. Apps help us track traffic real time, check the barcodes of products to see their social and environmental impact and more. The trend called "The Internet of Things" is pointing towards making objects smarter as they connect to the internet. Shared data generated by sensors can contribute to smarter individuals and collectives. Imagine that you could see the shared resources that your neighbors have decided to share among them, or you could each see who needs other resources not in the list. A single flow made visible can create so much beauty. To create a Life-Affirming Economy we need to communicate as fluidly as our cells. *Holopticism* is the capacity of any part to be able to see the whole, and this will play a significant role in our pursuit to become a social organism as intelligent as our bodies.

Imagine you could work with organizations that understand wealth as multidimensional; where money is not the primary directive, nor the sole way of measuring success. With effective ways to account for and value performance, reputation, skills, and innovation that allow each individual to be more creative, free and response-able. Intelligently interconnected sovereign entities engaging in clear and more collaborative based decision-making processes, social agreements and conscious contracts are evolving the very structure of the organizations we are part of. Wisdom-driven organizations are built around peer-to-peer economies, trust, and hierarchies of value, instead of command and control. Collaborative technologies for example enable services like Couchsurfing and Lyft show that **when people participate with reputation being explicitly valued and made visible, that trust and honesty is encouraged in the system.** Gore-Tex and other organizations are innovating in the ways they distribute compensation based upon performance, trust, and freedom.

Yet again, consciousness has influenced and been enabled by technology; as we are able to recognize value where it was previously invisible behavior changes. **Consciousness, creativity, and resourcefulness can be rewarded and encouraged in innovative ways through wise use of technologies.**

Our intention is to live and act on behalf of all life and the well-being of all forms on the planet, recognizing that all living things in all their diversity are interconnected and are one. The Fuji Declaration

Many of us know of the limitations of capitalism and learn how to shift our relationship with it. Interestingly, although the current monetary system with compounding interest has encouraged competitive behavior, for some people periods of economic downturn actually encourage more creative cooperation and many acts of generosity. So what is seen by many as a destructive economic system, is actually pushing us to evolve into organizing in more resilient, intelligent, and mutually-beneficial organizations and networks.

Every paradigm shift in history unfolds with a specific technology. Our first technology, our body/mind, enabled us to interact, invent language, explore the material world, and tap into the mystery. Money is our oldest information system; even writing was invented in Mesopotamia as a method of bookkeeping. Another technology that shifted culture, the printing press, gave ideas the capacity to transcend time and space and transformed culture. In our times, the internet has become humanity's nervous system, it connects us to large flows of communication and enables us to align our intentions and coordinate our actions by sharing and processing information with amazing speed. But let's go further. Think of automation and the phenomenon of mechanical minds replacing jobs. Is this part of a life-affirming economy? What will happen when many jobs are lost because machines are replacing people? *Dystopias* emerge from contexts like this one; it's not futuristic, it's happening right now.

Yet, there is hope. Who says that endlessly more jobs and work is taking us towards prosperity? What if we create communities, virtual and terrestrial villages where people can meet their survival needs efficiently with minimum of energy, thus leaving sufficient time for leisure and pursuing passion, creativity, and spirituality instead of being consumed by the pursuit of money?

In a world where artificial intelligence, automation, and biotechnologies are shaping the economy we need to think about more harmonious ways of living together. This may be a utopian dream, where interrelated communities improve their quality of living with mutualism, sharing, and creativity. And perhaps intelligent community wealth systems, to trade, measure and make visible all the wealth we create together. Not growing bank accounts, but human relationships. Whether we participate in communities of choice or chance, these connections are foundational to the individual's full spectrum flourishing.

"Communities can create urban environments by focusing not on growth in outputs but instead on minimizing their costs and the total work needed to achieve the outcomes they desire. Thinking of the city, town or village as a thermodynamic system, the aim should be to minimize energy losses, so local solutions are the ones to seek." Steven Liaros - Rethinking Cities

242

So, what can you do? Participate with life in ways that build your wealth not only your bank account. Generate lasting value within your community, local and global. Be receptive to what is emerging in the technological field. Use your good, true, and beautiful compass to discern what is in service of evolution and learn what inspires you. Pursue your passions, purpose, and deepest calling. Find out how technology can be an ally that supports your creations, whether it is a garden or 3D fashion made of recycling materials and share it to the commons. Yet be aware of the hype; this is a transitionary time and not all the technologies and emerging trends create well-being for all.

We require wisdom and discernment to observe, co-sense, and support the forms of exchange that revitalize individuals and communities. Together we can participate in systems that build value through relationships, where individuals are gladly contributing generously and receiving sufficiently. **For at the intersection of technology and consciousness is the human spirit, and what a magnificent and humbling place this is. As we recognize the Truth, Goodness, and Beauty that exists and is possible, elegant forms arise, intelligent action occurs, and vibrant pathways are illuminated.**

Ferananda Ibarra

Ferananda is an internationally recognized speaker, designer and pioneer in the field of New Economy. Her work in collective intelligence and metacurrency has given her a unique perspective on how to address the design challenges and architecture of our economic system. She is a TEDx speaker and through her work in VillageLab, participates in pioneering a new paradigm in economics. She works with intentional communities of all sorts and finds her bliss in creating new paths for prosperity, harmony and abundance.

Crystal Arnold

Crystal Arnold holds a BS in International Economics and is dedicated to creating a resilient local economy. In 2000, she began developing an expanded social perspective from her cultural studies in the Netherlands and Guatemala. Then, while working at an American bank, she began to understand how money systems actually work. She then discovered complementary currency systems, and realized their power to both generate enhanced freedom of exchange and revolutionize our quality of life.

Free Your Mind and Your Stuff Will Follow

Post Occupy, there was a collection of people in Vancouver that formed deep bonds and wanted to continue their relationships. The creation of a bi-monthly Free Market evolved. Food, music, books, art and more all given away freely. I was inspired to do the same further up the coast. So creating my own signs from recycled material, partnering with both the local recycling company and community association, I set up a table behind the library once a month for the summer. From five people on the first day to 150 on the second, I had some wonderful experiences! I gave away a $1500 piece of art to a family that loved it and had the perfect place for it. I watched one family light up with discovering garden gnomes on offer, another exude liberation with giving away two arm chairs and a TV. I made new friends, extended my community, engaged the local newspaper and began a conversation about local currency.

At the core of this inspiration is Sacred Economics.

Healing wounds that begin with money, specifically its creation and scarcity.

Bringing the act of exchange back to offerings and rituals that explore each of our gifts.

Reviving a flowing currency where giving and receiving exactly what is needed and community support runs freely.

Dana Wilson

Dana has lived and worked internationally for nearly a decade as an independent photojournalist and media producer. She co-founded Mortal Coil Media in 2007 and produced a series of creative documentaries - telling stories from the margins with sensitivity and integrity. Though still focused on themes of social justice, environmental reverence and peace building, she has shifted energy toward organizing events and documenting the efforts of soulutionaries as we transition to a more resilient society. This includes active engagement in the coastal Transition Town collectives, board membership with Salish Sea Productions, and stewardship with Gaiacraft - an international collective of permaculturists engaged in ecological outreach.

She believes deeply in community council using non-violent communication techniques, grounding in with nature, and the healing power of regular meditation, love and compassion for all our relations.

She is currently completing an advanced Permaculture Design Diploma in media through Permaculture Institute USA, while co-creating an ecovillage on the Sunshine Coast, BC.

Media & Storytelling

CHANGE YOUR THOUGHTS AND YOU CHANGE THE WORLD. Norman Vincent Peale

Before there was language, we began telling stories. Through drawing images of the natural world around us - plants, animals, and stars in the night sky - we longed to share our experiences. When we cultivated the use of fire we also began cultivating our language and our culture, gathering at the meeting place of the Tribe, Clan, or Village, creating the center and hearth to share stories, wisdom, experience, and celebration. As we have evolved, mythos and cultural stories have served as a primary method for humanity to understand itself and its place in the great and mysterious vastness all around us.

So what stories are we telling about ourselves? Everyday much of the world's population is interfacing with media and through the incredible advancement of technology, we have become inundated with a constant stream of messages and information. The access to this volume of information has sparked a renaissance of human potential and created a global connectivity in a way that we have never known. Our comprehension of the universe around us is advancing with each generation. And yet many of the stories being shared are stories of destruction, disease, of greed, of betrayal, of violence, of the worst aspects of humanity. The nightly news, blockbuster movies, broadcast programming mostly focus on devastation. We are saturated with these stories, both horrified and fascinated, drawn like moths to the flame, not realizing its potential to singe the very essence of our hearts.

So what of the stories of compassion, stories of elation, of innovation, of catharsis, of beauty, of hope? What of the stories of generosity in the face of struggle, solutions in the face of despair, stories that teach us, stories of people uniting to make a difference, to help each other, stories that compel curiosity, that showcase our genius, that share the best of humanity? They are few and far between in mainstream. While we idealize stories of love, of peace, of championship, of triumph they remain in a fantasy realm, not often do we see real life reflections in our media. Around the world, these stories ARE happening in everyday lives, not as isolated incidents, but as a sweeping tide of change. So why does our mainstream media not reflect this to us?

There are several theories about our current media and whether it is feeding an agenda or perhaps it's feeding us what we seem to want? Do we secretly love stories of destruction or perhaps are we hungry for a new story?

There is a rise in popularity and accessibility to a form of media called Conscious Media. Pamela Jaye Smith describes Conscious Media as being "consciously crafted using timeless tools to create a specific uplifting, expansive, transformative effect upon the consciousness of the audience."

Conscious Media is endeavoring to be intentional of the messages it shares, focusing on principles that transcend any one perspective and offer reflection on the inherent fabric of the human story and of the Universal story of life. By transforming the narrative that feeds the underlying emotional nourishment we crave so that we imbibe an authentically empowering message, we have the power to transform the story of ourselves. The story is not always pretty or perfect, yet there is always a lesson to discover, informing our choices now by integrating our past learning.

Media has a responsibility to tell the truth, to share the whole story, so that we may learn together, sharing our experiences in order to shape our future. With the advent of viral and social media, we are all becoming the storytellers, holding wide-scale witness and accountability to the transgressions that could be previously hidden and featuring the everyday journey of individual lives that offer a window into soul of another.

As Joseph Campbell coined the term

THOSE WHO TELL THE STORIES RULE SOCIETY. Plato

"the Hero's Journey," there is a pattern to many stories, tracing a consistently experienced archetypical journey through the various rite of passage in life. We are the Hero of our own lives, leaving behind what we know, setting out to face the unknown, encountering allies who guide us and being tested by challenges until we are faced with the ultimate pinnacle moment - will we fall or rise? Will we find the power within us to triumph, being made wiser and more humble on the other side of our journey, either continuing on to new journeys or returning to our origins as changed people? This mythos is seen throughout our human history and is one of the key storylines still playing out through our media, our movies, our images today.

What I am curious about is what happens after the Hero arrives home? What happens after we play out the Hero's journey and leave behind the great struggle of violence, of war, of claiming victory over life and proving that we are powerful? What if we arrived home to take up the path of the builder, the gardener of Eden, creating art and food, sharing abundance, the Master who has no need to triumph over nature, but instead recognizes that we ARE nature, aligning to become a part of creation, returning to innocence and wonder imbued with intelligence and capacity to create along with life instead of struggling against it.

That's a story I strive for and long to tell. What story do you want to tell?

Jamaica Stevens

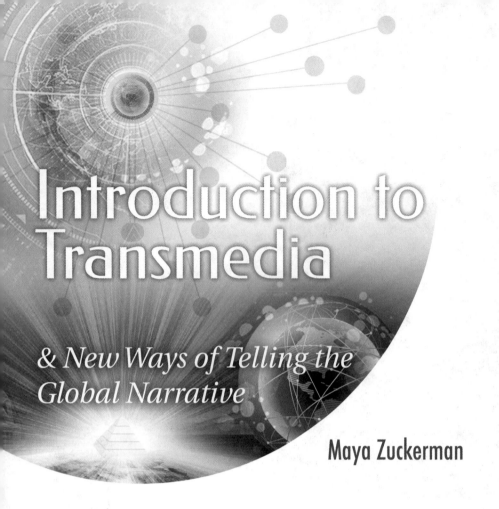

Introduction to Transmedia

& New Ways of Telling the Global Narrative

Maya Zuckerman

We used to tell stories, seated around a roaring fire, our elders captivating us with amazing tales of heroes long gone; tales of how we conquered the night with fire, went on hunting expeditions and how the world came to be. Fables which taught us lessons in how to be part of the tribe, how to belong; myths giving us purpose, meaning and explanation of the magical and mysterious world around us.

Millenniums have gone by since these ancient times and again we find ourselves in an era where we are back around the campfire—only this time, it's a global one.

The information age has brought with it a true connectedness to the global village.

News breaks on one side of the earth and within seconds, the social media swarms with information, commentary, truth and falsehoods.

We are immersed and oversaturated with information.
How do we decipher truth from fiction, and how do we create an intentional holistic narrative for the human race and our planet?

What Is The Problem?

As modern society evolved, we destroyed our ancient mythologies and created new ones. In the 20th century, the power that we unleashed on our planet in the form of global war and the atomic bomb married with our space exploration, transformed our human myths and created a myth gap: we became Gods—destroyers of worlds, as Oppenheimer, the father of the atomic bomb said, quoting the Bhagavad Gita, the ancient Hindu scripture.

In the gap that was left where our myths used to live, we welcomed marketing myths and inadequacy **memes**, forged by the geniuses of marketing and Public Relations.

Meme - An idea, behavior, or style that spreads from person to person within a culture.

Many of these memes came from a well-oiled marketing and media machine which, for most of the 20th century and now 21st century, has been focused on quite a repetitive model of the hero's journey, leaving no room for any other journeys. The Heroine's Journey, the Indigenous Journey and the Collective Journey were never really discussed in the mainstream media. Hollywood's formula has been quite specific and non-inclusive of any other narrative besides the one that sells more products, concepts and conformity.

But amidst the powerful stories and mythologies of Hollywood and Madison Avenue, there has always been another voice, another mythology. One that talks about a global unity of the planet; a holistic one that takes into consideration that the earth is one organism that we are all a part of. A story of all beings that inhabit this earth.

So how do we even start telling these stories and what methods can we use?

Solution

One school of thought, when approaching a large undertaking like this one, is to use transmedia storytelling, multiplatform narratives and world-building techniques.

Transmedia Storytelling - Is a technique used to tell a holistic story or engage in a storyworld across multiple media formats. The idea is that the world we are creating can be explored down many rabbit holes, like in life. We can choose to follow one character's point of view on film, play a game from a different character's perspective, follow other characters on social media, and so forth. Knowing where our audience is getting their information and interaction is key, so we can engage with them in smarter ways and lure them to further explore our storyworld.

Transmedia storytelling is essentially a complex ecosystem of story, narrative and technology. In order to tell a cohesive narrative we need to engage our global audience with great storytelling, immersive experiences and inspire them to participate and co-create.

To further understand these techniques and how to engage with them, I will use the global narrative as our storyworld and give examples of different platforms we can use to engage the audience.

Specifics

First, here are some concepts:

Storyworld - The universe where the transmedia narrative can exist; in our case, our world!

In it is the real world; geography, people, animals, climate, and the digital world; the internet, media, games, virtual reality, augmented reality and more. Those can be considered "platforms"- media as a platform, social media as a platform, technology as a platform, and space as a platform. The platform is where our stories and overarching narratives are going to take place.

Nonlinear Storytelling

Linear video narratives have been at the forefront of our narrative consumption for the past 100 years. We can now engage with stories and narratives wherever we are and whenever we choose to engage with them.

The days of sitting in front of one screen are over. We are in the days of media everywhere and thus we snack, binge and engage with media in different ways.

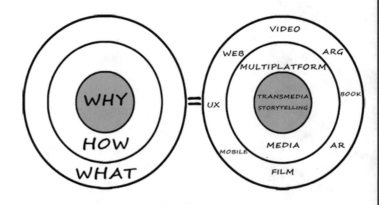

Nonlinear storytelling is at the forefront of how we engage with a lot of media now.

Stories that are segmented are non-chronological, with multiple rabbit holes and audience touch points, that can be immersive and interactive.

We need to start understanding and looking at these concepts of transmedia and multiplatform, nonlinear storytelling as ecosystems of narratives. We are moving away from simplistic narratives to all-encompassing storyworlds that do not stop when the movie is over—where reality and fiction are starting to blend. It is both exciting and dangerous, but we can be empowered to create and tell empowering narratives for our shared future.

Story versus Narrative

It is important to distinguish between these two, as they are often mistaken for the same concept:

Story - A sequence of events, with a beginning, middle and end, that are contained. The human species is a storytelling one; this is how we make sense and bring meaning into the chaos of our experiences. And great stories are well-edited and wonderfully told.

Narrative - Is open-ended. It is based on the voice of the narrator to conclude an outcome. It is the overarching connection between all the stories.

How do we start engaging our audience in this storyworld?

The first step is a compelling story that will attract audience to engage. That is the heart, the why of how we begin. It's the messaging that is so important and, in our case, a heart-based message that will wake up our audience and entice them to engage with us.

When building a narrative to entice participation, whether it's fictional or nonfictional, it is important that we build a set of values we truly believe in—but also values that are global and encompassing enough—that people outside of our choir can identify with and be activated by our stories and want to engage with our storyworld.

In the media world, a story bible where all of the rules of the world we are creating (again, even within a nonfiction story) is written and developed. This bible may include your main character's background, information about your values and culture, big plot lines and arcs.

In the story bible is also where we focus on our language, jargon and reframing of language.

Language plays a huge part in developing an intentional global narrative.

We are moving from using inadequacy-based language to empowering and encouraging language in our Transformational culture. Our stories need to be imbued in that language as well, but in a way that is translated for mainstream media to understand it, or be intrigued with it enough, so that we are creating

an inviting narrative for people to engage with and not creating a barrier to entry by using jargon that is noninclusive.

Our use of specific words needs to be intentional and explained to our audience at times, to help them grasp the depth of meaning we may hold for specific concepts.

Once we have a comprehensive narrative we wish to deploy, one area visionaries and storytellers sometimes neglect to explore and nourish is their audience. We are creating a global narrative for our communities to engage with, so it's quite important to engage them from the get-go.

To start understanding who the audience is, we need to understand who our diverse communities are. From our inner circles to mass circles—who is playing in our sandbox and who do we want to attract to play? Where do they hang out—geographically and online?

We need to create an easy touch point for them to enter our story-world. For example: a well-produced video spread on social media that is easy to share and connect to. A call to action that is inspiring and easy to be activated by and then a way for our audience—who just got activated—to act now.

Crowdfunding is a great way to interact with your audience as you are beginning the journey. Like this very book which started in that way, it is a way to get a bona fide response from our audience as to whether they are interested in playing. However, developing a successful crowdfunding campaign is extremely time-consuming and should be considered only

with a very specific marketing and PR campaign and with a community that is ready to engage and support.

Tools and Technology

There are so many new platforms to engage with our audience and lead them down the rabbit hole.

Media and tech everywhere

This is our status quo and "Moore's Law" predictions support that this is only going to exponentially grow in the world. ("Moore's Law" is the observation that, over the history of computing hardware, the number of transistors in a dense integrated circuit has doubled approximately every two years.) So knowing the tools and creating an engaging global narrative is what we can do as a community to tell a different and more empowering story from the one that is currently told in the mainstream media, and help create a bridge between the narratives to include everyone.

The information age has brought with it a true connectedness to the global village.

Social Media as a platform

The many platforms we all use are a great space for evolving the global narrative. We share memes, videos and ideas that support our own values, ideas and belief systems. We are broadcasting our own personal brand to the world.

There is an emergent global narrative showing up in our social media and we can observe and decipher it.

But what we can also do is intentionally curate our feeds to bring this narrative forward. Whether we are actively working on a media project or curating our own personal feed, bringing an awareness to what we put out in the social media ecosystem will help shape our global narrative to an attainable and desirable one.

Games

Games and play are a natural part of human and animal evolution.

We learn through playing when we are kids and increasingly more has been written about the benefit of play and gaming as teachable tools.

There are so many ways to develop games and incorporate them into the storyworld.

From simple tangible games people can play together face-to-face, to complex video games that need years of development and money to produce, the mainstream video game world is engrossed with a similar narrative as the media world.

But there are some games hailing from the "games for change" movement that are creating games that are impactful and educational.

Using game mechanics and tools we can create on going engagement with our audience. And beyond that, we can create teachable "leveling up" experiences that will educate and lead our audience through the creation and building of the global narrative.

Virtual and Augmented worlds

A quick explanation of the difference between virtual reality and augmented reality:

Virtual reality - Is a digitally simulated, fully immersive digital world, where sensory experiences can be recreated - and these can include virtual sound, touch, smell and even taste.

Augmented reality - Is a live direct or indirect view of the real world, through technology that adds or augments the view with usually more information.

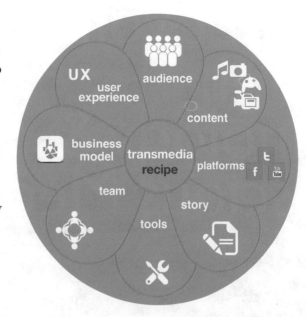

In virtual reality, our real world is replaced with a digitally simulated one. In augmented reality, our real world is enhanced with a digital one. There are going to be further technological advances in this field in years to come. We can use these technologies to our advantage and tell stories that simulate the world we want to create, and show how the world can evolve by viewing an enhanced augmented view of what is possible.

Conclusion

Many platforms and tools can be used to tell a very immersive story. There are so many ways to deliver the story, engage our audience and immerse them.

The most important question to ask is: What do we want our audience to do and how do we inspire them to get involved in creating the global narrative? There is no "one size fit all" in these concepts but ongoing rapid prototyping, testing, exploring and experimenting will shed light on how we can continue evolving the human narrative.

Using these tools, techniques and methods, we can start creating a human meta-narrative that imbues all of the different stories of humanity within it and lets all voices be heard.

Global Evolution through Technology & Conscious Media

Maya Zuckerman

"The medium is the message," said Marshall McLuhan, one of the 20th century's brightest communication theorists. But these days there is a new message and a new medium. Our advanced technology has created a media ecosystem within the media ecology.

The term *"media ecology"* can be defined as "the study of media environments, the idea that technology and techniques, modes of information and codes of communication play a leading role in human affairs." This emergence came in the form of conscious media and was transmitted through new technology channels and platforms and has supported the global evolution.

To dive into such a rich subject as this, we will need to define and introduce a few concepts:

Global Evolution

There are a lot of theories about the evolution of our species--especially dealing with evolution with our coping mechanism which in practice translates into the evolution of our consciousness. There are many wonderful theories and models around these concepts.

The one I always return to is Clare Graves' "The Emergent Cyclical Levels of Existence Theory" (ECLET), which was the biggest influencer on the "Spiral Dynamics" theory and Integral theories.

The basis of the theory is that our nature is an emergent, ever-evolving open system that reacts and transforms as a result of changing external stimuli that causes it to develop its coping mechanisms on many levels.

Another concept we should introduce is *"Conscious Media"*:

In their white paper named "The Explosion of conscious media," Gaiamtv.com defined it as, *Content that uses alternative modes of thinking--integrating spiritual, experiential and contemplative idas with methodologies drawn from modern science--to explore and understand the human condition and ways in which individuals can align their spiritual, emotional and physical beings to live in harmony with themselves, others and the planet.*

They looked at different topics which were covered by this moniker, such as: Holistic Healing and Integrative Health, Social Connections and Relationships, Harmony of Life,

With the media ecosystem being controlled by 6 corporations, it leads to 90% of all media in the control of these conglomerates. "The illusion of choice" and an open media ecosystem shines a light as to why our human narrative is not evolving.

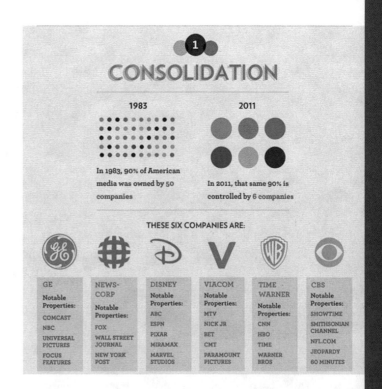

CONSOLIDATION

1983   2011

In 1983, 90% of American media was owned by 50 companies

In 2011, that same 90% is controlled by 6 companies

THESE SIX COMPANIES ARE:

GE	NEWS-CORP	DISNEY	VIACOM	TIME WARNER	CBS
Notable Properties:	Notable Properties:	Notable Properties:	Notable Properties:	Notable Properties:	Notable Properties:
COMCAST	FOX	ABC	MTV	CNN	SHOWTIME
NBC	WALL STREET JOURNAL	ESPN	NICK JR	HBO	SMITHSONIAN CHANNEL
UNIVERSAL PICTURES	NEW YORK POST	PIXAR	BET	TIME	NFL.COM
FOCUS FEATURES		MIRAMAX	CMT	WARNER BROS	JEOPARDY
		MARVEL STUDIOS	PARAMOUNT PICTURES		60 MINUTES

Energy and Oneness, Science and Spirituality, Paradigm Shifts, States of Consciousness, and much more.

So how do we harness the power of the amazing technology to create a global evolution through conscious media?

We start with the immortal words of *Hamlet*–"Words, words, words"– or in our case "*memes*":

Understanding global memes and creating compelling stories that evolve ideas and creates new narrative structures to make older ones obsolete is a necessity for us to develop our human narrative and our species.

What makes one meme go viral and another lay fallow has been researched by many smart memetic and swarm scientists. Unfortunately, it seems that joke and cat memes have been wildly successful while global warming is a failed meme, as Joe Brewer, a change strategist for humanity, states: "It does a terrible job of spreading. It is really hard to get people to think about it and act upon it, it is really hard to get people on their own to feel compelled to tell stories about it, or to bring it up at cocktail parties."

So instead of humanity coming together to work on revitalizing our ailing environment we chose to share cat videos. Or are we really choosing?

This is where the place of conscious media is vital. According to a Business Insider article, with the media ecosystem being controlled by 6 corporations, it leads to 90% of all media in the control of these conglomerates.

"The illusion of choice" and an open media ecosystem shines a light as to why our human narrative is not evolving.

The "programing preferred" narratives is the status quo in the content delivery systems and is controlled by the biggest corporations and governments.

Beyond the control over what is delivered to our screens, the messaging is tweaked to create behavioral and psychological habits and specific consuming patterns that are stagnating our evolution as a species and also destroying our world.

Beyond that, our internet and other machines have been developed for years without taking into consideration the natural human interaction, thus creating work and jobs that are unnatural to our development and natural tendencies. Even the newest tech, such as the IPAD, which have been designed with human-machine inter-

action in mind, still dumb down our cognitive abilities and create dangerous addictive behaviors.

So how do we de-program ourselves from the diminishing, inadequacy narratives that are controlling our stories?

We do that by choosing to use technology to deliver empowering, holistic and educational narratives which help evolve our species.

We do it by choosing to unplug and cut the cord of mainstream media and drop into channels that are broadcasting and creating more holistic and empowering messages.

Some channels to consider: Participant Media, Upworthy, Films For Action, Elevate, and the slew of independent filmmakers that are out in the world, bringing forward new narratives.

We do need to remember that like the mainstream media, even in the feel-good space of conscious media, we need to put our skeptic hats on and decipher truth from faux, as this has become a huge challenge in the information deluge we currently live in.

Lowering the barrier of entry to create technology and content with the prices on tech, hardware, cameras, computers and so forth, reducing year by year and the exponential developments in the different fields, has opened up the floodgate for all of us to take back the ownership on how we communicate and what we communicate, so that essentially we are now free to take back our information streams.

But what technology can support that?

The Noetic Web is one of these instances:

Mark Heley, author and conscious media strategist, has coined this term and defines it as, "The evolution of our current online architecture from a top-down, mostly single-server based model to a radically decentralized peer-to-peer network. This evolution parallels both nature and intelligence in that its architecture is essentially fractal. The technologies which are making this possible include the blockchain, next generation fully-distributed peer-to-peer networks, and the semantic web. Essentially data is becoming freed from the interface."

Mark sees it as important for many reasons:

"The current internet is essentially a tracking system where we trade our data for supposedly 'free' functionality. As the limitations of that model become ever more apparent, the only real alternative is to free the power of the network. The social utility that the growing online ecology of peer-to-peer services will provide has the potential to fundamentally restructure society. For example, think of the value ecological versions of Uber and Airbnb could provide to make communities more sustainable. It could leverage the empowerment of a global ecovillage network to the point at which alternative and sustainable economies, run on alternative **cryptocurrencies**, could rival state-sponsored economic and financial systems. Also, the computing efficiencies created by decentralised networks mean less power is needed to run them, massive server farms rapidly become obsolete and even e-waste is reduced by extending the useful functioning lives of computers that can serve the new networks even when we are not actively using them."

256

How can we start building it?

As Mark states, "Not only is it already being built, but its widespread adoption is inevitable. The synergy of these technologies maturing will provide a platform that will be beyond the control of any governmental or **corporatocratic** system. It will also be beyond the internet--in the near future we will have many nets: **meshnets, outernet, parallel nets**, all potentially part of the communications toolkit in the same way wifi and bluetooth are. The user will have control not only over their 'identity,' but over the way that they share, perceive, and shape their online experience. Each person will have their own internet."

In summary, the advent of technology and conscious media is giving us back the power and empowering us, the communities, who wish to build thriving and resilient communities and regenerate the earth, to take back our information, our narratives and knowledge.

Like this very book, this is a call to all of us to be mindful and intentional when interacting with media and technology and attempt to broadcast, consume and interact with both media and technology that supports us in our evolution as fully realized earthlings.

Maya Zuckerman

Maya is a transmedia producer and emerging technology aficionado. She brings a wealth of experience from different media silos: visual effects, film, production, gaming, startups, product management and brand narratives. She has worked with big feature films and on game cinematics such as "Prince of Persia," "Ghostbusters" and "Star Trek Online," and for companies such as Ubisoft, EA, Vivendi, Sega and Activision. In her career she's also worked in product development for software and interactive products for companies such as the Chopra Center, Harpo Productions and Salesforce.com. She co-founded TransmediaSF in 2012 and has co-produced over a dozen monthly meetups, including two weekend events: a Transmedia Jam and a StartupWeekend Transmedia. She is currently the Marketing Manager for Keyframe-Entertainment.

*Maya started exploring the fledgling world of electronic music in the late 90's in Tel Aviv, Israel. She explored the early rave scenes, the house music and mega club explosion that was happening in TLV and also started going to outdoor parties in the deserts and forests of the Holy Land.

*It was after her move to Vancouver, Canada that she really engaged with the Psychedelic Trance community after experimenting with the local rave scene. Her regular night out during school was Friday night at the legendary Organix trance night at 23 water street club.

*She first went to Burning Man in 2001 and has been going ever since – engrossing herself with the community on and off playa. She's ran theme camps, created installations, made movies on and off the playa and has been an active participant in the local SF community. One of her community highlights was interviewing Harley Dubois – one of the BM founders – for her YouTube channel WorldWideGoodNews in 2010.

*In the Bay Area and California she has also frequented small to large outdoor parties and festivals, including: Gemini, Pulse, Symbiosis, Moontribe and most street festivals in SF.

*She is extremely passionate to see the evolution of the Electronic Music and Transformational culture into a thriving community and humanity.

Elevate

A Little Production Company on a Big Mission

http://elevate.us/
http://facebook.com/elevatefilms

We are ELEVATE, a little production company on a big mission. We live to the fullest in Ojai CA, pop. 8000. Positioned one hour North of Los Angeles we're just close enough to play in Hollywood, yet far away enough to create our own game. The game we love most is producing filmed works of art that make us feel good about being human.

Our portfolio includes feature films, record setting network TV specials, commercials, and music videos for multi-platinum artists. Our clients include some of the world's foremost scientists, futurists, evolutionary experts, and best selling authors. We are totally committed to using the awesome power of movies and viral media to activate real and lasting change – to bridge ancient wisdom with modern technology – to share stories that deserve a voice – to champion worthy causes – to amplify the good news of the world – and of course… to ELEVATE!

ELEVATE is a tribe of creatives who stand for the power of art and media as a tool to inspire the masses and activate change.

The Bloom
A Journey through Transformational Festivals

http://thebloomseries.com/

A Film Series by Jeet-Kei Leung and Akira Chan in association with Elevate Films, Key-frame-Entertainment, Muti Music & Grounded TV.

THE BLOOM is a documentary webseries illuminating the emerging culture of tansformational festivals, immersive participatory realities that are having profound life-changing effects on hundreds of thousands of lives.

Amidst the global crisis of a dysfunctional old paradigm, a new renaissance of human culture is underway. Over the course of 4 episodes and 23 transformational festivals around the globe, THE BLOOM: A JOURNEY THROUGH TRANSFORMATIONAL FESTIVALS explores the alchemy of themes that weave a true story of genuine hope for our times: a new blooming of human consciousness emerging through creativity, love and joy and an emerging culture pointing the way to a bright and promising future.

The BLOOM's Mission and Vision

THE BLOOM tells the vibrant, compelling and colorful story of a cultural renaissance in progress with the artistic sensibility and inspired creativity from which the culture has been birthed.

THE BLOOM promotes the sustainability and evolution of transformational festival culture by creating a shared vocabulary and understanding of essential issues, empowering participants to contribute towards the integrity of the culture and be a part of collectively navigating its course.

THE BLOOM builds a bridge of understanding and creates an invitation to communities and allies with similar values who may find resonance with the transformational aspects of festival culture.

THE BLOOM contributes to the creation of a better world by disseminating the model created in transformational festivals to communities and audiences in many contexts.

Leveraging Digital Networks & Communities

David Casey

Networks are changing the way we live. Humans are more connected than ever before, and we are utilizing the power of the internet to meet each other, organize and assemble, exchange resources and information, finance each others' endeavors, and interact in many other ways that would have been unthinkable less than 25 years ago.

The purpose of this essay is to leave you, the reader, with tools that will empower you to learn, exchange, find meaningful work, travel, and bring projects into the world.

This essay provides an overview of the current landscape and trends of digital networks and communities, the underlying mechanisms that make them tick, and how they can be leveraged by you to live your dream life.

There are many different kinds of networks. Merrium-Webster provides us with a few definitions for networks:

- A system of interconnected lines or channels

- An interconnected or interrelated chain, group, or system

- A system of computers and other devices that are connected to each other

- A group of people or organizations that are closely connected and that work with each other

- A usually informally interconnected group or association of persons

Networks are nothing new; in fact they have been an organizing principle for humans over thousands of years before the invention of the internet. For example, the hawala trust network has been functioning for over 1000 years as a global remittance system to move money across the Islamic world. The most interesting network for our purposes is the peer-to-peer ("P2P") digital community, a new variation on the last definition of network. P2P networks are platforms that facilitate the exchange of resources, information, and other forms of value directly between individuals, often strangers living on opposite sides of the planet.

Digital networks are online platforms that facilitate connections in the real world, which in turn fuel the growth of the network. Digital networks often grow through *feedback loops*, virtuous cycles of growth driven by individuals interacting with each other through sharing, providing value, and building trust. The dance between "online" and "offline" is critical to the function of a digital network. These networks are interesting because they are openly available tools that give any individual the power to change their lives and change the world around them, because they are accelerating a number of trends towards individual empowerment, openness, innovation, trust, *decentralization*, sharing, access to resources and information.

260

TO RAISE NEW QUESTIONS, NEW POSSIBILITIES, TO REGARD OLD PROBLEMS FROM A NEW ANGLE, REQUIRES CREATIVE IMAGINATION AND MARKS REAL ADVANCE IN SCIENCE. Albert Einstein

The Trends. With the capacity to share information instantly with the rest of the world, humans truly have manifested one collective consciousness. The past five years have witnessed an explosion in growth of "sharing economy" networks that facilitate the exchange of value between individuals. The trend is towards increasing individual empowerment through access to resources, informations, and peer networks. "Off-the-shelf" software gives any one of us the ability to create their own platforms, websites, social networks, and many other tools to connect with each other and exchange information. By the same token, we can now use "crowdsourcing" to educate ourselves, connect to work and travel opportunities, finance our projects, exchange commodities and services. Increased access to information and resources have led to a lower barrier of entry into many markets, driving innovation in nearly every sector of our society. This has led to a renaissance in entrepreneurship that includes independent artists and storytellers, makers, digital nomads, and others empowered by access to create their own realities and share their creations with the world.

Thanks to networks, many large systems that underpin our society are trending from hierarchy towards decentralization, including inside of the world's largest corporations. Flexibility, adaptability, responsiveness, autonomy, and "crowdsourcing" are being integrated into formerly rigid hierarchical structures. In the financial sector, P2P decentralized *cryptocurrency* platforms like Bitcoin are beginning to compete with the established international global banking system. Communications networks are headed in a similar direction towards a type of P2P network called a mesh network, where each individual cell phone, modem, and device can become an internet wifi hotspot for public access. In the production of music, art, and manufactured goods, access to tools and software has kicked off a revolution. A decentralized energy production and distribution network is soon to follow, based on renewable sources like wind and solar.

How They Function.

Digital networks and communities share common underpinnings and structures that enable them to thrive. Many networks are structured as multi-sided platforms, places where a provider and a seeker can connect. For example, Couchsurfing is a platform that connects a traveler to a host with a couch. In order to function, this system relies on both sides of the platform being populated. In addition to a populated database, the key feature that makes many platforms work is the reputation trail. Trust is the key currency, and an individual paves a reputation trail through positive interactions with other individuals. There is often a common passion or hobby that bonds users together. In the case of couchsurfers, it is the passion for authentic travel experiences and cultural exchange. When these elements are present, people can share resources with each other and co-create a collective pool of *social capital* - the value of networks. These pools can be thought of as a sort of digital commons where all participants of the network can add value to the collective wealth and enjoy access to it. Sometimes, trust is the only currency and exchange lubricant, and fiat currency (money) is not always needed.

To grow the digital commons, networks crowd-source information and resource availability, utilize. For a network to be truly effective, a useful (and under-used) resource is identified, and the platform provides a method to facilitate its exchange or access. As a network is formed and strengthened by like-minded individuals around the world, it achieves the network effect. Each additional member of the network adds to the total value. Each resource provider makes the network more attractive for seekers, and visa versa. This is the engine that fuels growth.

Many **Sharing Economy** platforms operate under these principles. The sharing economy model enables humans to find, trust, and share resources with each other without ever having met in the physical world. Within the sharing economy, it's important to distinguish between gift-exchange models like Couchsurfing, and profit-driven models like Airbnb.

Why They're Useful.

Networks can be leveraged for just about anything. Three incredibly useful applications of networks are (1) gaining skills and access to education, (2) obtaining access to basic human needs like food and shelter, and (3) building projects and attracting **resources** to them. Through these networks, you can live the life of your dreams by learning the ancestral tools of self-sufficiency (growing food, building with earth), gaining income-generating skills, finding a job or apprenticeship, tapping international support networks for food and lodging, or launching a project and attract resources to it.

The Networks.

Like people, networks come in many shapes, sizes, and colors. One kind of network is the work-exchange model. Work-exchange networks like WWOOF, Workaway, Worldpackers, and Moving Worlds allow you to exchange your time and skills for a place to stay, food, and an experience that ideally provides you with some real-world education. Money is also sometimes exchanged, to balance differences in value being offered by either side. WWOOF is an organic farming network connecting volunteers to hosts. Workaway connects travelers to a broad range of hosts. Worldpackers helps backpackers connect to opportunities at hostels. Moving Worlds connects skilled professionals to meaningful volunteer work with non-government organizations. The best-case scenario will provide you with food and housing, while training you in a valuable skill-set. Often, hosts require a longer-term commitment and some pre-existing skills in order to make this a fair exchange. On the flip-side, Villagecraft is a platform for community skillshares and related events. A related model includes professional networks like Idealist and LinkedIn, platforms that allow you to leverage existing skill sets to search for jobs and professional development opportunities.

The **gift-exchange model** is unique from work-exchange in that value often flows in only one direction. Individuals can gift things of value to one another, and accumulate "karma points" or increase their reputation and trustworthiness within the community. Couchsurfing operates on the gift-exchange model, as well as a number of platforms based on gifting and points (Freecycle, Listia, Freegan.info), digital versions of the "free stores" of 1960's Digger fame. Other platforms (Swap) digitalize the ancient art of barter. Home Exchange allows you to swap homes with people across the world. One of my favorites, Shared Earth, connects farmers and gardeners to farmable land, birthing "the largest community garden on the planet." (Shared Earth)

Dhamma, the Vipassana Meditation network, is a gift-based organization that offers room, board, and education in an ancient meditation technique to individuals. Individuals voluntarily donate time, money, and other gifts to the organization.

Dhamma has received gifts of entire properties to build new retreat centers, steadily growing their organization and serving as a living demonstration of the validity of the gift-exchange business model.

Activation.
When you are ready, you can leverage networks to build your project, organization, business, or non-profit. Bringing your dreams to life is a long and rewarding path. It requires determination, clarity of vision, and purity of intention. With these elements present, there are not many limits to what you can build by leveraging networks. The following platforms will provide you with support, guidance, collaborators, workspace, and interest-based communities.

Meetups are a great way to find the people who live in your geographic area and share your interests. Joining an interest-based community is a great way to meet collaborators, supporters, enthusiasts, hobbyists, and business owners who will comprise your local network. The Meetup.com platform allows you to find local interest-based groups and events, and discover what resources exist in your area. For example, organic gardeners convene to transform each others' backyards into productive garden spaces, while socializing and sharing fresh home-grown produce in potluck-style garden parties. Building your local network and getting connected to local resources is vital, and provides ample opportunities for you to discuss your ideas and ask for feedback and support for your creations.

Networks of physical spaces for co-working, co-living, and co-producing are blooming around the world. *Coworking* networks like Impact Hub offer membership to community workspaces around the world.

Europe-based Hoffice allows anyone to transform their living room into a public coworking space, where others can drop in to work and network. Coliving networks like the Embassy Network and Nomad House are providing live/work space and community to nomads. The Nomadlist is a resource for digital nomads to let each other know best places to live and work around the world. Likewise, hackerspaces and makerspaces are empowering artists and creators with access to workshops stocked full of production tools for manufacturing, welding, carpentry, 3D-printing, and more. *Hackerspaces wiki* provides a global map, and membership networks like Techshop provide low-cost subscription access to these spaces. The freespace movement is a gift-based variation of a civic innovation space. Many of these venues offer great educational and networking events in the evenings.

Project NuMundo

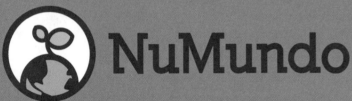

NuMundo is a web platform that interlinks impact centers and individuals into a thriving ecosystem with the potential to regenerate the earth and empower her people through the dissemination of skills, models, templates, techniques and blueprints that are the foundation of an earth-restoring economic system.

What is an impact center?

An impact center is a land-based project that offers individual transformation, regenerative living education and strives to leave a positive local impact. An impact center could be an ecovillage, organic farm, yoga retreat center, or even a hostel, as long as they meet the criteria below.

We vision a network of interlinked centers

We connect people to meaningful experiences to facilitate personal transformation and accelerate the development of impact centers.

Our Story

NuMundo began in 2013 when 4 passionate nomads bound together to create a platform that would be a transparent and trusted resource for travelers all over the world to find more enriching ways to travel, work, live and learn.

Our Team

Our staff has experience living and working at impact centers in North, Central and South America, and shares a love of Latin culture, pupusas, and cacao-fueled dance parties.
To find out more about us please visit

http://numundo.org

Finding a place to live and work and having access to the tools you need is a big step. Mentorship and financing can help you get your project off the ground quickly. Micromentor provides a free mentor-matching service. Mycelium is a network-based education platform designed to accelerate personal and professional development, and structured to mimic an underground fungal network. Financing is also becoming more feasible through leveraging networks. Crowdfunding platforms like Kickstarter and Indiegogo are tools designed for individuals to raise money and build community through small donations from large numbers of people. Angel investor networks like AngelList, CircleUp, and CuttingEdgeX allow investors to purchase small amounts of shares. Incubator and accelerator programs can provide you with mentorship, workspace, and financing to make your idea real. The F6S platform provides a searchable directory of programs that you can apply for.

My Gift.
After using many of these networks for years, I decided to build a network for individual transformation and planetary regeneration. Numundo is is a transformational travel network, connecting individuals to transformational experiences at impact centers: retreat centers, ecovillages, Indigenous communities, organic farms and other land-based education centers. *Impact centers* are the experimental gardens which are growing roots and bearing seeds for a new world that we know is possible. Discover your passion, match passion to purpose, and live a meaningful life. At impact centers, you can grow, share, serve, learn and teach, connect with community, and take the seed of the new world with you as you move through space and time. Through the power of this network and others, my wish for you is to leverage the resources available to you to transform your own life and the world.

Conclusions.
The trust-based network is a technology that can be employed to expand human altruism beyond the traditional family and "clan" to an interconnected global community. Through the power of networks, we are co-creating webs of support with the power to uplift individuals, communities, and the well-being of all who inhabit our home planet.

David Casey

David Casey is a graduate of UC Berkeley in Political Economy of Industrial Societies, with concentrations in Global Poverty & Practice and Energy & Resources. David has lived and worked in seven countries in Latin America, where he has built an extensive network of connections to individuals and organizations. His work has focused on marketing, social networks, experience design, and event production. He is the co-founder and CEO of Project Nuevo Mundo, a platform that connects people and impact centers, encouraging resource sharing on the web and on the ground to catalyze planetary regeneration and individual transformation. PNM has recently been accepted into Startup Chile, a prestigious business accelerator facilitated by the Chilean government. With his strong vision and clear articulation and conviction, he is able to attract wide support from multiple sources in the form of donated labor, land, and financial capital. Among other things, David is also the founder and co-producer of Cosmic Convergence, an annual gathering of art, music, tribal technology, education for conscious living, and Mayan culture on Lake Atitlan, Guatemala, and a co-founder of Tribal Alliance Retreat in Costa Rica. David successfully bridges the worlds of technology startups, permaculture, international development, and retreat and festival production.

http://numundo.org

The Modern Fire Circle

Community-Building through "Collective Viewing'" Experiences

Brad Nye

Using the Primitive Tool of Gathering in the Digital Age

Since the dawn of time, cultures and communities have been built through their stories and myths. And members of those early civilizations gathered at night around the fire telling stories that transmitted the codes of culture for generations to come.

Stories were told in virtually all hunter-gatherer societies and nighttime around the fire was a universally experienced "medium" for sharing information, a time for bonding with one another, for entertaining one another, and for shared emotion. These storytelling sessions were, in essence, the original social media, as ideas were spread from person-to-person. And those stories carried ideas and values therein that became the glue that bound us deeper to our families, communities and to each other.

The fire circle was indeed a powerful tool in our early development.

It's not hard to see then that today's "media screens" are the ancient fire circles of yesterday. We sit for hours each day in front of our television or computer screens, consume and "virtually" share story after story with our communities. We continue to evolve in this way.

In his book "Understanding Media," the great Marshall McLuhan shared his famous "the medium is the message" idea. He theorized that it's not the content that is most relevant, but more the "medium that shapes and controls the scale and form of human association and action." This is huge to apply to today's thinking.

We might therefore begin to look at the ancient form of "gathering" itself as one of the most important technologies of our times, as it had been since early man walked the earth.

Architecting a Gathering

Throughout my career, I have used events to weave community. This form of social architecture traveled with me across my various areas of interest - arts, entertainment, technology, social enterprise, even Burning Man. But no matter who the intended audience was, the objective was always the same: to create an environment for participants to connect and transform.

When I began producing screenings and discussions of documentary films around the Bay Area (eg. Zeitgeist, Thrive, Happy, etc.) years ago, I noticed how these films were attended by like-minded people excited to validate their worldviews, learn something new and meet other individuals in their communities. Through these experiences, it became obvious to me that media could be used as a powerful alignment tool.

So I invite you to consider this as well for your community building efforts. And as this might be the case, let us look at some steps that will enable you to design a powerful "collective viewing" gathering or series of gatherings for your community.

Purpose – Why Are You Gathering?

Assuming that there's motivation for you to bring your community together on a regular basis, ask yourself a few questions to architect the blueprint of how you can use a "collective viewing" experience to build or evolve your community.

- For what purpose are you bringing your community together? (eg. celebration, fundraising, membership development, revenue generation, coherence building)
- What venue will serve as a "clubhouse" for those experiences?
- What type of media will advance community values, awareness or understanding?

There are other questions to consider, but for now, these will put you in motion and serve as the foundation for which you build your "collective viewing" series for your community.

Venue Selection – Where Will You Gather?

I consider this to be one of the most important aspects of your planning. The venue will help you create a strong container for catalyzing your community together. So consider these points in your selection:

1. Convenience of location
2. Warmth and comfort
3. Available networking space
4. Emotional connection for your community

Curation – What Type of Media?

There is no shortage of possibilities here. You may choose from feature films, television programs, internet videos or anything on DVD. What's most important here is to curate something that possesses some of the following qualities:

1. Compelling story and narrative
2. Relevancy to your community
3. Conversation-provoking
4. Action-causing, purpose-driven

Promotion
Create the Anticipation!

From the moment that first email invitation is sent, the field of possibilities is being created. It sends out a flare, a signal, or a frequency if you will. The potential audience member immediately connects, or dismisses the opportunity based on their many filters of perception. The anticipation takes over! As a result, whoever comes to the gathering is already pre-selected as like-minded.

Post-Film Discussion
Facilitate Your Objectives

Discussions do not have to have the director present. In fact, having someone with a strong grasp of the subject matter and/or a good facilitator will help provide for a vibrant discussion even more. The goal is to actively engage and inspire the audience, as they have much to share when invited. Lead them to a call-to-action through the discussion so when they walk out of the venue they are activated to do something.

Of course, there are many other things to consider in putting your gatherings together, but these will help you to get started.

Our Shared Experience

What's most important with any collective viewing experience is to remember that watching media in a group changes the perception of that media. In essence, viewers cross the "threshold" into another consciousness and shared experience. And in that darkened room, their sense of identity dissolves to some extent, bringing about some disorientation, but also the possibility of acquiring new perspectives. This is called a "liminal experience." The individuation of the self is blurred and social hierarchies may be reversed or temporarily dissolved. It is from there that a new view of the world may come forth.

Tribes are a very simple concept that goes back 50 million years.
It's about leading and connecting people and ideas.
And it's something that people have wanted forever.   Seth Godin, 'Tribes'

Everyone walks through the door with their own "Me View," and through the collective viewing and group discussion that ensues, they emerge with a more informed "We View." It's the natural principles of collective intelligence at work.

Societies need shared myths to evolve. But in our modern era where the media we consume comes from so many diverse sources, it's impossible to have one "shared" story of our times any longer. By employing the principles of McLuhan, we can see that the "medium" of gathering can be more valuable than anything. For it is through gathering that we experience our shared humanity, our collective sense of oneness and our care and compassion for the others who sit across the circle from us with our hearts wide open.

Since the prolific creation and consumption of media content is not going away, let us focus on the form and intention of our media fire circle being the glue that holds our communities together over the course of our evolution.

Brad Nye

Brad Nye is a social entrepreneur and founder/CEO of WeVu, a web-based service that helps anyone organize a 'collective viewing' gathering around their favorite films, television shows or online content.

As a seasoned events producer, Nye has produced more than 700 networking and educational events in the arts, media and technology sectors, including over 100 film screenings and discussions.

As a social architect/community builder, Nye has founded and operated two successful organizations: VIC, a 3,000-member technology trade association in Southern California and Artsfest, which advanced the arts by curating over one hundred events annually in multiple locations across the San Francisco Bay Area. In addition, Nye was the founder of the Burning Man theme camp, 'Red Lightning'; launched the event space, 'The Nexus', in Richmond; on the founding teams of TEDxMarin and TEDxBlackRockCity; and founded and produces the annual LA tech industry holiday party, Digital Family Reunion.

As a marketing professional, Nye spent fifteen years in entertainment industry-related marketing, promotion, licensing and merchandising, including management roles at Twentieth Century Fox and Universal Pictures.

Nye is a graduate of UC Berkeley.

http://wevu.com

The Evolver Network

http://evolvernetwork.org/

The Evolver Network (TEN) facilitates knowledge sharing and grassroots activism to catalyze the growth of the global transformational movement. TEN's global collective of local groups, called Evolver Spores, connect together the people and initiatives in their city that are creating a regenerative society, from visionary healing to public policy to ecological restoration.

10 Principles of TEN

- Educate Minds
- Build Community
- Restore the Environment
- Promote Permaculture
- Celebrate Art & Creativity
- Advocate Social Justice
- Highlight Holistic Healing
- Design Better Systems
- Transform Money
- Dream Big

Through TEN, you might learn about beekeeping, alternative currency, innovative governance models, herbal medicine, water cycle restoration, gender and sexuality, Indigenous rights and digital privacy. You'll also likely find peer support and networking for your conscious business or healing arts. Often people attend an Evolver Spore because they are tuning in to a deeper spiritual reality and want community that is oriented toward practical action and creative liberation. We encourage a culture of celebration, inclusion and autonomous self-organizing.

Every quarter, Evolver Network hosts a media campaign focused on a specific issue. Local Evolver Spores host events for community to connect, learn, and take hands-on action. We partner with organizations, companies, and media producers focused on the theme, to expose their work to a global audience. Locally and globally, the themes generate shared resources and collective impact across organizations.

Network-Level Programs

At the international level, Evolver Spores coordinate to share solutions and innovations. We host online webinars, a traveling lounge for festivals and conferences, and strategic media partnerships.

Relationship with Evolver.net & Reality Sandwich

The Evolver Network is the nonprofit side of the Evolver Social Movement. We have independent operations from, but work in solidarity with, Evolver.net, the Evolver Learning Labs, and Reality Sandwich. Evolver.net is an e-commerce website which sells products such as herbal medicine, art and books. The Learning Lab offers online interactive webinars, where you can learn from pioneering thinkers and practitioners in a live global classroom. Reality Sandwich is an online magazine for transformational culture, and reports on similar topics as what the Evolver Network covers. TEN has its own section on Reality Sandwich.

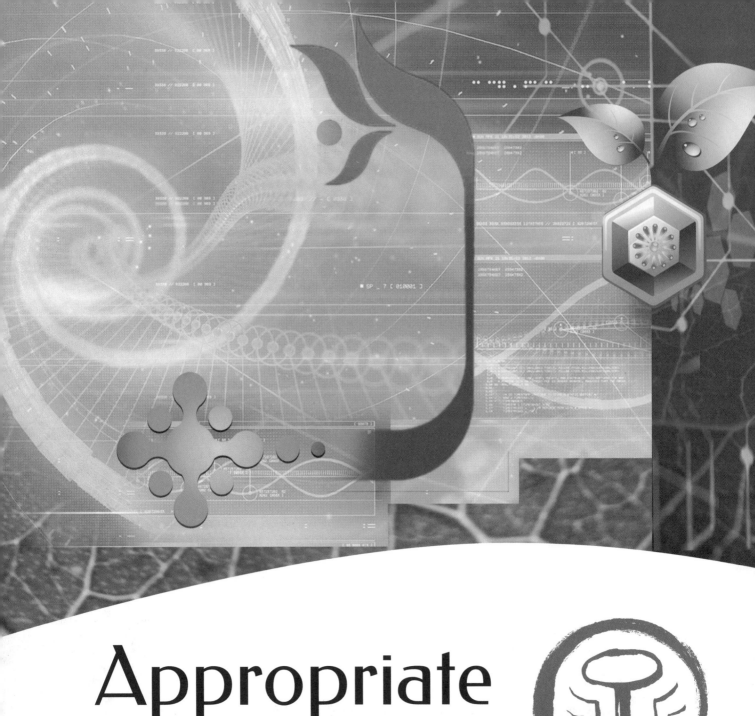

Appropriate Technology

TO RAISE NEW QUESTIONS, NEW POSSIBILITIES, TO REGARD OLD

PROBLEMS FROM A NEW ANGLE, REQUIRES CREATIVE IMAGINATION

AND MARKS REAL ADVANCE IN SCIENCE. Albert Einstein

If we looked at the timeline of human evolution, humans have been around for almost 2.2 million years. We have spent most of that time deeply interwoven with our natural environment, considering ourselves intricately bound, connected to the cycles of the seasons, adapting to the changes of bioregions, our life hanging in balance based on our capacity to intelligently work with nature to meet our needs and progress as a species. Yet it has only been in the last 250 years since Industrial Revolution and the renaissance of technology that we have seen the rise of our global interdependence on systems that now have us on the precipice of ecological failure. What happened?

We have to be mindful of over idealizing a "simpler time." Our past was wrought with challenge and struggle and science and technology have created the conditions for the human species to exponentiate our growth, and evolve in our understanding of the world through the wonders of innovation. Yet somewhere along

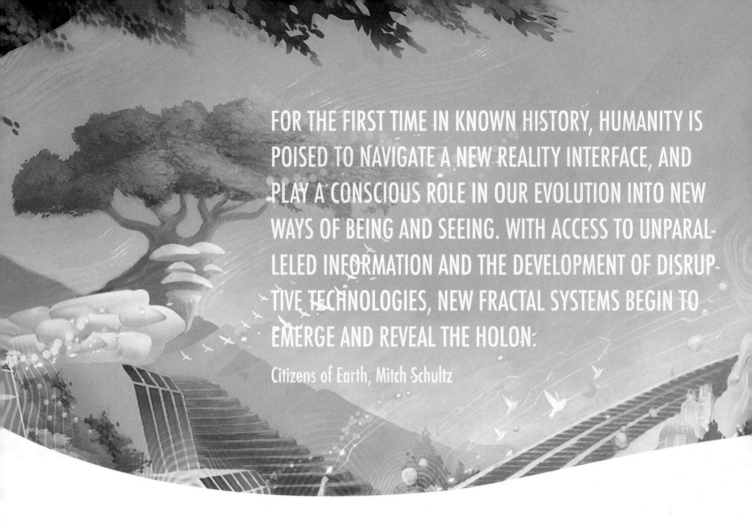

FOR THE FIRST TIME IN KNOWN HISTORY, HUMANITY IS POISED TO NAVIGATE A NEW REALITY INTERFACE, AND PLAY A CONSCIOUS ROLE IN OUR EVOLUTION INTO NEW WAYS OF BEING AND SEEING. WITH ACCESS TO UNPARALLELED INFORMATION AND THE DEVELOPMENT OF DISRUPTIVE TECHNOLOGIES, NEW FRACTAL SYSTEMS BEGIN TO EMERGE AND REVEAL THE HOLON.

Citizens of Earth, Mitch Schultz

the way we became disconnected. We forgot that the parameters haven't changed. We are still deeply interwoven to our natural environment, still have a need to adapt to the changes, and our life, now more than ever, hangs in the balance of our capacity, intelligence, and willingness to work WITH nature, not against it.

It is time to utilize the dwindling resources available to us now in order to shift our dependence on outdated technology that is no longer serving the long-term viability of our environment. The goal is to improve the standard of living without environmental damage or economic exclusion.

Becoming adaptable based on responses and changes within an environment, we have all that we need in this moment to use our intelligence to create the ripe conditions for

human and all species to flourish not only now, but for all of the generations to come.

Growing in popularity as a response to the economic crisis in the developed world in the 1970's, Appropriate Technology is defined as "technology encompassing technological choice and application that is small-scale, decentralized, labor-intensive, energy-efficient, environmentally sound, and locally controlled." However, the practices of appropriate technologies have been adopted by developing countries for some time. Using low cost materials easily sourced, requiring fewer resources, considerate of the local culture, easy to use and maintain, inclusive of local collaboration and usually open-source, these technologies are meant to be accessible for developing countries and provide easeful implementation that resolves a need.

Some shining examples of Appropriate Technologies are solar power panels and street lights, rocket stoves, passive solar building designs, bike and hand powered water pumps, concrete canvas that can be set quickly into permanent low-cost housing in disaster relief areas, Lifestraw personal water filtration device, compostable toilets that reduce water waste and provide organic matter for non-edible farming, rolling water barrels to transport water in areas where people have to travel long distances to find potable water, to name a few. These "soft technologies" create behaviors that bring about change and involve human motivations and interactions without requiring mass production or engineering complexity.

However, localized appropriate technologies are not the only solution

274

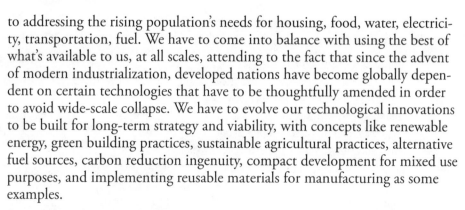

to addressing the rising population's needs for housing, food, water, electricity, transportation, fuel. We have to come into balance with using the best of what's available to us, at all scales, attending to the fact that since the advent of modern industrialization, developed nations have become globally dependent on certain technologies that have to be thoughtfully amended in order to avoid wide-scale collapse. We have to evolve our technological innovations to be built for long-term strategy and viability, with concepts like renewable energy, green building practices, sustainable agricultural practices, alternative fuel sources, carbon reduction ingenuity, compact development for mixed use purposes, and implementing reusable materials for manufacturing as some examples.

Science has always probed into the uncharted territories, reaching beyond our perceived limitations. History has shown us that often - even when we are skeptical at first - the least expected innovations in thought or application have proven to be truths, leading to revelations in our thinking of what is possible.

How is technology in service to the continuation of life in its optimum expression? How can we leverage all of the genius available to us to create the conditions to thrive? Everything that exists in this world is made of this world. If we made it, we can remake it, reforging our creations to reflect benevolence, wisdom, healing our disconnection from nature to create advancement that truly evolves humanity to be in balance as co-creators of our future in service to life.

Jamaica Stevens

Tranformational Technologies

There are countless modalities that can be consciously employed to enhance our lives. Technology is no exception. By using newly-developed Transformative Technologies and implementing **Appropriate Technology**, we can create a better life for future generations.

According to Transformative Technology Lab, "Transformative Technologies are actual hardware and/or software devices, developed based on credible science, that can produce a reliable and positive change in human experience. These technologies might be consumer devices, medical devices, interactive exhibits or art pieces."

Examples of Transformative Technology include wearables such as Google Glass and FitBit (a wearable data tracker that tracks information like steps taken and sleep quality to help users reach their desired goals).

Heartmath, an app that uses **em-Waves** technology to determine mood to assist the user in guiding them to emotional place they wish to be, and Spire, a breath-tracking app, are other examples of health-focused technology currently in use. These are but a few examples of the conscious, beneficial technologies that highlight the industry's current thriving status.

The clear progress that our technology has already made helps paint the future in a positive light. As the range of emerging technologies--fields of technology that broach new territory in some significant way--shows, there are also many advances in future tech that we should be aware of.

For instance, **Neuromorphic engineering**, *a new interdisciplinary subject, takes inspiration from biology, physics, mathematics, computer science and electronic engineering to design artificial neural systems, such as vision systems, head-eye systems, auditory processors, and autonomous robots.* This will enhance the progress of **artificial intelligence (AI)**; *the field of study that explores how to create computers and computer software that are capable of intelligent behavior.* Our real-world experiences will also undergo significant changes through **augmented reality (AR);** *defined as a live direct or indirect view of a physical, real-world environment whose elements are augmented (or supplemented) by computer-generated sensory input such as sound, video, graphics or GPS data.*

Future technologies will also yield benefits in energy and economy. Progress in **Energy harvesting**, a process by which energy is derived from external sources (e.g. solar power, thermal energy, wind energy, salinity gradients, and kinetic energy), is captured, and stored for small, wireless autonomous devices, will provide alternate energy sources. In addition, using alternative modes of currency – such as **Bitcoin**, bartering or **cryptocurrency** – may altogether change our monetary system as we know it.

These examples showcase the many paths our technology will take. As the future of technology unfolds, being aware of Permaculture principles and Appropriate Technology is vital to the prosperity of our society and health of the planet.

The term Appropriate Technology emerged from the original term

"Intermediate Technology" envisioned by Dr. Ernst Friedrich "Fritz" Schumacher in his influential work, **Small is Beautiful.**

Appropriate technology looks at the technological choices that focuses on small-scale, decentralized production (moving away from mass production) and is energy-efficient, is people-centered, environmentally friendly, and locally controlled.

In the recent years, it's been used in conjunction with **open source** principles and this has led to the creation of **open-source appropriate technology (OSAT)** and in its methodology takes into account peer review and transparency of process. It has been proposed as a new model of innovation in sustainable development.

In 1972, the **Club of Rome**, a global think tank that deals with a variety of international political issues, commissioned 4 researchers to conduct a computer simulation experiment of how would exponential economic and population growth interact with finite resources.

They examined 5 variables of pollution, population, industrialization, food production and resource depletion. **"Limits to Growth"** was the book published with their finding.

The gist of these predictions give us the bleak scenarios of what might happen if we, as a species, do not change our ways and act responsibly and swiftly. The *Appropriate Technology* gives us a model to start learning how to change these predictions, one decentralized group at a time.

Globally there are many groups, non-profits, startups, for-benefit groups and other concerned citizens

Appropriate Technology Characteristics to consider:

- Addresses a need of the people in a developing country.
- Promotes "a free and healthy life in a safe environment"
- Socially and culturally acceptable
- Raw materials availability
- Labor (including skilled) availability
- Can be made, maintained, and repaired in the country
- Has a reasonable cost and price relative to the country
- Is attractive to the end user (by their definition)
- Can lead to decentralized production
- Is a viable replacement for the current approach
- Is sustainable
- Does not harm the environment
- Microfinancing could be used to develop the business

of the world who are harnessing technology in new and old ways to start reversing and regenerating the topsoils, the forest, our transportation systems, our food systems-- the whole system redesign.

In many ways, this very book you are reading reminds us of lessons from a pre-industrialized world while providing sustainable options for our future. There is an understanding amongst its writers that in order to move forward, we need to examine our ancient tribal and village structure, mimic nature, decipher best practices and evolve them into our current and futuristic states.

In order for humanity and the planet to regenerate and become resilient to the massive climate changes, over-population and the collapse of old economic and social structures, we need to consider new and old technologies and methodologies.

Appropriate technology, amongst many other modalities, are offering us solutions to these ecological problems, where individuals choose roles and tasks for themselves and execute them accordingly. As such, responsibilities are connected to people who do the work, rather than elected or selected officials. These processes create tangible results and help communities take care of their food system, create resilience, invite new governance models and alternative economy.

When reflecting upon technology, an important question comes to mind: how can we say that our lives are truly enhanced by technology if it only further contributes to the destruction of our planet? As such, we must remain active participants in taking care of our planet since we are part of it. The technology we build in the future should focus on conserving and protecting the oceans, the animals, our soil, our food supply, and be centered on making our life healthier. By staying conscious of the current status of technology, as well as the already flourishing, promising future possibilities, we can decide how to best implement and contribute to these opportunities in our lives. It is up to us as a species to implement them.

Co-written by Julian Reyes, Natacha Pavlov and Maya Zuckerman

Founded by CEO Julian Reyes, Keyframe-Entertainment is a media network that creates global positive change by inspiring, informing, and entertaining through Transformational films, Visionary Art, and Electronic Music. Keyframe produces, finances, and distributes cutting edge projects in order to generate growth for its partners and strengthen community-building worldwide. Keyframe was launched in 2004 as a music label and artist management company and has since expanded its scope to offer Marketing, Social Media strategy, and PR services to DJs, festivals, and EDM companies.

Julian serves on the Board of Directors of the Electronic Music Alliance (EMA), a collaboration that cultivates social responsibility, environmental stewardship, community-building and volunteerism.

Keyframe is the Executive Producer of "Electronic Awakening," "The Bloom Series" Episode 3, and "ReInhabiting the Village."

http://keyframe-entertainment.com

I, Human

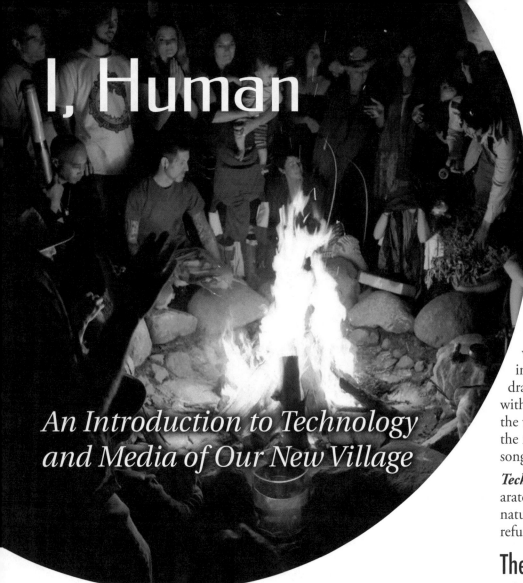

An Introduction to Technology and Media of Our New Village

we were otherwise unable to accomplish. This includes smartphones and the code which drives its software. It includes pens and the symbols we draw with them to communicate with one another. In fact, it includes the very intonation of sound beyond the most animalistic of grunts and songs, what we call language.

Technology is everything that separates us from our most primitive nature; it is the vehicle by which we refuse to be what we are.

The Technomythology of Man (Prometheus Absolved)

First, there was a body. With it, legs and limitations, opposable thumbs and an instinct for survival. The corporeal experience was a deep inner knowledge of foraging and fighting. Displays of dominance intermingled with the discovery of cooperation.

The uneasy balance of consequence and community elevated our bodies from a vessel of cells into an expression of a complex system of movement and sound, dance and song.

We learned to communicate our needs, and fulfill them in ourselves and those in our nearest surroundings. In time, those needs became wants and desires. These luxuries became staples of the human experience: shelter, warmth, and companionship.

Wesley Wolfbear Pinkham

MAN IS THE ONLY CREATURE WHO REFUSES TO BE WHAT HE IS. ALBERT CAMUS

To write an introduction to Technology for the purposes of re-entering our communal village is to write a history of society from our humble roots. No other animal has tapped into the sacred knowledge of its surroundings to create such an exponential impact on its environment. Wildly creative and devastatingly destructive, the long march of progress is our genetic and manufactured legacy. But if we are, as Camus claims,

the singular creature who refuses to remain true to our nature, then is that not the very essence of man?

Before we jump into the meal, let's set the table with an attempt at defining Technology. I propose the following: Technology is any means of invention harnessed to express that which was previously unable to be expressed.

That is, it is any creation along the timeline of existence that has helped us to accomplish something that

278

Touch and noise were not enough for our genetic ancestors, they sought understanding. His Holiness, the Dalai Lama, takes this one step further, saying that, "Love and compassion are necessities, not luxuries. Without them humanity cannot survive."

From body language, symbols. From noises, expressions. We carved our identity into our surroundings, painting in caves, drawing in sand. We sought connection and a shorthand way to express knowledge. Our time was short and our legacies were always in danger. We utilized the technologies of primitive tools and language to make our lives easier. Convenience was a survival tactic.

How complex thought evolved from language is, at this moment, a chicken and egg conundrum for cognitive historians. When man began to name the animals, the trees and their children, they began utilizing a nodal, pattern-based cognition in a way that few creatures can claim.

I should state that I am not a linguistic anthropologist. But as the question of how language originated remains one of the most confounding evolutionary questions remaining, I think it is safe to posit the following: it seems likely that through language, we evolved into consciousness and self-awareness. Sounds turned to symbols, or perhaps they turned into words, or perhaps symbols turned into sounds and back into symbols.

In the biblical creation myth, man names the animals and only much later does he feast on the fruit of knowledge, able to determine good from bad. But the fruit must be a superfluous symbol at this point; we had already begun our need to understand.

By identifying the relationships between us and not-us, the vastness of the Universe began its true unfolding. All technological progress has been an unwinding of the sacred truths in the mysteries of the physical world. We have always found the next level of technology in the discovery of new traits in the natural world: fire, wheel, paper… All invention is inherently organic, coming from the complex fabric of the chemical, electrical and energetic intricacies of our surroundings.

But the mythological components of invention—the aspects of discovery that linked man to the spirit—have been replaced by a faith in the ingenuity of human cleverness. Because our technological advances are shrouded in complexity, in biotech or microprocessors or satellite systems, we have lost the appreciation for the grand mystery of it all.

Since the 17th Century's Age of Reason, we have absolved Prometheus of his sin—delivering fire to humans from the hands of Zeus—and unnecessarily stripped the sacrament from science.

If there's one tangible synergy that the Village can bring to Technology, it is to rekindle the flame of coexistence between the unnecessary bifurcation between spirit and science. The answers we seek must bridge the gap between the holy paradox of knowing and not-knowing.

Synchromysticism

All of these technological advances have been mired by the sheer force of distance. Across vast continents, we developed an inability to speak the same language. We built systems too complex, too full of proprietary nuance for one person to understand and we became inextricably interdependent on each other's specializations.

Every step of the way, our collective intelligence has been the method of discovery. This is what I'm referring to as **synchromysticism**: the ability of our shared experiences to lead to the "Eureka" moments that make inventors famous. How could there ever be a Thomas Edison without the universe from which he came? How many people may have simultaneously discovered fire or the wheel or a toothbrush but had no

Revealed truths became codified and canonized in secret tongues, held by kings and mystics. Mythology became a seasonal experience, called upon for harvest, for rain, for teaching children their obligations and initiate them into adulthood, to move the soul from the physical realm to the one after life.

Was it necessary for the mystic, the shaman, the priest(ess) to separate themselves from the general population? Or for religion to hide knowledge in the catacombs of bureaucracy? In the *New Village*, we see a need for universal training in mystical thinking, infusing healing, prayer, and medicine into every corner of life. We desire open source spirituality and an interface for sharing sacred knowledge, plugging in variables from our own experience and connecting to a larger database of generational understanding.

IF THERE'S ONE TANGIBLE SYNERGY THAT THE VILLAGE CAN BRING TO TECHNOLOGY, IT IS TO REKINDLE THE FLAME OF COEXISTENCE BETWEEN THE UNNECESSARY BIFURCATION BETWEEN SPIRIT AND SCIENCE.

And we too became more complex, more proprietary and more nuanced. We said, "I cannot understand myself, let alone my brother," and we sought a creator mysterious and vast enough to know us.

We have accomplished so much of what we could imagine: space flight, telepathy in the form of smartphones, solar power. Exponential growths in science and technology are closing the gaps between reality and our wildest dreams.

Freud conceptualized the distance between man and his gods as the distance between what he could do and what was unattainable. "Now he has himself approached very near to realizing this ideal, he has nearly become a god himself," he writes. "Future ages will produce further great advances in this realm of culture, probably inconceivable now, and will increase man's likeness to a god still more."

way to share it with the world? What are the core truths hidden amongst all of the world's religions?

Many of these systems—trademark, copyright, Digital Rights Management—will come to a close within time. It is so rare to truly "discover" anything in this world without borrowing: from another culture, era or field. All technology is a remix of what came before it.

How often have you thought of a brilliant idea, only to think, "someone must have already thought of that"? Before the internet age, you may have still executed that idea, done it differently, or at least experienced the act of simultaneous creation. Now, though you may be put off to doing something that's already been done, you are also able to connect with those who share in the same line of thinking. "Great minds think alike," as they say.

For centuries, the means of transmitting information had become increasingly proprietary.

Technology as Universal Language

In getting back to the history of our humble little planet… Just when Nationalism had scarred our surfaces… Just when World War had hardened our hearts… Just when nihilism seemed to explain it all… We built a computer.

In this neutral instrument of war, we saw a glimmer of ourselves in its mainframe. Its electrical pulses and analytical thought showed us what was unique in the human condition. The analytical components of computation shined a light on our states of both Calculator and Being.

The millennia of ambiguous metaphor was sliced apart, and a cross-section of humanity exposed its true nature. We began to analyze The Code. Data. DNA. Algorithmic regularity. In a long stream of 1's and

0's, reality revealed itself like a chess game with basic rules but a playbook of infinite possibilities.

We began to read the code and we took the new language and wrote it into pictures and phones and medicine. We played and warred and danced to it. And we turned our consciousness back into itself, discovering the possibilities and limitations in encapsulating an entire human being into a digital language.

Wind through your hair. Boredom. Orgasm. The taste of water. Any and all experience have an underlying electrical/chemical reaction. We simultaneously became both infinitesimally small and quantumly expansive. As we become more bionic, we begin to better understand and appreciate the sensory experiences this world has to offer.

Living Life by the Scientific Method

Perhaps the greatest joy in incorporating all of these advances into our New Village is finding the synergy between what has always worked in communal experiences and what is new and exciting to the senses. By infusing the scientific method into the sacred art of living, we are capable of building systems that address the needs of our time in the most tangible and accessible way possible.

In the 20th century, many villages were conceived and created based on back-to-nature principles that excluded themselves from the modern conveniences of the era. As previously mentioned, technology and convenience are a means of survival. They make repetitive tasks easier, allowing us to focus on what the next opportunities could be in our evolutionary growth.

In the 21st century and beyond, if we can embrace modernity and reimagine the tools of mass culture for the purposes of building connection as opposed to ego-driven self-interests, we can synthesize the need for living as a village and enjoy the fruits of our technological advances. By utilizing the scientific method, we create experiments in living that do not limit the parameters of success to whether an idea does or does not work. Instead, we can play out our ideas and learn from the real-life examples as a means of growth and deeper understanding. By sharing our experiences, through words, videos and travel, we accept our role in the outcomes of our own generation.

We feel gratitude towards the progress which brought us to this point. We humbly embrace the challenge to leave the world better for future generations.

Wesley Wolfbear Pinkham

Wesley Wolfbear Pinkham is traveling at almost 30 km/s on the only celestial body known to harbor life. He has spent the last two years leading up to the 1st Edition traveling to 35 states and 4 countries, visiting festivals, emerging art communities, protests, and communal spaces to better understand how people gather for transformation.

With a background in non-profit arts administration, festival production, and digital marketing in the private sector, Wesley launched Wolfbear, a creative consultancy that seeks to help brands tell stories of a future filled with laughter and optimism. A photographer, writer and community leader, Wesley is helping the worlds of art and business merge mythos into a constant conversation of commerce and creativity.

Wesley earned a B.A. in Cultural Studies from the UCLA Department of World Arts and Cultures and studied abroad at the Hebrew University of Jerusalem. He was previously Director of Marketing and Programming at Yiddishkayt, the home for alternative Jewish culture in Los Angeles.

http://wolfbearconsulting.com
http://wesleypinkham.com
http://linkedin.com/in/wesleypinkham

Recode
Changing Building Codes to Legalize Sustainable Design

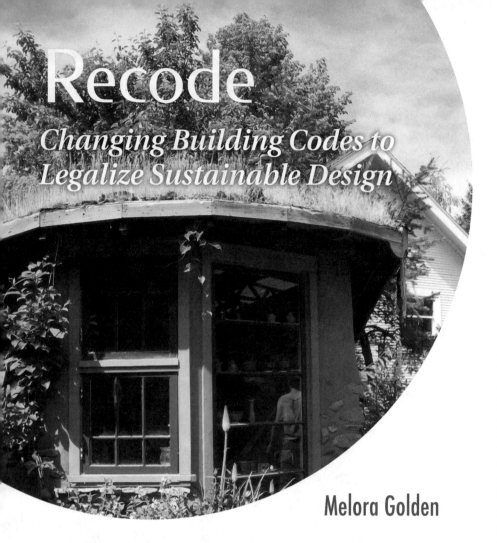

Melora Golden

We believe that human beings can use effective design to meet our needs while living in harmony with the earth. When our infrastructure was designed we didn't understand the needs of the environment and the limitations of natural resourc-

For sustainable design to be used on a scale that will make a difference for the environment, it has to be legal.

es. By consciously redesigning our built environment, which requires changing our codes, we can protect the natural world and create a sustainable, regenerative future for ourselves.

How Recode does its work is just as important as the code change work itself. We believe that to build a more resilient culture we also have be conscientious about the way we interact with each other. Recode uses a collaborative, relationship-focused approach.

Recode was formed in 2007 by Tryon Life Community Farm. When starting a sustainability education center, they found that many of the technologies they wanted to implement were illegal. Rather than just

maneuvering their project through the code process, they formed Recode, to create an accessible legal pathway for sustainable technologies.

For sustainable design to be used on a scale that will make a difference for the environment, it has to be legal. Since Recode formed, we have focused primarily on sustainable technologies that conserve water, protect human and environmental health and create beneficial reuse opportunities. We began by legalizing greywater, have expanded the types of composting toilets that can be used in Oregon, have written a national composting toilet code for the International Association of Plumbers and Mechanical Officials, and have advocated for a transition to a performance-based code regulatory framework for on-site sanitation.

Our current campaign focuses on creating legal pathways for ecological sanitation options like composting toilets, urine diverting systems and greywater systems to be used as part of a sanitation alternative in Oregon. We want a composting toilet and greywater system to fulfill a building's requirement for a waste water treatment system. Currently only

septic tanks, alternative treatment technologies (septic tanks with additional filters), and a sewer hook-up meet the requirement for a wastewater treatment system. This means that people can put in a composting toilet, but that one of the three above redundant systems will be required
as well.

In addition to catalyzing code changes, Recode creates educational materials and gives trainings to support our code change efforts. Below is more information on our approach to creating collaborative relationships with government.

Relationship

- Whoever you are sitting across from at a table is a person, just like you. They care about people, they care about the earth and they're looking for solutions too. If they seem resistant it's much more likely that they are actually overwhelmed.

- Their jobs are hard already, the changes we are asking for require that an already overburdened system does even more work.

- Be an ally not another critic.

- Try to form a personal relationship. Recode often starts meetings with a personal check-in.

Approach

- Be positive: we can do this!

- Be solution-oriented. Never come to the table with just a problem, always come with a solution as well.

- Learn about the technical aspects of your subject, do extensive research. Knowledge is power and creates respect.

- Be ready to do the work. Don't show up to the table to tell government officials that they are doing things wrong. Figure out who has jurisdiction over the issue you're interested in, find the code or rule that is a barrier, identify a technical solution and be ready to continue to do the work with them as they make the changes you want.

Messaging

- Find your shared goal. Sometimes this is close by and sometimes you have to really extend out towards the big picture to find a place of commonality. This is the big 'Why?' This is the reason the change needs to be made, the big picture benefit. For Recode it is protecting water.

Respect

- Be open to other perspectives, listen and learn.

- Respect the knowledge and experience of the officials that you're working with; your proposals will benefit from incorporating their insights.

- These policymakers and regulators have used the current codes to successfully protect human health in this country. Much of the world does not have reliable access to clean water.

- For issues other than sanitation, try to look at the regulation and

understand why it exists. Regulations are generally created in response to the concerns of a certain time or a certain group. As we learn more and our values change, our rules also need to change. This is a natural pattern.

Power

- Be a catalyst, bring energy and inspiration to an overworked, financially stressed system. You don't have the same accountability to a bureaucratic system that policymakers and regulators have. Their reality is that change moves slowly; you come from a different reality, invigorate them with your truth.

- Stay committed. Always retain your large-scale vision as you're incorporating technical facts into the solution needed to create the environmental benefits you are seeking.

- Because of the way that power can be used in society it is easy to feel powerless, frustrated, and angry when interacting with a person who is in a role that

holds embodied power. Try to perceive the innate humanity in the person sitting across the table from you. When we are able to see through the stress of these emotions and connect with the individual in front of us, even a little bit, it can really help move our cause forward more quickly and thoroughly.

- Everyone, to some degree, has experienced the oppressive use of power. Many people have to deal with multiple layers of societal oppression. Engaging in personal healing and processing around these traumatic, and often ongoing, experiences is a way to help activate your power. It is true that power is held by groups of people and structures. It is also true that we can take power and use it to shape our world. Please try to find the tools that help you heal and connect with your power, share those tools with others, and make personal work part of the conversation about social change and cultural transformation.

Recoding Ourselves

Every decision about the physical world comes through a human process. The most impactful aspect of creating change is improving the way we communicate and interact with each other. A significant piece of interaction has to do with the stress and protection we carry around based on past negative interactions with others including our families, partners, and institutions that utilize societal power. Old stress and contraction gets triggered and we can react in a way that is disproportionate to what's actually happening in the current moment.

That's why personal work is so important in creating a world where humanity shifts the way we treat each other and the planet to become more compassionate and aware. There are many practices that can teach us to observe our reactions, therefore creating some space where our cognitive brains and our hearts can participate in the decision-making about how we react to the people around us and the situations we're in. The way we act and the choices we make in the current moment create reality and shape the world.

Take a moment to connect with why you want to change the world. Let your mind travel down the pathways of your vision for the future. Reclaim a deep awareness of the parts of your vision that exist today. What personal contribution are you making to the manifestation of the long-term vision and to supporting and maintaining the building blocks that exist today? Remember that one of your most important actions is to take care of yourself, increasing our resilience gives us more capacity and compassion to deal with the inevitable human friction that occurs as we try to work together to make things better. Keep up the good work, you're doing great!

Melora Golden

Since 2008 Melora has provided vision, strategic planning and organizational development to Recode, serving as a community organizer, project manager, executive director and now as an advisory council member. Before becoming an environmental activist, Melora was a social worker for 12 years. She has traveled to 37 different countries learning about the variety of cultural approaches to life and about universally shared human values and needs. She attended the International Permaculture Convergences in Brazil and Africa and was one of the organizers of the second and third Women's Permaculture Gatherings. Melora believes that bringing more intention and tools to group interaction will significantly increase the effectiveness of those groups. She consults with individuals and groups to increase the productivity and enjoyment of their work through holistic organizational design. She has lived in Portland OR since 1996, and is dedicated to her wonderful city and to the regenerative evolution of the world.

http://meloragolden.com

http://recodenow.org

http://tryonfarm.org

Techn:ecology

Reframing Technology in Light of our Ecology

It is now highly feasible to take care of everybody on Earth at a higher standard of living than any have ever known. War is obsolete. It is a matter of converting the high technology from weaponry to livingry.   Buckminster Fuller

Techn:ecology is a **meme** born of the experience of living in a built environment in a world where technique informs at the foundations of our every innovation, yet often as if in a void disconnected from our ecology. Even in this era when we make the course correction from the mechanistic worldview into a new and deeper understanding of the biological and living systems approach, there is a need to practice designing with the top-line in mind and achieving the higher level of awareness required for overcoming our current challenges.

TECH

A basic use of the word "tech," as shorthand for technique or technology is about knowing the most efficient way to do things, and in a healthy system, a natural evolution through learning directly from the iterative process. This is another opening for evolutionary processes, as we are presented with the opportunity to engage the collaborative value of **open sourcing** things rather than to develop proprietary branded techniques and innovations. This has been a long journey through history and pre-history in terms of the various ways to approach the earning of a living or the collecting of our food for survival and the apparent cultural conflict over whether to share or not or by which system of measurement to define exchange.

Bringing awareness to our own worldview, and sometimes unconscious cultural bias, sheds light of perspective around the nature of sharing, having, taking, buying, renting, or owning anything. The dominant cultural assumptions in our world are built on relatively recent colonialistic structures and it can become very dynamic to understand how to change that as we peel back the layers. What is the system of ownership? Is there value in access over ownership? Once one owns, what are the costs of ownership, how does one earn in order to maintain? How is one's value measured in their microclimate and larger ecosystem? Where is the feedback received and acknowledged in a healthy system?

ECOLOGY

Amidst what some evidence suggests is a 6th mass extinction on our planet, it is a challenging time to witness the loss of biological diversity and is a call to question our very purpose on the cusp of a Planetary Era for humanity. What it means for us at this point is to be choosing cultural and biological diversity and even looking deeper at what programs of "success" are running within our evolution as a species that have brought us to this point where we are seemingly in need of a course correction to avoid driving out of existence the diversity of life which has supported us to thrive in the first place. If we choose to consciously evolve toward rather than away from this greater diversity, how will we do it? How will we repair the harms humanity has caused in our Global Village, not only the harms against our human family, but to all of our relations?

So then we ask, what would be another structure and another way to make a difference that can affect for change at a larger scale on existing systems? We can look at the changes we might want to see happen in terms of shifting the focus from a human-centric world to expanding and encouraging natural ecosystems and biological and cultural diversity. These are significant shifts in worldview and it is humbling to imagine the complexity of communicating to the masses of people who are reproducing in large numbers driven by

their inherent DNA and drive to thrive and survive. So what is the answer? Is there a way to send out a wakeup call through the nervous system of humanity as a whole? What would that look and feel like? In most accounts from our recent cultural perspective it looks like a celebration of approaches and engagement of positive potentials and optimistic life-affirming choices and shifting perspective toward a greater identity within a cooperative ecosystem.

INNOVATION

With such complex questions to answer, we engage here through encouraging the creative human spirit in the realms of being proactive, *regenerative*, and creatively solutionary. Our inherent human capacities include a dimension that is able to find solutions within the problems, to find products and resources within the waste streams, to thrive in the pulse of this Ecological Renaissance. Along with redesigning use we also have innovations in the realm of Maker Spaces and Creative Commons. There is even a project called Open Source Ecology developing "The Global Village Construction Set." Clearly innovation is born of ingenuity and it is that aspect of the human spirit that we are calling to inspire and encourage. Our monetary and financial systems will also naturally need to evolve from a foundation on debt to a foundation on quadruple bottom line values and ethical foundations. We have an opportunity to transcend and include the best of the old systems, from business startups to banking; we are ready for the rising waves of eco-entrepreneurialism to make a total culture shift toward enlightened leadership.

Knowing the difference between technology as an industry and technology as a flowering of human potential realized through collaboration and cocreation is a hallmark of the Planetary Era. As we deepen our understanding that technology is our approach to enhancing ourselves and our environment toward greater comfort and success, we simultaneously are awakening to an expanded sense of self. As we integrate the opportunity to define our success in the context of Life in a larger way, we can embody that through the development and discovery of a synthesis and a *new techno-economic mode*. Grounded in simple mindfulness practices such as honoring the trees that are providing essential oxygen for our every breath, we can further define our success by how well we are able to steward and support the biological and cultural diversity of our beautiful, living, Spaceship Earth.

A. Keala Young

A. Keala Young is a whole system designer, regenerative practitioner and permaculture teacher with a background in the healing arts. His work on ReInhabiting the Village as editor and contributing author is an opportunity to further his life's mission to support the integration of living systems coherence for individuals and communities in the Global Village.

Since 2005, Keala has been consciously committed to the path of co-creating living & learning communities and has studied far and wide, gathering the seeds and skills to undertake the adventure of realizing models for the future now. He has over a decade of experience presenting and curating educational content for festivals and is a cofounder of the budding Atlan Ecovillage in the Columbia Gorge. A steward of Gaiacraft, Keala also serves on the board of CultureSeed, an educational nonprofit with the mission to inspire and support thriving global communities through experiential learning programs and transformational events.

EcoCentro

IPEC Institute of Permaculture and Ecovillage of Cerrado

IPEC (Institute of Permaculture and Ecovillage of the Cerrado) is a non-governmental non-profit organization which has its office in Ecocentro, located in Pirenópolis, Goiás. The IPEC was founded in 1998 in order to establish appropriate solutions to problems in society, promote the viability of a sustainable culture, provide educational experiences and disseminate models in the cerrado and Brazil.

In this context, the permaculturist André Soares and educator and writer Lucy Legan ministered Permaculture courses all over the country and abroad, enabling many permaculturists in the evolution of a proposed change. In 1999 they began building a space to demonstrate the viability of the principles of Permaculture and Bioconstruction, the Ecocentro.

Water

The Ecocentro IPEC is fully responsible for all the water it consumes. We will not return contaminated water into the environment; all wastewater is bioremediated and only when clean again will it be returned to the environment or used for irrigation.

Infiltration trenches or Swales

Water is power! In permaculture design we try to plan the system in order to avoid as much of any form of loss or waste of energy. Rainwater that falls on the site is a fabulous source of energy being injected into the system. What we don't recycle is simple being wasted.

Solar Water Heating

The Low Cost Solar Heater (ASBC) is responsible for the warming of all water used in showers and a few taps of the Ecocentro. The ASBC consists of a set of pvc or polypropylene plates through which water passes and is heated, then stored in a thermally insulated box.

Water Tanks (ferrocement tanks)

The ferrocement technique is widely used in the Ecocentro IPEC. It can be applied in various functions, such as cisterns to capture and rainwater storage tanks for aquaculture, ponds and even swimming pools. In regions with long dry seasons, as here in the Cerrado, this technology becomes a vital strategy.

Natural Water Treatment Plant

1. The first step of bioremediation is the grease trap, septic tank located near the kitchen, below ground, an anaerobic environment where the solids begin to be broken down by microorganisms which feed on the organic matter remains.

2. Next, the digester is fully closed or anaerobic reactor where solids are separated.

Electricity

In Brazil most of the electricity used comes from hydroelectric plants installed in gigantic dams, which, in addition to causing tremendous environmental and social impact, consume valuable resources. Despite using a renewable resource, water, concentrated form of energy-generation creates spending on distribution, high technology and construction of large and complex installations.

The flooding caused by a hydroelectric plant, in most cases, covers vast areas of forest plant species and destroying the habitat of many other species of animals increases the risk of extinction. The power lines necessary for transmission of energy, besides causing tremendous visual pollution, cost fortunes and often requires the felling of forest areas for crossing lines.

Renewable does not mean eco-friendly! The larger the scale, the larger the impact on the environment..

Photovoltaic Energy

Sunlight can become a source of extremely abundant, renewable and free electricity through the photovoltaic plates. Unlike collectors for water heating, taking advantage of the sun's heat, solar panels convert light into electrical current. The plates are constituted by cells fabricated with silica crystals.

Fossil Fuels

Ecocentro IPEC's use of fossil fuels is reduced to almost zero! All vehicles and machines powered by Diesel are converted to run on recycled vegetable oil from pastry shops and restaurants in the area. Soybean oil, corn, sunflower, among others, which were once part of restaurant waste, are now collected for free and then recycled.

Water Electricity

Water is a natural abundant resource on our planet. Moving water means energy and water energy can be converted to electricity cleanly and efficiently. The turbine generators are driven by flowing water from a river or even a small stream where there is a gap.

Wind Energy

From the earliest days the wind has been used in the service of humanity. From the sailing ships of the first explorers to the mill farms, the wind has been present in every step of human evolution, and the Ecocentro IPEC has now taken it to the next level!

Solar Energy

Ecocentro IPEC harnesses energy from the sun in many different ways. Our houses are built with insulating materials and do not require spending cooling or heating. We make food in solar ovens that take advantage of the greenhouse principle avoiding higher energy expenses.

Sanitation

In Permaculture we classify the wastewater into two types; gray water and black water. The gray water is water that comes from kitchen, showers, rinses in general; water containing only food scraps, grease and soap and that is not contaminated with feces. The black water is water containing human excrement and urine.

The Ecocentro IPEC develops environmentally friendly and economically viable solutions for the treatment of wastewater, whether in large or small-scale, adaptable to different climates and regions, and above all respecting the natural cycle.

Whoever comes to visit us knows how seriously we take our disposed "materials" ...

One thing is certain: IPEC's nearest recycling centers of black and grey waters never flow directly...

Check out these technologies, and apply them whenever possible.

Treatment Beds

One of the simplest and most efficient ways to treat gray water is with biological treatment beds; in which all Ecocentro IPEC shower, kitchen and laundry waters are treated. The beds can handle both large volumes of water in communities, neighborhoods, housing projects, schools, etc.

Ecosystems Alive

Revolutionizing the sewage treatment systems, a "machine" that incorporates plants and animals, any kind of sewage can become clean water and be used again in the garden, shower, bathroom, etc. The system consists of a series of tanks densely populated by filtering plants composed in water treatment stages.

Septic Infiltrator

Even the waste that comes from conventional toilets can be treated in an environmentally friendly manner. Unlike septic tanks widely used in most homes that just break down solids instead of freeing the water from bacteria, staphillococcos, etc. contamination, septic infiltrator's all organic matter is digested by plants with filters.

Compostable Toilets

What in large urban centers becomes the biggest headache, Ecocentro IPEC transformed into indispensable resource: human waste. When a conventional toilet is used, about twenty liters of drinking water are polluted and simply wasted in the discharge, and what is worse, without solving the problem.

Biolytic Treatment

Gray water that is coming out of sinks, showers and washing clothes. This water is soap, fiber and other waste but has no human feces, so it can be treated and reused with relative ease. After going through a very simple biological filter gray water can be reused for irrigation.

This article is a brief overview of the technolgies leveraged by EcoCentro. Please go their website for more information and details.

http://ecocentro.org/

Whole Systems Design

WHEN PEOPLE GO WITHIN AND CONNECT WITH THEMSELVES, THEY REALIZE THEY ARE CONNECTED TO THE UNIVERSE AND THEY ARE CONNECTED TO ALL LIVING THINGS.
Armand Dimele

Throughout this book we have attempted to look through the lens of varying perspectives while acknowledging our place in nature, our inter-connectivity, embracing that in truth we are woven within the context of a web of relationships. From microcosm to macrocosm, we exist inside a pattern that is made of individual parts, interrelated and nested within a whole. We are individual and collective, one organism with distinction in our purpose and yet only functional when we synergize and act in accordance with the all. This is not theory, this is the blueprint of life. And with the blueprint accessible to us, we can now become the social architects designing our reality in symbiosis with creation, as an aspect of creation.

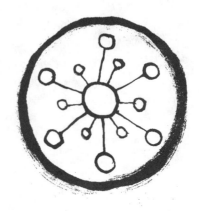

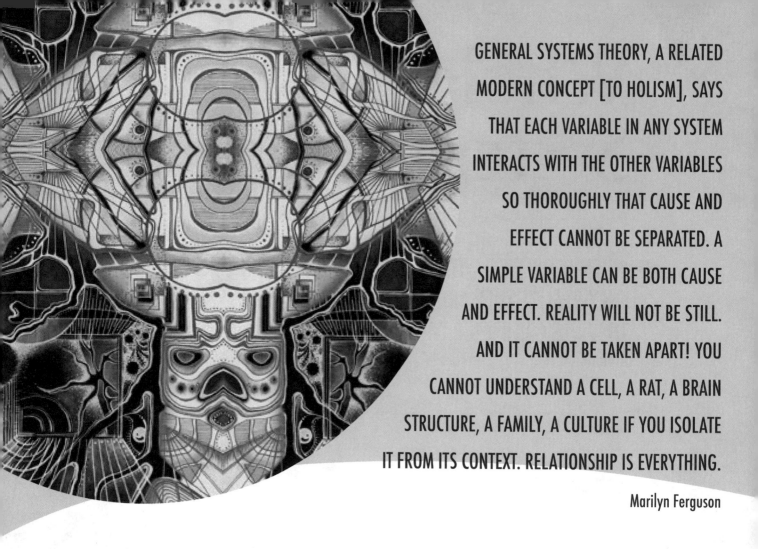

GENERAL SYSTEMS THEORY, A RELATED MODERN CONCEPT [TO HOLISM], SAYS THAT EACH VARIABLE IN ANY SYSTEM INTERACTS WITH THE OTHER VARIABLES SO THOROUGHLY THAT CAUSE AND EFFECT CANNOT BE SEPARATED. A SIMPLE VARIABLE CAN BE BOTH CAUSE AND EFFECT. REALITY WILL NOT BE STILL. AND IT CANNOT BE TAKEN APART! YOU CANNOT UNDERSTAND A CELL, A RAT, A BRAIN STRUCTURE, A FAMILY, A CULTURE IF YOU ISOLATE IT FROM ITS CONTEXT. RELATIONSHIP IS EVERYTHING.

Marilyn Ferguson

Modelling after the geometry and patterns within nature - or Bio-mimicry - Whole Systems Design principles generate adaptable and yet constant structures and forms that function to work from the simplest framework while scalable to the complexity of multifaceted considerations. We honor the wholeness within the parts, fortifying the points of connection where things, people, or systems intersect, and through clear agreements and intelligent design, we cultivate a collaborative space within the connection for all parts to act in alignment and harmony.

Elegant, practical, intuitive, there is a way to work in flow with life that creates exponential results.

Leveraging multiple approaches creates resilient solutions, organizing elements into a series of dynamic interactions. When we isolate and compartmentalize the world into parts, our strategies for engaging our challenges diminishes. We discover that our capacity to resolve issues is inefficient at best and detrimental at worst. We must see the relevance that each part has to the other and yet still value the intrinsic nature of a single component.

Examples of Whole Systems Design in the domain of governance include models of decision making such as "Sociocracy," "Dynamic Governance," "Resonance," or "Consensus Decision Making" to name a few. In the domain of farming, some examples are "Permaculture" or "Bio-Dynamic." In energy it's "Regenerative Design." Cross sectorally it's "Ecolog-

ical Design," Alan Savory's "Holistic Management," Ken Wilbur's "Integral Theory" is the all-inclusive theory which seeks to weave coherence between all schools of thought.

Just as our nervous system, cardiovascular system, musculoskeletal system all have to function together for our body to thrive and achieve homeostasis in response to the constantly changing factors that our organism experiences, we too have to work with nature in order to ensure not only the survival of all species, but to truly know and fully realize our human potential. As pioneers constantly delving into the uncharted territory of the infinitely vast Universe, by embodying Whole Systems Principles, we unlock the mysteries of this exquisite design all around us.

Jamaica Stevens

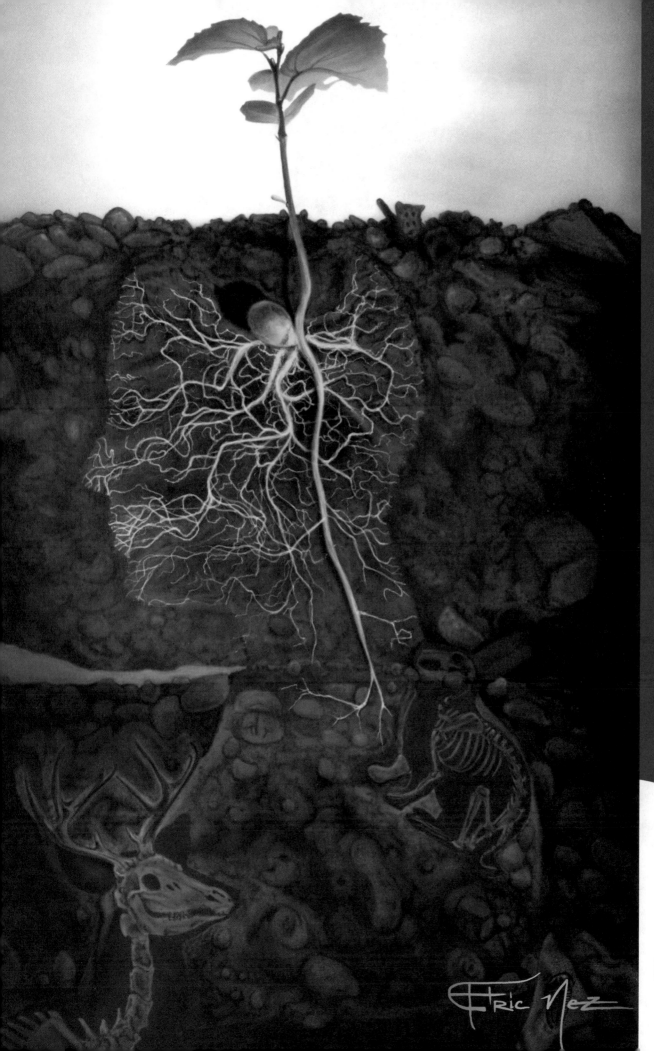

Mapping Whole Systems

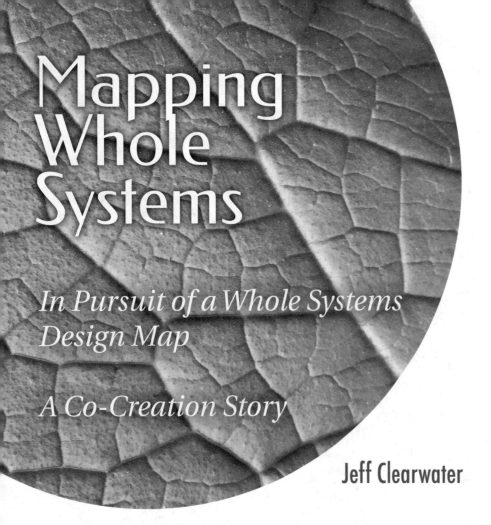

In Pursuit of a Whole Systems Design Map

A Co-Creation Story

Jeff Clearwater

Every society is based on a story. It's the underlying narrative, the most essential agreement field–the mythos–the essence of the culture. For Indigenous cultures the rich web of stories that made up their collective wisdom had, at its heart, a Creation Story.

The Creation Stories of the Indigenous cultures have one thing in common. For those able to hear the subtle divine wisdom therein, Indigenous creation stories reveal a deep understanding of Humanity's place in Nature. In fact this is perhaps why they are continually retold: to remind its people of that place.

Modern society is no different–though it might want to think it is–as conventional science seeks to replace "mythos" with "fact." Yet if we look deeper, it is clear that modern science is just one contribution to the rich interplay of our collective sense of Humanity's place in Nature, taking its place alongside art and spirituality.

Western society's story has at its core a long journey–one beginning with the separation of Nature and Humanity–attempting to assert a "conquering of Nature" and to prove that all the gods and goddesses are dead, that consciousness arises as an accident of physical forces, only to now come to begin understanding that all is alive–that consciousness is intrinsic in the Universe–that separation is an illusion and that Humanity, in order to survive, must seek once again its rightful place in Nature.

The Importance of Mapping Whole Systems

Whole Systems Design is essentially a journey; a journey that attempts to show our individualistic and fragmented Western society its way back to a reintegration with Nature and with each other–a synthesis of modern science, ancient art and the essence of Spirit–a retelling of our collective story.

The mapping of Whole Systems–as a tool for **Whole Systems Design**–then becomes an essential part of our collective participatory design journey towards a new paradigm society.

What is a Map?

It is an analysis and a synthesis. It's a taking apart so that we can see the whole. It's an art piece showing one lens of reality. It's a capturing of one or more clear moments of seeing, so that we may find our way again. The map is not the territory. Each map holds the danger that the user may forget this; may fail to see reality because it doesn't match what they expected from the map.

Maps can be powerful pattern languages that evoke deep knowing and sensing, or they can mislead us, forcing unnatural structures upon natural flows; trapping us in our mind's old grooves and ways of seeing–or not seeing.

There are myriad types of maps, each serving a different purpose. They

294

I believe that the community - in the fullest sense: a place and all its creatures - is the smallest unit of health and that to speak of the health of an isolated individual is a contradiction in terms. Wendell Berry

include process maps, framework maps, and *mind maps* and they often mix more than one modality. This article primarily deals with framework maps that serve the design process—the subject of design here being no less than the redesign of human society. The particular map we are developing is a ***Regenerative*** Whole Systems map that serves as a tool for designing human systems in general and villages of all kinds; what we call communities of intent.

Analysis and Synthesis - Toward an Ontology of Regenerative Design

All maps categorize—they isolate and highlight parts of the whole—often based on a categorization criteria that arises out of the worldview of the story behind the map. The purpose of a map is to find our way, to categorize only so one can find the whole again.

The power of a map to bring us back to the whole is only as strong as the story the map evokes, either intuitively upon sight or from association with a story that has already touched us. Some maps are able to convey the story with the graphics, others are not energized without the associative narrative.

The map we seek to develop is one where human society's systems are seen in their entirety, with the inherent evolutionary and dynamic relationships between the parts conveyed in a context where they can be rewoven back to a whole—a whole which affirms and uplifts life as an organic and interdependent process in which we are conscious participants and stewards. We want a design guide that can show us the major areas—the broad strokes—and then provide a framework for diving in. This framework map becomes a gateway to a deep engagement with each category, where the sub categories become the entry to detailed analysis and synthesis. The map becomes the beginning of an Ontology of Regenerative Design.

Ontology - Ontology is an organizational system designed to categorize and help explain the relationships between various concepts in the same area of knowledge and research.
(For a more complete discussion of ontology see:
http://www-ksl.stanford.edu/kst/what-is-an-ontology.html

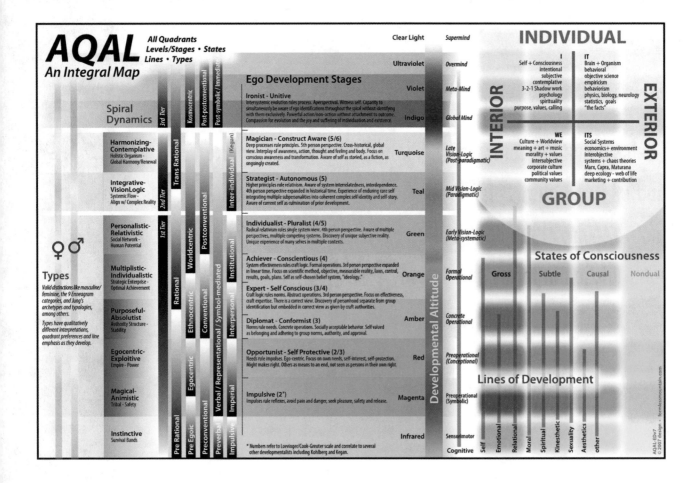

Invitation

I'd like to invite you on a journey–exploring various whole system maps–with the hope that we may co-develop better maps for Regenerative Whole Systems Design. This invitation goes far beyond this article; it's a process and a dialogue, it is a song and a dream, it's collective telling of a new story. It's an invitation to design a new society.

Existing System Maps

First let's take a quick look at some powerful existing whole systems design maps that we draw from in our work.

Integral Theory - Ken Wilber

Perhaps the most comprehensive whole systems school of thought and mapping system comes from Ken Wilber's inspired Integral model as conveyed in his book, *A Theory of Everything* (2001). It really is that; a grounded deep understanding in the nature of reality, the evolution of consciousness, and of human and social development.

Integral is a multidimensional map, integrating what is called "all Quadrants, all Levels or Stages, all States, and all Lines of Development–or the AQAL map. It integrates the Spiral Dynamics system of Don Beck and Chris Cowan and includes and transcends Maslow's hierarchy of needs.

Its primary insights are the development and place of the "I" and the "We" and our relation to the "It" and the "Its"; the stages and lines of development of the ego and social systems, and the various states of consciousness. Integral theory is grounded through the teachings of Integral practice; a profoundly deep guide to personal development. This understanding of personal development is also being applied to every field of human and social development.

We find AQAL to be a powerful insight into evolution and whole systems understanding. Its utility as a design tool comes from its ability to guide us to include all dimensions and facets of a system as well as understand the evolution and nature of the worldviews inherent in any system. All parts of the map are useful in village design, especially when integrating personal well-being with collective well-being and co-sensing a new emergent worldview.

In 2007, Ferananda Ibarra and Juan Pablo Rico formed a group called ICON–the Integral Communities Network–where they applied Integral theory to the design of intentional communities. VillageLab's Whole Systems Design Map draws on that work. You can access the results of that work here: http://iconglobal.wordpress.com

Wheel of Co-Creation / Tribal Convergence Network

Barbara Marx Hubbard was one of the first visionaries to offer an inspired co-creative map of whole systems. Barbara's Wheel of Co-Creation is based on a powerful story of the evolution of Humanity, a story she received in a transcendent vision in 1964 and refined over the many years of her fabulous work inviting all fields of human society to evolve into its next level of evolution. The strength of the Wheel of Co-Creation is in that story and its ability to bridge from the old world paradigm and way of seeing society into the new; to invite every "existing field of human endeavor" to connect through their heart and for each person to find their place through their own "vocational arisal," and bring forth the "Golden Innovations" in every field of human endeavor.

Barbara's story behind the Wheel is one of the emergence of collective consciousness by the synergistic effect of each field's evolution combined with the "Global Communication Hub" that can bring forth these golden innovations, creating a new planetary DNA for Humanity. Barbara's powerful work and delightful message has inspired millions to imagine a new humanity, a wonderful new positive future.

The Wheel of Creation categorizes human endeavors along lines all can understand using the traditional sectors of our existing society. The focus here is to invite all existing "vocations"–all spheres of influence into their next evolutionary spiral.

Tribal Convergence Network Guilds

The Tribal Convergence Network uses a close adaptation of the Wheel of Co-Creation in the design of its famous tribal gatherings and in the selection of its Guild structure. TCN has modified and/or expanded some of the sector descriptions with an eye toward grounding village and movement design. From this wheel they have formed Guilds to move forward each area of Humanity's evolution–seeking to fulfill Barbara's vision of each field of human endeavor spiraling to its next level of evolution. For more see http://tribalconvergence.com/projects-offerings/1190-2/

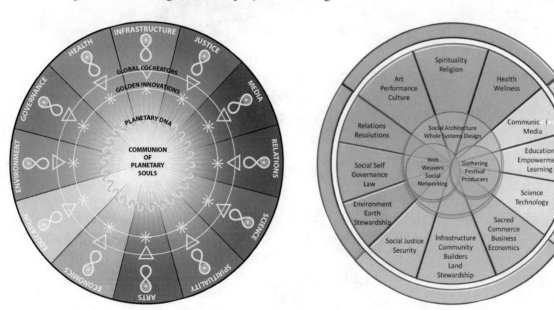

...only to now come to begin understanding that all is alive–that consciousness is intrinsic in the Universe–that separation is an illusion and that Humanity, in order to survive, must seek once again its rightful place in Nature.

Global Ecovillage Network

The Global Ecovillage Network has used a simple Four Quadrant map for over 20 years in describing whole systems sustainable design. The map was designed by GEN's educational initiative–Gaia Education– to map out a whole systems design curriculum according to 4 quadrants: Social, Worldview, Economic and Ecological. It stands today as the outline for Gaia Education's Ecovillage Design Education certificate program, as well as the organizational basis for the 4Keys set of whole system design books produced by GEN. See the 4Keys - http://gaiaeducation.net/index.php/en/publications/4key.

Recent evolutions of this map have been inspired from GEN's work in Africa and other cultures where "worldview" was not received so well as a category, and where "culture" means everything. GEN's new whole systems map now uses these 4 categories: Social, Cultural, Ecological, Economic. The new map is used in their description of the dimensions of sustainability and is used in their new online interactive "Solutions Library." (http://solution.ecovillage.org/)

VillageLab's Whole System Design Map
Purpose

- VillageLab has several interlocking purposes for developing our Regenerative Whole Systems Design Map

- Create an easily graspable evolutionary understanding of human systems development

- Create an intuitively accessible overview of all elements of human systems design

- Differentiate clearly between whole systems frameworks and design processes

- Create the main and subcategories we will use for both whole systems design and as a guide in our support and activation of various community technologies, processes and practices

- Suggest an organizational structure and process for the Guild Boards for VillageLab's Research, Development, Demonstration, Replication and Dissemination Program

Creation Story

As we touched on earlier, the power of a map is only as good as its immediate intuitive grasp, coupled with the narrative or story it evokes. The following is the story behind the VillageLab Whole Systems Design map.

It starts at the bottom of the map with its root in Nature–in the Ecology sphere–recognizing 4.5 billion years of evolutionary wisdom in life on Earth as the most integrated whole systems design model we have. We wholeheartedly embrace what some call *biomimicry* as the basis for whole systems design, as Nature creates ever-increasing spheres of ecological integrity and collective intelligence over time. Like the Wheel of

The figure above shows the Whole Systems Design map with the following sections:

Social — Personal Skills in Community
- Health & Wellness
- Spiritual Growth
- Communication
- Shadow Work
- Leadership
- Autonomy

Culture
- Art, Ritual & Play
- Gathering & Celebration
- Nuturing Community
- Tradition & Culture Shifting
- Youth & Elders
- Story & Worldview

Political
- Group Process
- Decision Making
- Governance
- Conflict Resolution
- Participatory Design
- Local > Planetary Politics

Whole Systems Design
- Integral Theory
- Pattern Language
- Collective Intelligence
- Pedagogy & Peeragogy
- Eco & Bio Mimicry
- Permaculture Design
- Deep Wealth Design

- Water
- Food
- Energy
- Shelter
- "Waste" Cycling
- Appropriate Tech

- Access to Wealth
- Currencies & Holopticism
- Collaborative Economy
- Local > Global Economy

Ecology
- Watersheds
- Ecosystems
- Earth Stewardship
- Bioregion > Biosphere

Economy
- Access to Resources
- Generating Value
- Gifting & Generosity
- Resource Sharing
- Ownership & Equity

VillageLab

Co-Creation and Integral Theory, the VillageLab map draws on the evolutionary wisdom of the Universe itself as we evolve to ever higher states of consciousness as expressed in human systems integrated in evolving natural systems.

From that base root at the bottom of the map human society evolves upwards. First in tribal cultures with little technology, humans are integrated fully into Nature, occupying a niche like any other animal; drawing our sustenance directly from the Earth with few "human systems" differentiating it from other animals save for the technology of its hunting and gathering tools (eco-tech). With an ever-increasing sophistication of tool use–and especially with the control of fire–human tribes start to create extra stores of food and goods. As we create storable and tradable value, our economic system is born, even if rudimentary and not recognized as one. As soon as there is excess food to store, share or trade, then we have an economy. There is no mistake that "eco" is the root of both ecology and economy: they both mean the taking care of the "home." In ecology we study our larger home of Nature; in economy we take the value we've created from nature and share it in our immediate home.

These Ecology (Eco-tech) and Economy sections of the map make up the lower hemisphere of the whole map symbolizing that they are the physical plane root that all other systems evolve from. They are a natural grounded result of Humanity's first relationships with the earth and each other on the physical plane. The evolution from Ecology to Economy is represented as going to the right side of Humanity's "systems body," as we evolve from the bottom of the map. This symbolizes the development of our collective "left brain" evolution (the left brain controls the right side of the body), as we begin to develop a rudimentary economic system that arises directly from our relationship with the ecology.

The combination of the Ecology (or Eco-Tech section, including our water, food, and shelter technology) and the Economy sectors of the map represent what will become the "production system," producing and distributing goods and

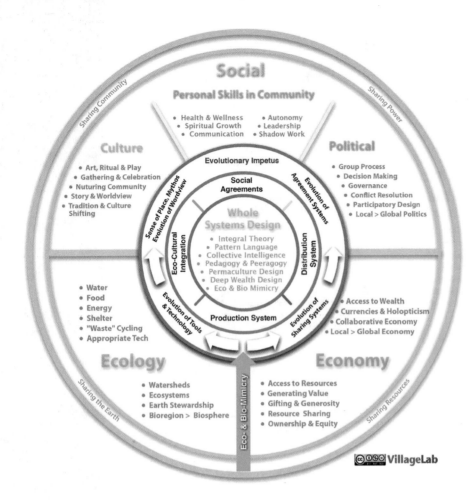

The diagram contains the following text:

Social
Personal Skills in Community
- Health & Wellness
- Spiritual Growth
- Communication
- Autonomy
- Leadership
- Shadow Work

Sharing Community

Sharing Power

Culture
- Art, Ritual & Play
- Gathering & Celebration
- Nuturing Community
- Story & Worldview
- Tradition & Culture Shifting

Political
- Group Process
- Decision Making
- Governance
- Conflict Resolution
- Participatory Design
- Local > Global Politics

Evolutionary Impetus

Social Agreements

Sense of Place Mythos · Evolution of Worldview

Evolution of Agreement Systems

Whole Systems Design
- Integral Theory
- Pattern Language
- Collective Intelligence
- Pedagogy & Peeragogy
- Permaculture Design
- Deep Wealth Design
- Eco & Bio Mimicry

Eco-Cultural Integration

Distribution System

Production System

Evolution of Tools & Technology

Evolution of Sharing Systems

- Water
- Food
- Energy
- Shelter
- "Waste" Cycling
- Appropriate Tech

- Access to Wealth
- Currencies & Holopticism
- Collaborative Economy
- Local > Global Economy

Ecology
- Watersheds
- Ecosystems
- Earth Stewardship
- Bioregion > Biosphere

Economy
- Access to Resources
- Generating Value
- Gifting & Generosity
- Resource Sharing
- Ownership & Equity

Eco- & Bio-Mimicry

Sharing the Earth

Sharing Resources

VillageLab

---

services to the culture–the integration of adjacent sectors of the map showing their resultant human system–as shown on the smaller circle on the inside of the map.

If we go up to the left side of the map starting at the Ecology field, we see the evolution of human systems analogous to the "right hemisphere" of the brain (left side of collective "body"), as we collectively embrace the emergence of Culture from Nature. In early tribal society Culture and Nature had no perceived difference, as we filled a niche like any other species. As we evolved our sense of collective self, our cultural awareness grew, evolving into a rich array of earth-based mythos and spirituality communicated through story. In essence our Culture is a society's "Nature"–the collective ecosystem of belief, values, ritual, art and play–intricately linked to the sense of place, climate, food, materials, shelter and

energy that nature provides. Hence Culture is adjacent to Nature and their combined energy is one that there is no single word for in English, save perhaps "Eco-Cultural Integration" or Cultural Integration.

Returning to our evolutionary trip up the right side of the map (in the "left-brain" of human systems), as soon as our economic "system"–as rudimentary as it was in early tribal societies–created an excess in food and goods, we also created the need to decide how to distribute that excess. Who controlled that decision and how it was arrived at–whether by brute force and "power-over" or by sharing–represents our early political systems; how we share (or don't share) power, and how we decide things. We've entered the social sphere of society now (the upper hemisphere of the map). Economy, like Ecology is a huge invisible architecture for the Social sphere, but it is not until our Economic system evolves into a conscious system that it becomes visible

as part of our social identity. Economy emerges as soon as you pick a berry and share it. Economic systems emerge as soon as we start to set up systems to share and trade the wealth. Politics emerge the moment someone seeks to control the stash of berries. Political systems emerge as society starts to consciously function under a certain acquiesced to or chosen power relationship.

The intersection between the Economy field and the Political field is the field of the Distribution System of a society; the system that takes the goods and decides how to distribute them, based on individual or collective values.

The top hemisphere of our map–the Social–symbolizes the emotional and mental planes of human systems design. The Eco-Cultural impetus arising on the left is met at the top by the Political-Economy impetus from the right. The Social Realm expressed as one human meta-field (cultural, individual and political combined as Social) completes the trine of the Ecology-Economy-Society trine found in many maps.

At the very top we find the individual and the issues involved in individual development and integration of the individual into society (a realm that the AQAL map excels at). Though Wendell Berry's quote at the beginning of this article states that individual health is meaningless outside the context of a community, it is also true that communities can only be as strong as the health and well-being of the individuals that make up the community. That said, the individual is symbolized at the top of the map–not because we uphold individuality as somehow ruling community–but because the individual stands tall only when fully supported by the healthy architectures of all it depends on below.

The Subcategories as Gateways to Research & Development

A quick study of each section shows that the subcategories are crafted to uphold this basic evolutionary model, as well as designed to be entry points into the next level categorization of the fields of study and elements of each section. With this care of subcategory headings, the map becomes the beginning of an Ontology of Regenerative Design.

Whole Systems Design Tools

At the center of the map are the design tools we then use to synthesize all of this differentiation of the sections and their subcategories. Remember that analysis is ultimately for re-synthesis. Represented in the center are the best processes we've found to make sure that the design of each element fits into the design of the whole that we are intending.

True Whole Systems Design - Integrating the Sectors

Whole systems design does NOT mean just having all elements "in mind" or present while designing one element in isolation. It means designing each element so that if fits into the whole. This is an essential differentiation between old and new paradigm design that many practitioners miss. This requires a sophisticated linking between the guild members and decision-making processes of each guild area. This is why VillageLab has chosen 4 primary Guild Areas–Ecology, Economy, Social (cultural and personal) and Political–so that we have a manageable number of guilds that we link together through our sociocratic design process.

Regenerative Whole Systems Design is an art and science that is just in its infancy and a subject worthy of its own article and much more. The Whole Systems Design schools of thought and design tools listed above are a great start on our journey back to Human/Nature reintegration. But we have a long way to go. It is our hope that this exploration of the mapping of whole systems and processes and our contribution to the field will help all along the way. Please accept our invitation to come map and design and play with us! The world is waiting!

Jeff Clearwater

Jeff Clearwater has enjoyed a 36-year career in community innovation, appropriate technology, business start-ups, not-for-profit development, and new paradigm economics. Jeff has served on 7 NGO Boards and founded 5 successful appropriate technology businesses as entrepreneur, engineer, designer, contractor, consultant, advocate, and teacher.

Jeff has provided consulting, design and installation of solar, wind, microhydro, micro-grid and integrated systems, water and waste systems, and electric vehicles for businesses, governments and organizations in over 15 countries.

Jeff was a founding Council member of the Ecovillage Network of the Americas, was Ecovillage Office Director at Sirius Ecovillage for 6 years, and co-founded Living Routes. Jeff now provides consulting in Whole Systems and Ecovillage Design - integrating his experience in technology, social architecture, permaculture, and a new paradigm in economics. Jeff has visited over 60 intentional communities and provided whole systems design to dozens.

Jeff and his Beloved, Ferananda Ibarra are co-founders of VillageLab - a social catalyst devoted to energizing the village meme in all aspects of society using the principles of Collective Intelligence and Metacurrency. Jeff also offers a workshop called Economics as a Path of the Heart. Jeff's blog can be found at The Visionary Commons.

http://villagelab.net

The 3 Phases of Project Management

Bringing the Vision into Fruition

Jessica Plancich

Description

Taking your Big Idea into reality through nurturing the 3 phases of project development and outlining your vision to most effectively engage your team and keep them inspired.

Like producing a feast for the royal court, creativity meets practical strategy with action steps leading up to the big day in a methodical and inspired way.

At each step, the health of the individual, the larger group and community, exchange with the earth and networks are imperative for this to be executed in an integral way.

Let's map out the steps involved in nurturing your Big Idea from ***Development to Implementation to Completion.***

Development Phase

This phase is dedicated to doing the legwork to set the foundational pieces upon which to build your platform. Though passion and enthusiasm would often drive us into the Implementation Phase quickly, this can be premature and create a shaky, weak infrastructure for the entire effort. It's best to take time to thoroughly define the questions posed below before initiating greater resources of time, energy and money.

What are we doing?

Defining the outcome and the target you're after is critical. What would a successful outcome look and feel like? Share this in clear and descriptive language. Outline the vision in writing as well as through the use of engaging visual layouts, a map or storyboard not only for the project/event itself, but how this fits into the larger, broader organization's mission. Here's where you craft your mission statement and core values.

Why are we doing it?

Create a compelling reason for participants to engage in a meaningful way that speaks to the group's core values, concerns and shared vision for the future. Here's where you craft your vision statement.

Where is it going to happen?

Putting the proverbial stake in the ground will help shape your teams and the resources available to create your desired outcome.

Who is going to make it happen?

Enrolling key leadership is essential in order for them to have the authority and support to build their teams. Clearly spell out a Job Description (include key roles and responsibilities, expectations and agreements around reciprocity and exchange), this will give your prospects what they need to make an informed choice. Of course, you get to sprinkle your charm and persuasive skills into the mix; however, they will be more apt to accept your seductive offer if they have full disclosure about what they are signing up for.

When is the deadline?

Before this is created, ensure that you have the spaciousness to see your team through these phases without undue rush or pressure. There is such a thing called ***eustress***—a useful form of stress and healthy framework to move a project to completion. The leadership team is wise to take all angles into consideration before driving hard and fast deadlines that create distress—the disruptive kind of stress that undermines integrity and team cohesion.

Implementation Phase

This is dedicated to putting the previous stage's research and development into action, where feedback loops are generated to inform the projects' progress. This will allow for micro-adjustments along the way to keep the team informed, focused, adaptive and inspired.

What

Shape your team's divisions or sub-committees by taking a meta-perspective on the kind of people power required to pull it off. Craft an organization chart that maps the organization/ project's divisions or teams and how they intersect and communicate with one another. Utilize user-friendly project management software (like Asana) to keep team members apprised of progress, updates and changes. Test the software's capacity and applicability before rolling it out to the team.

Why

Give cohesive messages in clear, understandable language that conveys the mission and motivation behind the efforts and the compelling reason behind your vision. Avoid verbiage that is obtuse or so forward-thinking visionary that it loses relate-ability. Utilize pictures and video to feed enthusiasm and inspiration to your team to convey the feeling state of what kind of outcome you're enrolling them in.

Where

Once a location is determined, it's imperative to create a line of steady and consistent communication with those who tend to the land. Assign a team dedicated to liaisoning with the land stewards (ideally some of the stewards would be on this team) to ensure that consideration is made for questions such as:

- What are the land's special needs? Water concerns? Delicate ecology?

- What is the projected impact of project on the land and how can you best address it for a more positive outcome?

- How can leadership best take these into consideration and support these needs, what would a win-win look like?

- How could the project or organization's presence actually heal or enhance this piece of the planet (think Leaving a Positive Trace)?

Who

Who are all the parties involved in the effort? Before you can effectively invite this cast of characters, first take your divisions/ sectors identified above and clearly identify who fits into them:

- Leadership

- Supportive team members

- Target demographic

- Surrounding neighbors

- Larger surrounding community

- National or multinational networks

- What are their special needs?

- How will they be best engaged?

Engaging and Empowering Leadership – Identifying Leads

When crafting a map of the organization/project, consider these as a place to start:

- Visionary + keeper of the meta-perspective (aka CEO)
- Production (aka CTO)
- Strategist, detail master + time tracker (aka COO)
- Sharing the vision to the outside world (aka Public Relations)
- Keeping the vision and cohesion for the team (aka Human Relations)
- Number master + logistician (aka CFO)

Before you invite people into a role that you'd like them to fulfill, consider not just their strengths, but how to best garner their aspirations, dreams and how to use this endeavor to support who they're becoming, not just who they've been. People will be much more likely to give of themselves, especially during a startup phase, if there are additional ways they will benefit from their efforts. Since at an early state finances are usually tight, consider some of the secondary gains and perks that your team could gain by choosing to come on board. This could include: camaraderie and the benefit of group creative synergy, training, expanding their personal portfolio and resume, creative freedom to experiment with new ideas and leadership development.

Clear Agreements

Enthused visionaries often engage their supportive friends and networks based upon their loyalty and desire to support, but without outlining what they can expect, relationships can falter and quickly deteriorate. Clear agreements are especially imperative with those that you are friendly with to preserve the sanctity of your relationships. Throughout the course of the project, relations can get difficult and strained when emotions are high and logic is low. Hence the importance of creating a contract that addresses the following *in writing*:

- Outline of roles and duties
- What is the exchange or reciprocity involved? What's in it for them?
- What kind of authority do they have? (i.e. executive decision-making or consensus decision-making)
- Who will they lead?
- What kind of support will they get? (training, guidance, fiscal)
- What kind of budget will they work with?
- What kind of accountability can they expect/how will their progress be measured? (i.e. performance reviews, key productivity indicators, quarterly reports and outcome measures)

Team Cohesion

After determining your divisions/sub-committees, inviting your leadership to participate and clarifying agreements, it's time to ensure that you put systems in place to ensure that the team you've built will last and weather the waters ahead. Close collaboration with your Human Relations lead is imperative to determine the following:

- What kind of communication container will you create? Meetings: in person, telephonic or web based? Frequency and focus of meetings: financial, operations, production, visioning? What kind of platform or software will you use to track your project management? What kind of container will you create to ensure clear communication and clear conflict?
- How will you encourage and inspire them into their greatness?
- How will you ensure balanced energy output and prevent burnout and promote self-care?

Importance of Flexibility and Adaptability

It's important to note that in the implementation phase, the systems you put in place will give valuable feedback on the health and harmony of the team and the progress of the project. Throughout this process, remember to strive toward that juncture between retaining focus on the target and adapting your expectations based on incoming information and emerging needs. An innovative leader rides the wave between and uses the genius of the team to take incoming data and course correct in flexible and creative ways to keep making forward progress.

Gripping rigidly to a fixed outcome can create distress in your physical body, create discord in your emotional state and impact your behavior and interactions with others. This can ripple outwards in multiple directions and have a negative impact on your team morale and degrade the integrity of the project. Taking on a project and leadership of a visionary organization is a deep and potent spiritual journey that few of us are prepared for in advance. Expect it to challenge and confront you, to stretch and ask you to evolve. Knowing this in advance can ease the discomfort that can and will arise as you pioneer into new territory. By surrounding yourself with trusted confidantes and those willing to do this inner work, you will receive the many gifts along this Road Less Traveled.

When

Along the way, be willing to adapt your timelines and adjust milestone expectations on emergent information. If you find that your team is consistently not meeting these, it may be time to reconsider whether they have the resources or capacity to meet the expectations and make adjustments accordingly. This may mean more training or a new leader is required or that more support (team members or financial) may be necessary.

Completion Phase

Projects have various timelines and milestones. Whether the duration is one week or an indefinite period, giving your team a sense of completion within the life of the project is important for measuring and celebrating success as well as making informed micro-corrections along the way.

Here are some considerations when a project or phase is coming to its completion:

- How will you measure progress and outcomes? (There are many resources on the internet that you can adapt for your project on Key Productivity Indicators as well as Outcome Measures)

- How often will you measure progress?

- How will you gain input from your audience/participants/customers? (satisfaction surveys, suggestion boxes, town hall meetings)

- How will you gain input from your team? (leadership performance reviews, anonymous satisfaction surveys, focus groups)

- How will this information be used to inform and shape the direction for the future? (feedback to leadership, new outcome measures, a plan of correction)

- How will you present this information to your team and to your contingency? (annual report, stakeholders meeting, strategy session)

- How will you celebrate milestone completions? (*This is not to be overlooked to keep the team's morale and inspiration high - get creative and innovative!)

On a Final Note...

Remember that deciding to birth a Big Idea into the world is a tremendous undertaking and requires both tremendous preparation and patience. This is an arena for massive growth on a personal, professional and spiritual level. Be gentle with your expectations of yourself and your team, as anything with lasting potential will be built with integrity from the foundation, even if it means shifting focus, timelines and roles. Remember to keep things light and playful as the game that it is!

Jessica Plancich Shinners

Jessica Plancich Shinners is a social architect and connector of people. As a reformed mainstreamist turned integrative psychotherapist, she brings over 10 years of individual, family and group facilitation, nonprofit management, leadership and program development, corporate consultation, and integrative systems experience. Deeply passionate about creating practical pathways to returning to all things wholesome and real, she has practiced these tenets as a founding member of The Emerald Village, an Eco-Village and intentional community in Southern California and is a founding member of the Tribal Convergence Network. Jessica supports grassroots, non and for-profit organizations to build morale, learn how to communicate in heartfelt ways and establish a corporate culture that supports the thrival of the mission and vision. She operates somewhere between practical priestess-ing and sexy strategy creation to bring the invisible into integral form. She thrives in nature, keeps herself balanced through movement (especially dance!), connection to animals and loves to create beautiful sanctuary spaces for people to experience nourishment from the inside out.

http://innerfinity.com

http://theemeraldvillage.com

Tribal Convergence Network

TRIBAL CONVERGENCE NETWORK

Tribal Convergence Network (TCN) is a synergistic web of social architects aligning with conscious individuals and organizations. We develop sustainable and generative whole systems design templates to cultivate harmonious alliances and collaborative interdependent communities. We envision a globally accessible system of education, resources and tribal technologies which bring people and organizations together into a network of trust for the sake of a thriving, symbiotic, sustainable, and continuous cultural evolution. We are no longer fragmented in our efforts to create change in the world we live in. When we feel that we are part of something greater than ourselves we can activate the potency of emerging as a global community connected to the whole system. Therefore, we are building community!

Through grassroots community leadership events, trainings, regional community development, open source models and templates, and a monthly newsletter and events calendar we foster participation at local levels while bridging to organizations and projects focused on efforts to cultivate healthy people and planet.

We value deep connection with natural systems, the deepening of community relationships, and putting our skills to use by building something of lasting value. We are inspired by envisioning and implementing new ways of living and caring for each other and continuously evolving our perspectives on society and our agreements.

Our dedicated mission is to discover, model and transmit the best practices for group and community inter-

actions to empower thriving local collaborative communities and to act as a beacon, connecting and facilitating the alliances best suited to bring forth regenerative global partnerships for the betterment of all.

Through our council we work with principals such as Code of Kin, The Way of Council, our Unifying Principle, developing Evolving Leadership Structures based on Whole Systems Design, including the Guild Model for Community Organizing all rooted in our commitment to our Core Values.

Based on the "Wheel of Co-Creation" developed by Barbara Marx Hubbard and Buckminster Fuller, the Guild Model is a community organizing tool that we remixed with permission by the originator. A Guild is a group of community members

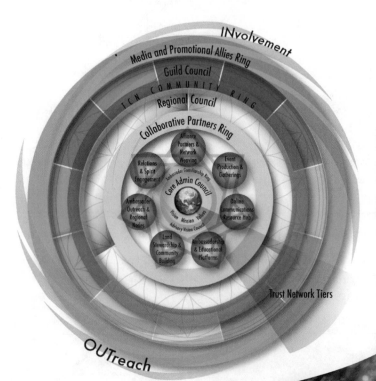

whose focus is on a particular sector of engagement in society, aligning their resources to support the vitality, impact, and success of all members of the Guild as they offer the gift of their purpose as part of a holistic community. This sector is their area of expertise or the field of their greatest passions and their choice to participate in this collective Guild interaction can elevate their own individual efforts as well as creating a synergistic elevation of the entire guild. Tribal Convergence Network stewards 15 Guild Group Facebook community pages, networking people with others to share resources with those who share similar interests.

http://tribalconvergence.com

7 Stages to Sustainability

Melanie St. James

A Framework for Locally-led Global Collaboration

7 Stages to Sustainability (7SS) is a universal "Asset-based Community-driven Development (ABCD)" framework.

While not explicitly named, the logical, ground up pattern in the 7 Stages to Sustainability (7SS) framework is inherent within the most successful social innovation and community development efforts and is the same pattern that nature itself uses to create and sustain life.

Identifying, assessing and thus revealing the impact of this under-recognized shared strategy can provide needed cohesion in a sea of chaos.

As a neutral ABCD framework that can address any issue a community or region faces, 7SS provides a context bridge between local and global perspectives, and aligns diverse development actors - NGOs, voluntary partners, local beneficiary groups and donors - around local needs, thereby reducing duplication and synchronizing efforts for greatest impact.

By enabling diverse partners to overcome technical, logistical, language and cultural barriers, 7SS is a catalyst for Public-Private Partnerships, cross-sector collaboration and increased civic participation.

7SS harnesses the power of the private sector to forge holistic solutions. Deliberately inclusive of six mutu-

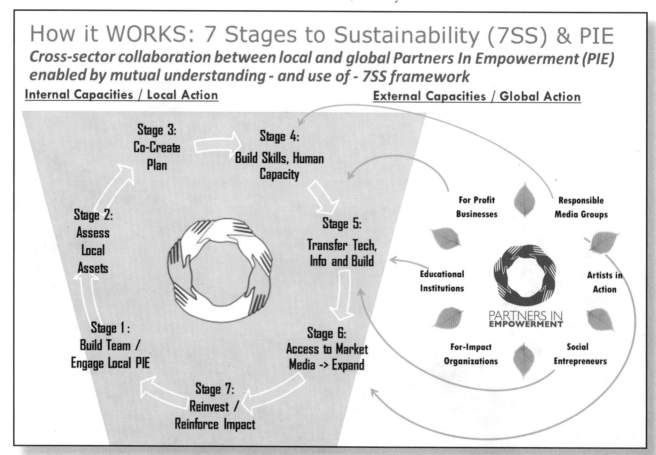

How it WORKS: 7 Stages to Sustainability (7SS) & PIE
Cross-sector collaboration between local and global Partners In Empowerment (PIE) enabled by mutual understanding - and use of - 7SS framework

Internal Capacities / Local Action — External Capacities / Global Action

Stage 3: Co-Create Plan
Stage 4: Build Skills, Human Capacity
Stage 2: Assess Local Assets
Stage 5: Transfer Tech, Info and Build
Stage 1: Build Team / Engage Local PIE
Stage 6: Access to Market Media -> Expand
Stage 7: Reinvest / Reinforce Impact

For Profit Businesses
Responsible Media Groups
Educational Institutions
Artists in Action
For-Impact Organizations
Social Entrepreneurs

PARTNERS IN EMPOWERMENT

ally-supportive social and economic sectors–including for-profit businesses, responsible media groups, artists in action, social entrepreneurs, for-impact organizations, and educational institutions–the 7SS system engages a full array of human and economic resources.

In creating a sustainable, repeatable, globally scalable, locally-led system for 7SS education delivery, particularly in the world's least developed countries, the 7SS Global Education Initiative is an opportunity to transform international development from top down to whole-system empowerment.

7SS - Summarized Below Internal/Local

Identify partners in empowerment: community members, local and international organizations, socially conscious entrepreneurs, and others who are working to address critical challenges to a healthier future, locally and globally.

Identify economic opportunities through collaborative assessment of local human and natural assets that can be developed to create sustainable livelihoods.

Together with community partners, co-develop a community development plan setting forth goals and measurable objectives where responsibility is shared between stakeholders.

External/Global

Support vocational skill development and other practical education.

Facilitate global transfer of critical information and appropriate technologies.

Increase access to local and global markets and use market knowledge to develop products and services that value and protect local heritage.

The Culmination

Deepen impact and self-reliance. Build creative Public-Private Partnerships and other sustainable, locally-based income streams supporting ongoing impact.

While not explicit, the 7SS framework is already widely in practice across the globe. It is inherent within the most successful social innovation and community development efforts and is the same pattern that nature itself uses to create and sustain life.

http://empowermentworks.org

http://theglobalsummit.org

http://bionova.org

Melanie St. James

Dedicated to building a thriving world from the ground up, Melanie's social change journey began at 20 in 1994 with a semester abroad in mainland China. Through field studies in Senegal and Zimbabwe, and being inspired by local social entrepreneurs, Melanie identified Empowerment WORKS' flagship approach to turning local resources into solutions, now called, "7 Stages to Sustainability (7SS)." In 2001, Melanie formed "Empowerment WORKS" as a global sustainability think-tank in action to advance 7SS and connect the world's most culturally rich, yet economically challenged communities with access to ethical markets, appropriate technologies and partners.

In 2007 after participating in the World Social Forum in Kenya, Melanie launched The Global Summit event series to unite eco, social and economic movements.

Melanie holds a Masters of Public Administration in International Management from the Monterey Institute of International Studies' Graduate School of International Policy Studies, and a BA in International Relations and Diplomacy from Schiller International University in Madrid, Spain. In 2013 she publicly released a medicinal plant based AIDS patent, laying the foundation for BioNova Health (http://BioNova.org). Melanie is trained in transformative mediation, Reiki, loves animals and is fluent in French and Spanish.

http://linkedin.com/in/melaniestjames

Moving Toward a Greater Evolution

The Emergence Process

Co-written and edited by
Patricia Ellsberg
and Sheri Herndon

The Emergence Process is an evolutionary and incarnational path guiding us from within to make the fundamental shift of identity from ego, "local self" to Essence, from our egoic separated self to our essential spiritual self, our True Nature. This is the healing of separation on the planet and is a fundamental step toward the evolution of our species. We discover that underneath all specific symptoms which feel so personal and unique, there is usually one fundamental source from which the particular problem springs. That is ego's separation from essence, or in traditional language, the human separation from God. Therefore, the prime solution to almost all our problems is the reunion of ego and essence. This is the shortcut to human transformation.

This process is based on the process that Barbara Marx Hubbard discovered through her own developmental template and embodiment practice. This story is shared in her inspiring book, ***Emergence: The Shift from Ego to Essence***. The process illuminates a sequence and set of deepening practices for overcoming the fundamental human illusion that we are separate from each other, from Nature and from Spirit. Ultimately, through practice to mastery, we connect to our Higher or Essential Self, the radiant core of our being, in an ongoing and authentically consistent way. This is our destiny. While there are many paths for awakening, this Emergence process, where we activate and embody our Divine-Human capacities, is a path that integrates with what we are already doing to heal ourselves and community. It is also a profound sacred technology for accelerating the evolution of our species. It is a lifelong journey.

This is the natural birth process of Universal Humans as we cross the threshold from our self-centered, self-conscious state of being into the next phase of our development.

In the course of ten progressive steps, we learn to gradually and authentically embody and become one with our

The coming of a spiritual age must be preceded by the appearance of an increasing number of individuals who are no longer satisfied with the normal intellectual, vital, and physical existence of man, but perceive that a greater evolution is the real goal of humanity and attempt to effect it in themselves, to lead others to it, and to make it the recognized goal of the race, in proportion as they succeed and to the degree to which they carry this evolution, the yet unrealized potentiality which they represent will becoming an actual possibility of the future. Sri Aurobindo

Essence, the Inner Beloved, so fully that we actually become the source of love and guidance we seek. This is perhaps the greatest and most nourishing love affair of our life.

Here are the Ten Steps of the Emergence Process expressed in very simple terms so you can get a feel for the journey:

1. *Entering the Inner Sanctuary* – Creating a safe and sacred inner space where you are protected and can contemplate the glory of your Essential Self; here is where you cultivate an inner receptivity to listen in deep contemplation on a regular basis. Here it is as if the patterns of the local self – the compulsive thoughts – are erased quickly by the vibrational field of the Beloved.

2. *Contemplating the Glory of the Beloved* – Here we begin to fully contemplate the Essential Self in its more personal form, the inner Beloved, and bring our focus to the specific qualities and wisdom of this all-knowing inner presence; we become the director of our attention.

3. *Incarnating* – Remembering our deeper identity and physically feeling this Essential Self as a vibrational field penetrating our body. The illusion of separation is dissolving. Falling in love with our Essential Self, an inner love affair is taking root.

4. *Inviting the Beloved to Take Dominion* – In which you as the local self invite the Beloved to become one with you and to allow the new vibrational pattern to nurture and support you. Here we cross the great divide from unconscious to conscious self-evolution and embark fully on the Way of Love.

5. *The Bliss of Union of the Human and the Divine* – In which you experience the ecstasy of union with your own divinity. Our divine essence is experienced personally as the creator within, the source and life pulse of the great creating process. The bliss of union accelerates the alchemy of our transubstantiation and incarnation.

6. *Shifting Our Identity* – n which you literally experience a new identity of your ego infused with the radiant love of your Essence. Here we are becoming a self-governing, sovereign person in the world. Be the Beloved!

7. Transferring Authority – In which you transfer authority and responsibility to your Essential Self and gain you own authentic power. As we learn to govern ourselves through inner authority, we become sovereign persons grounded in a knowing that we are an aspect of the infinite, non-dual reality. This authority opens us to our authentic power, a power which flows from the Essential Self and leads to power from within – true empowerment of self and others. In this stage, we are naturally maturing our own inner masculine and feminine aspects of ourselves.

8. Educating the Local Selves – In which you heal the sacred wounds and transform critical inner voices through the unconditional love of the Inner Beloved. We learn to create a new center of gravity in ourselves.

9. The Repatterning of Life – In which you let go of what no longer serves you and open to the new. Following the "compass of joy" through the darkness of our confusion and allowing our life to repattern itself to a higher order, one that is more resonant with our inner values. When our separated minds quiet down, our deeper self, which is one with the universal organizing intelligence, comes forth. This emergence is the purpose of meditation, and it is also the process of incarnation and co-creation. We learn to sustain the internal resonance of our inner state as we work in the world. We learn to maintain resonance with others, doing the work together. In fact, being and doing blend together seamlessly.

- *Allow your Life to Repattern*
- *Self-Presencing*
- *Radiate the Presence*
- *Realize that you are fully Response-Able*
- *Cultivate Resonance by Attuning with Others*
- *Nourish the Beloved*

10. Fulfilling the Promise – In which you become a whole and co-creative being living as Essence in Action. Like our forerunners in the cultures of the past who separated out youths from the tribe at puberty and initiated them in the ways of adulthood, so we initiate ourselves together in the ways of the evolving human growing in us now. We celebrate within us the shift from the creature human to the co-creator. Being the Essential Self in person is the ultimate service to humanity, because then we transmit that essence to others spontaneously, helping others to shift from ego to essence by our presence.

These seed patterns are stepping stones toward incarnating our divinity in an ongoing, ever-deepening process. We are all on a developmental path and we can choose how deep we want to go. Our invitation is to feel them and discover where you are on this spiral path and to step into a deeper relationship with the Emergence Process so we can embody our divinity here and now. These capacities explored in each of these ten steps are essential for the future of our species and our remembering of who we truly are.

The Essential Self education begins when we are no longer looking outward to external principalities and powers but are experiencing the Bliss

312

of Union, the Shift of Identity and the Transfer of Authority as an ongoing, ever-deepening process. In continues as we are willing to learn responsibility for our own emergence and fulfillment as Universal Humans. We are no longer victims. There is no one to blame. We become co-creators of our own reality, entering in an inspired dance between the local selves, the Essential Self, other people, and the larger Reality that is informing us all. The choreographer of this divine dance is the universal designing intelligence incarnating in and as us through the Emergence Process.

All of us now attracted to the path of our own emergence should realize that we do not do this for ourselves alone, nor do we do it as ourselves alone. We are actually an expression of the Great Creating Process itself, finally expressing consciously through and as Universal Humans on planet Earth. The Essential Self is the personal expression of that Process incarnating in each of us, as us. Our maturation as Adult Universal Humans holds within it the fulfillment of the promise given to the human species through its great avatars, seers, and visionaries. We are to be the promise fulfilled, the end of a long struggle of life itself in the first chapter in the history of the world and the beginning of the second chapter which will tell the story of the appearance of new forms of life, co-creative with the process of evolution itself, never before seen on planet earth.

Excerpts with permission from the book *Emergence: the Shift From Ego to Essence* by Barbara Marx Hubbard.

To learn more about the Emergence Process and the book, Emergence, visit http://evolve.org. To learn about courses, teleseminars and coaching around the Emergence Process, contact Patricia Ellsberg at http://PatriciaEllsberg.net. She is the lead teacher of this process worldwide.

Patrica Ellsberg

Patricia Ellsberg is a social change activist, meditation teacher, coach and public speaker. In their first year of marriage in 1971, she helped her husband, Daniel Ellsberg, release to the press the Pentagon Papers, a top secret history of US involvement in Vietnam, which helped end the Vietnam War. She teaches and works closely with her sister, Barbara Marx Hubbard, in furthering social transformation through Conscious Evolution and bridging spiritual experience with social activism.

Sheri Herndon

Sheri is an evolutionary social architect at the nexus of social networks, conscious evolution and co-creation. For the last 20 years, she has worked in the field of evolving consciousness and creating new forms for culture to be fully actualized, including founding a global media network, inspiring synergistic whole system design, curating evolutionary best practices, and providing high-level strategic consulting. In service to making visible the new emerging paradigm, Sheri produces and hosts a weekly podcast called "Heart of It All." She holds an MA in Interdisciplinary Studies from the University of Washington.

Inhabiting the Future

Robert Gilman

As we reinhabit the villages in our lives, we are surrounded by a deep sense of the world in crisis. The Chinese character for "crisis" is derived from the characters for danger and for opportunity. Which of these is your stronger motivation as you face the future?

In this chapter I'd like to share a perspective on our times that I've come to after being immersed in the sustainability movement and ecovillage development for over three decades. It may surprise you that I'm actually quite encouraged about the future for the Earth, for humanity and for all of life – "encourage" in the sense of filled with courage and drawn forward by the opportunities. Yes, I'm well aware of the dangers and I don't expect the process to be easy – especially over the next few decades – yet we must take a long view.

The quickest way to understand that long view starts with this "outline of history":

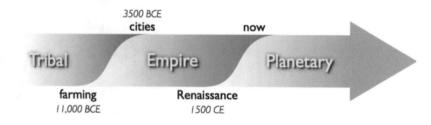

This image divides human history into three major "eras" with two transitions between them. Here are some of the major changes as we move from one Era to the next.

Characteristic	Tribal	Empire	Planetary
Main Livelihood	Hunting & Gathering	Agriculture	No more "main"; huge diversity of occupations
Basis For Social Organization	Kinship	Violence-enforced, Religiously-sanctioned Hierarchy	Self-organizing consensual collaboration
Communication Level of Development	Oral Stories, Songs and Verse	Elite Literacy	Widespread electronic multi-media

WE DON'T NEED TO WAIT.
WE CAN START INHABITING THE PLANETARY ERA
IN OUR CONSCIOUSNESS AND IN OUR EXPERIENCE NOW.

The Planetary Era will also need a sustainable relationship with the natural world. How do I get the characteristics for the Planetary Era? By looking at trends and by analogy with natural systems, as I explain in ***What Time Is It?***

Here's an example based on how an ecosystem, like a forest, develops over time:

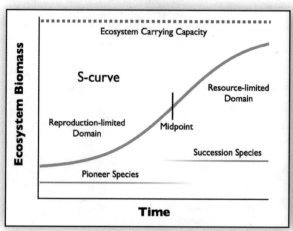
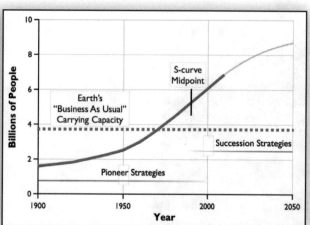

The change from the early, reproduction-limited domain to the later, resource limited domain, while gradual, is actually quite dramatic – and best seen in the attributes and success strategies of the pioneer and succession species. They are almost mirror opposites of each other:

Attribute	Pioneer	Succession
Resource Consumption	high	low
Resource Efficiency	low	high
Growth Rate	high	low
Longevity	short	long
Diversity	low	high
Complexity	low	high
Relationship With Others	isolated	cooperative

How does this apply to our world and our future? Around 1990, the human population shifted from a "pioneer species" time to a "succession species" time.

Old beliefs and pre-1990 experience create a cultural lag, but eventually we will have to adopt succession species' strategies.

What about climate change or nuclear war or some other threat? Won't one of these kill off humanity? It's possible. These are all serious issues. However, they all have well understood technical solutions. The obstacles to implementing those solutions are all human. It is in our hands – and in our power – to avoid catastrophes of our own making.

Rather than despairing about things that may or may not happen in the future, let us choose to work now for the best future possible.

Many of the human obstacles to moving fully into the Planetary Era are subtle and widespread. Our daily life is already more Planetary than Empire but our minds have been shaped by ideas and language that we've inherited from the Empire Era.

Here's a comparison of some of those subtler differences:

Empire	Planetary
Linguistic	Linguistic, visual, kinesthetic, etc
Categorical, polarized	Pie charts, gradients, continuous variation
Linear, hierarchical	Multi-dimensional
View life as a power struggle	View life as discoveries to make, puzzles to solve, mysteries to live
Moralistic, absolute, redress-oriented	Systemic, contextual, healing-oriented

For example, the Empire mindset is strongly oriented to categorical thinking – something is either A or B, black or white. So, when interpreting the Outline of History chart above, it is easy to think that each era completely replaces the one before it. The Planetary perspective is more like this:

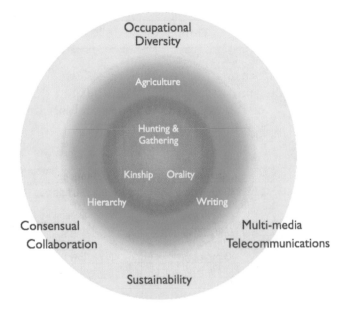

As each new era comes along, it goes beyond the characteristics of the old era but also includes and transforms the old characteristics. For example, kinship never went away but it lost its dominance in the transition to the Empire Era. Likewise, hierarchy doesn't go away in the Planetary Era, but it loses its dominance and is no longer violence-enforced. Functional hierarchies are just another form of consensual collaboration.

It will likely take decades, likely generations, for society as a whole to complete the transition into the Planetary Era. We will know we have truly arrived when we have ended warfare as an institution – something that many futurists, myself included, expect will happen in the 21st century.

But we don't need to wait. We can start inhabiting the Planetary Era in our consciousness and in our experience now.

Much of the material in Reinhabiting The Village points in the right direction. However, if your image of reinhabiting is filled with a sense of going back to a supposedly simpler, happier time in history, you may be quite surprised in the times ahead. Our world is vastly different than it was even a century ago, and even more so than when most of our culture's ideas were formed:

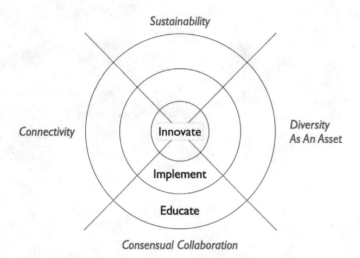

So how can you start inhabiting your own bright future now? Think like a succession species (see the table above) and then find a way to innovate, implement and/or educate about new cultural forms that increase sustainability, improve connectivity, treat diversity as an asset and/or increase consensual collaboration.

If we do this, we will be working with the momentum of the times, the future will lose it dread, we will be catalyzing a wonderful future for all life… and we will have a lot of fun doing it!

Robert Gilman

Dr. Robert C. Gilman, Ph.D. is a sustainability pioneer and cultural midwife. He is the President of Context Institute and the Founding Editor of IN CONTEXT, A Quarterly of Humane Sustainable Culture.

Trained as an astrophysicist, Robert decided in the mid-70s that "the stars could wait, but the planet couldn't." He turned his attention to the study of global sustainability, futures research, and strategies for positive cultural change.

He was instrumental in the founding of the Global Ecovillage Network in the early 1990s and lived for three years in Winslow CoHousing, one of the first cohousing projects in the US based on this Danish model for community living.

He is the co-author of the Household EcoTeam Workbook and Program, which became a national program in the Netherlands. He served as City Councilman in Langley, Washington, and worked with the American Institute of Architects on issues regarding sustainability and the built environment.

Dr. Gilman is currently immersed in applying the breadth of his knowledge to creating a training program (Bright Future Now) and core curriculum (Foundation Stones) for 21st Century change agents.

http://context.org
http://ecovillage.org

Conclusion

> ## I ALONE CANNOT CHANGE THE WORLD, BUT I CAN CAST A STONE ACROSS THE WATERS TO CREATE MANY RIPPLES.
> Mother Teresa

So here we are at the end, except for "ReInhabiting the Village: CoCreating our Future," this is just the beginning. Like the cycle of the seasons show us, some things end and yet others begin. And even after all of this time, we - Humanity - are still young in our journey. We have much to learn.

This book is an artifact. The stories shared here are a glimpse of a few voices, yet the "Voices of the Village" are many. Through the website we invite you to continue the journey with us. YOU are invited to participate in this movement. Peruse the interactive Ebook, write a blog on the community Blog Page, add an event on the Community Calendar, find mentors and consultants to help you with your project, feature your project/organization/business on the Resource Directory. As more information, more practices, more perspectives are shared, ReInhabiting the Village will continue to be a co-created space of discovery, a living library of wisdom built by community, for community. The Village belongs to us all and we will continue as stewards of this collective Vision.

What lies before us is determined by us. Each action, each step, each voice, each choice is part of the story we all share. What story will you tell? What story will your great-grandchildren tell? Let it be a story of healing, of truth, of responsibility, of trust, of transformation, of peace, of creativity, of dedication, of compassion, of freedom. Let it be a story of Love.

In Service,
Jamaica

Jamaica Stevens

Creator and Author

Jamaica Stevens is the Author and Project Manager for "ReInhabiting the Village: Co-Creating our Future". As an Organizational Design Consultant, Jamaica works with Whole Systems theories empowering people, projects and organizations to THRIVE! She is also an experienced event producer, workshop leader, group facilitator, community organizer, project manager, writer and life coach. She is an Associate Consultant with VillageLab, the Creator of Tribal Convergence Gatherings, Executive Producer of Awaken Visionary Leadership Summit, Co-Founder of Tribal Convergence Network, Organizational Consultant and Producer with Lucidity and other festivals.

With a focus on urban and rural community Village projects, Jamaica is devoted to revealing the genius of each human and sharing collective intelligence to co-create solutions that benefit all.

http://ReInhabitingthevillage.com

http://Tribalconvergence.com

htpp://Jamaicastevens.weebly.com

TRIBAL
CONVERGENCE
NETWORK

Julian Reyes

Executive Producer

Julian Reyes's experience reflects a lifelong dialogue with computer graphics, film, music, and technology. Julian formed Keyframe-Entertainment in 2004 as a music label and artist management company. Currently, Keyframe is a media network that creates global positive change by inspiring, informing, and entertaining through Transformational films, Visionary Art, electronic music and festival culture. Keyframe produces, finances, and distributes cutting edge projects in order to generate growth for its partners and strengthen community-building worldwide.

Keyframe is the Executive Producer of Andrew Johner's "Electronic Awakening" and Jeet-Kei Leung's "The Bloom Series Episode 3: New Ways Of The Sacred." Julian is honored to be joining "ReInhabiting the Village" as Producer and Contributor, and hopes that this work offers audiences a view of the potential of visionary culture.

Julian serves on the Board of Directors of the Electronic Music Alliance (EMA), a collaboration that cultivates social responsibility, environmental stewardship, community building and volunteerism.

http://keyframe-entertainment.com

A. Keala Young

Content Editor

A. Keala Young is a whole system designer, regenerative practitioner and permaculture teacher with a background in the healing arts. His work on ReInhabiting the Village as editor and contributing author is an opportunity to further his life's mission to support the integration of living systems coherence for individuals and communities in the Global Village.

Since 2005, Keala has been consciously committed to the path of co-creating living & learning communities and has studied far and wide, gathering the seeds and skills to undertake the adventure of realizing models for the future now. He has over a decade of experience presenting and curating educational content for festivals and is a cofounder of the budding Atlan Ecovillage in the Columbia Gorge. A steward of Gaiacraft, Keala also serves on the Board of Directors of CultureSeed, an educational nonprofit with the mission to inspire and support thriving global communities through experiential learning programs and transformational events.

http://atlancenter.org

http://cultureseed.org

http://gaiacraft.com

Natacha Pavlov

Copy Editor

Natacha is the Copywriter / Communications Assistant at Keyframe-Entertainment. She is also a writer whose literary style blends elements of creative nonfiction and historical fiction with doses of spirituality, magical realism and myth / folk tale elements.

She is honored to be using her literary background to support the "ReInhabiting the Village" book project as Copy Editor. As a being committed to lifelong growth and learning, she looks forward to learning about the multiple facets of the social movement directly from instrumental members and proponents of this vital community.

http://natachapavlov.com

Davin Skonberg

Graphic Curator

Davin Infinity is a Creative Director with 18 years of graphic design and book publishing experience. His life's work and creative play is at the intersection of art, social psychology and culture design. He has traveled the world visiting and studying sacred sites and cultures as well as visited over 50 eco-village and new paradigm societies like Damanhur & Auroville.

Davin has also spent the last 15 years training in several lineages of shamanism, martial arts and Mystery Schools. Davin is an experienced and powerful metaphysical guide into the Quest of human potential and transformation. This work is a major influence in his films, workshops and visionary art.

http://shamaneyes.net

KEYFRAME
ENTERTAINMENT

Dana Wilson

Graphic Curator

Dana has lived and worked internationally for nearly a decade as an independent photojournalist and media producer. She co-founded Mortal Coil Media in 2007 and produced a series of creative documentaries - telling stories from the margins with sensitivity and integrity. Though still focussed on themes of social justice, environmental reverence and peace building, she has shifted energy toward organizing events and documenting the efforts of soulutionaries as we transition to a more resilient society. This includes active engagement in the coastal Transition Town collectives, board membership with Salish Sea Productions, and stewardship with Gaiacraft - an international collective of permaculturists engaged in ecological outreach.

She believes deeply in community council using non-violent communication techniques, grounding in with nature, and the healing power of regular meditation, love and compassion for all our relations.

She is currently completing an advanced Permaculture Design Diploma in media throughPermaculture Institute USA, while co-creating an ecovillage on the Sunshine Coast, BC.

Jonah Haas

Social Media and Marketing

Jonah Gabriel Haas is a messenger, a cultural anthropologist, a lucid dreamer, and a successful entrepreneur. After completing a double BS in Marketing and Anthropology from the University of Maryland, College Park, he went on to earn an MA in Cultural Anthropology from the University of California, Santa Barbara. During this time he focused his studies on the anthropology of tourism, community-driven resource management, and the social theory of global capitalism. Currently, Jonah is the Marketing Director and Co-Founder of Lucidity Festivals LLC, which is now in its 4th year of transformational festival production in Santa Barbara, CA.

As an ambassador of personal, communal, and global transformation for the benefit of our One Human Family and Mother Earth, Jonah is committed to facilitating the emergence of activated land nodes. He envisions these thriving hotspots of human potential as interconnected conscious communities acting as educational institutes and sustainable models for New Earth village living.

Antje Martina Schaefer

Curation Support

Antje manages communication with contributors and the submission process for Re-Inhabiting the Village. She is honored and excited to collaborate with and support a network of change-makers and is dedicated to connecting people with valuable tools and resources in their interwoven paths to create a better world.

West Coast festival culture and the rich creative community of San Francisco provided Antje with the perfect playground to discover her passion and gift for bridging and cross-pollinating ideas between communities. She is the creator of Connexus Dance, a platform for transformational urban gatherings furthering the movement for authentic connection through conscious partner dance. She is based in Portland and teaches Cocréa - a holistic partnered dance and movement modality she co-creates with her partner, Wren LaFeet. Together they teach workshops and retreats throughout the U.S. as well as internationally.

http://connexusdance.org

http://dancecocrea.com

Kelli Rua Klein

Designer

Kelli is a storyteller. She has extensive experience over 20 years in the graphic design, brand marketing and business operations fields. After earning a BS in Graphic Design, she discovered she also held a love for story, process, and organization. Her professional path has led her through a diverse 18 year tenure at Nike Inc, where she held global and regional operations and management roles.

In 2013, she resigned from Nike to start Bridgewalkers LLC; a consulting business with a mission to support individuals and organizations with their organization and storytelling. She leverages visualization tools and techniques to deliver creative messages that bridge art, audience, culture and whole systems design.

As a world traveler and lifetime student of nature, art and culture; Kelli became interested and active in a variety of transformational festivals and intentional communities. She is dedicated to building and participating in communities that steward the land they reside on and helping build educational and storytelling methods and models for our future.

http://bridgewalkers.com

http://linkedin.com/in/kelliruaklein

Jessica Perlstein

Jessica Perlstein is a visual artist based out of San Francisco, California. Her work carries a strong message of taking action towards respecting a thriving planet and the resulting reality we can create when we choose to be aware of our symbiotic relationship with nature. Her latest work emphasizes a movement toward environmental sustainability, merging ecological technologies and innovations within our current society to illustrate a thriving community.

More work can be seen at

http://dreamstreamart.com

http://facebook.com/JessicaPerlsteinArt

Amanda Kay Creighton

Amanda is an explorer, writer and supporter of sustainable and intentional communities. After traveling around the US on a bicycle and visiting 100 communities, she co-created the film Within Reach, which documents ecovillages, cohousing, coops, Transition Towns and more throughout the United States. Since making the film, she has lived in either a teepee, cob house or tiny home at Hummingbird Community and Dancing Rabbit Ecovillage, and continues to visit and document communities worldwide.

Watch the film at

http://withinreachmovie.com.

Autumn Skye Morrison

"I feel a deep stirring, a potent blossoming of creativity. I believe this sacred duty to share inspiration is especially crucial at this challenging and exciting time of change in the world. As we follow our personal creative paths, we contribute vitally to the whole. We get out of our own way, and give permission to our unique expression. As in any form of art, there is a profound potency when we are able to release judgment. Because of this, I constantly strive to move beyond mind, breathe deep into the process, and find bliss in each step. I realize that I am supported in my authentic expression, and understand that the art is not born from me, but through me. In this knowing I am both humbled and empowered. Each canvas takes me on a journey, and as my paintbrush follows, each time I am lead back to my centre.

I offer my artwork as a mirror to reflect your divinity. May each painting be a reminder of the rich depths of inspiration available within your being and beyond your perception. My artwork is an invitation to allow your own expression to flow from the boundless wellspring of your spirit."

Autumn Skye has been painting since she was old enough to hold a brush, developing a deep wonder for the world around her. Amidst journeys to explore, teach, and share inspiration worldwide, Autumn Skye otherwise lives and paints in her home on the Northern Sunshine Coast of British Columbia, Canada. In her studio, she overlooks a sweeping expanse of ocean, islands, mountains, and sky. She considers herself among the blessed of the blessed.

http://autumnskyemorrison.com

Dana Wilson

Darren Minke

Dana has lived and worked internationally for nearly a decade as an independent photojournalist and media producer. She co-founded Mortal Coil Media in 2007 and produced a series of creative documentaries - telling stories from the margins with sensitivity and integrity. Though still focussed on themes of social justice, environmental reverence and peace building, she has shifted energy toward organizing events and documenting the efforts of soulutionaries as we transition to a more resilient society. This includes active engagement in the coastal Transition Town collectives, board membership with Salish Sea Productions, and stewardship with Gaiacraft - an international collective of permaculturists engaged in ecological outreach.

She believes deeply in community council using non-violent communication techniques, grounding in with nature, and the healing power of regular meditation, love and compassion for all our relations.

She is currently completing an advanced Permaculture Design Diploma in media throughPermaculture Institute USA, while co-creating an ecovillage on the Sunshine Coast, BC.

I create art in several different mediums. I make my stain paintings with wood panels that have a combination of stain, oil paint, and gold leaf applied to them in order to bring out the images I see in the wood grain. My digital art is created with photographs, often 100 or more, that have been extracted and collaged together with photoshop. I also have several oil painting that were done earlier in my career, as well as different series of photographs.

http://facebook.com/DarrenMinke-Art

http://holylove.tv/2014/05/darren-minke/

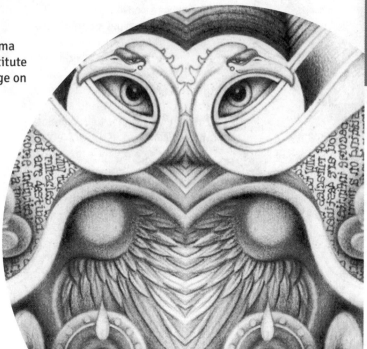

Dustin Engleskind

Dustin has been a self-employed Professional Touring Videographer and Photographer for over the last 6 years.

He has extensive experience in Directing, Engineering, Projection, LED Walls, and Media Servers. He provides advertising and speciality shoots for mostly corporate clients and award shows.

He is also a entertainer in the circus as a performer, event production, and photographer. This and the circus arts are his creative expressions.

http://obsidianproductions.com

Aaron Cyrus Dorr

Aaron founded ACDImaging in 2001. He specializes in all aspects of Photogrpahy and Film Production. He is a Film Maker, Photographer, Live Action Videographer, DVD Author, Editor and Artist.

http://acdimaging.com

Eric Nez

Painting was a gift given to Eric Nez at an early age. His parents (an artist and a Yogini, both lovers of life) ensured that he had plenty of tools to express his innermost realms. A devoted artist-mystic, Nez is both a visionary and a cultivator of the soul. Tending the garden of personal transformation offering to the sacred self: nurturing environments, wholesome nutrients, divine light, and love in order to grow the fruit of the spirit and share it with the Universe.

The Muses of yoga, plant medicines, alchemy, and magic lay their offerings at the feet of Nez. His creative process is deeply influenced by transpersonal psychology and the spiritual sciences, as well as intergalactic experiences. The spirits of the mystics all gather around Eric's studio to whisper in his ear and guide his hand as he paints.

The art of Eric Nez is both a stepping-stone and a chronicle of his evolution of consciousness. Explorations of distinctly different worlds, powerful life experiences, and teachings from ascended masters all of these find a home in the work of Nez. His every painting is offered up to the universal collective as a catalyst for transformation.

http://ericnez.com

George Atherton (GeoGlyphiks)

Art is a tool for transformation, on both a personal and planetary level. We can use the arts to express our realities, promote and reflect social change, and connect more deeply with ourselves, each other, and the Earth. From the Lascaux cave paintings, to the Egyptian hieroglyphics, to current-day infographics, it has ever been the role of the arts to preserve and promote knowledge.

In my art, I work with visions of a harmonious, ecologically thriving world and Awakened states of being, in the hope that viewers of my work will joyfully manifest these states of being in their own lives. A truly open Internet makes it possible to share creative inspiration with vast numbers of people through the arts, and to that end I tag my own with links to the information that inspires it.

I work primarily in the digital medium. After drawing concepts in my sketchbooks and dreamjournals, I fully realize those concepts using an electronic drawing pad and digital ink.

I draw inspiration from a variety of related subjects, including meditation & Yoga, mixed martial arts, lucid dreaming, comparative mythology, culture jamming, bright green innovation and permaculture. Through these practices, I find the empowered center where creativity surges.

http://geoglyphiks.com

Ishka Lha

Ishka Lha was born in 1982, and grew up on a ranch atop the rolling hills of northern California. She has been making art for as long as she can remember, and received significant artistic instruction and encouragement from her mother and father, both professional artists and architects. At university, she earned a Professional B.A. in Architecture, but promptly set her course for a career as a fine artist after graduation. She realized what she most wanted was to design new worlds on canvas. Ishka's art explores and celebrates the natural radiance inherent in all dimensions of life through the use of neo-traditional iconography, geometry, dreams, ritual, legend, and prophecy. She is especially interested in what archetypes and mythologies can most govern our passage through time into greater awareness and heartfelt livelihood. She is part of a growing visionary art movement that is dedicated to the happiness of all beings, and approaches her creative process as a spiritual practice. Her work is primarily featured in private collections, galleries, gift shops, and transformational art and music festivals throughout the U.S. and abroad.

http://ishkanexus.com

http://facebook.com/pages/
The-heART-Spirit-of-JAH-Ishka-
LHA/110087709023164

https://instagram.com/ishka_lha/

Ka Amorastreya

Kathryn June "Ka" Amorastreya, born in west Texas, creatively expressed herself from a young age. Her passion and purpose was ignited when she began studying healing arts, and practicing massage therapy. It was then she discovered the power of healing, and the beauty in helping people improve their quality of life. Weaving this love for healing into her artistic expressions, she began making tools and adornments from elements of nature. She started teaching herself to paint at the age of 22, and found herself revealing powerful visions, creating her own unique language of beauty. Meanwhile, she developed her expressive talents in the performance arts. Ka continues painting, dancing, and creating as much beauty as she possibly can. She has danced and offered her art at events and galleries internationally since 2002, and her work has been featured in a number of mainstream publications. Ka is also a founding director of the Visionary Arts Foundation.

"The aim of my art is to bring more beauty into this world. When I experience beauty, I feel my heart open. When I have an open heart, I experience more love, peace, joy and healing in my life. At times, I feel bombarded by the fear and violence propagated by a conflicted humanity, and the only way to truly find my balance and power again is by generating more beauty through art. It is my mode of prayer, and my devotional response back to the perfect beauty of nature, the love songs of Great Spirit. My hope, is that when people view my work, they feel their heart soften, even if just a little, so that they may feel more at ease. I wish for my art to be a mirror, to reflect the viewer's own Divinity and beauty. I believe we are all Divine and sovereign, and that we, collectively and individually, have the power to create the beauty we wish to see in the world. I hold great reverence for our collective potential to bridge Heaven and Earth through our love, imagination and creativity.

http://serpentfeathers.com

Mark Goerner

Drawn at an young age to conceptual design and problem solving, my sketchbooks and early projects reflected an industrial design sensibility that extended into fantasy environments with a futuristic mindset.

Exposure to manufacturing during this time helped fuel an interest in machinery, an interest in varied forms of fabrication, and a curiosity towards materials. This, coupled with a love of nature on both a macro and micro scale, archeology, architecture and automobiles, helped form my base of interest.

These interests were fed in later years by attending classes at the Art Institute of Chicago, Rhode Island School of Design, and later, Art Center College of Design in Pasadena where I received a Bachelor of Science with distinction in automotive design, with a focus on entertainment design. Internships included work for Coca-Cola, and BMW/Designworks contributing on various conceptual automotive and product design projects.

The freelance career began by providing design services, illustrations, storyboards, character and concept development to post and pre-production studios such as, New Line Cinema, Disney, Fox, Rhythm and Hues as well as industrial design work for BMW/Designworks, Intel, Toyota, and Honda. Then I dove into the feature film industry as a conceptual illustrator for: "Minority Report", "Constantine","X-men 2", "Superman", "Iron Man 2", "Thor", "Bunraku", "Fantastic Voyage", "Avitar" and "Battle Angel Alita". These experiences lead me to developing training DVDs on digital painting, co-authoring "Concept Design 1-2", and to teach seminars around the world on design methodologies along with classes in visual communication, production and entertainment design at Art Center, Gnomon and locally. Recently, I co-designed the world's largest wheel for Harrah's casino, Las Vegas which opened in 2013.

2012 marked the beginning of my venture into starting a design studio developing artful products of exceptional quality, venturing at times from the purely digital world in the process. The creation of this new studio and involvements in immersive installations for festivals is taking me back into the built world while still delving into conceptual projects and inspirational visions of spaces not yet known.

http://GRNR.com

Krystal Smith
(Krystaleyez)

Krystleyez communicates a clear vision of unity to the world through the gift of her sacred artwork. Through the synergy of imagination & information, spirituality & science, dream & reality, her paintings offer cosmic keys to access mystical realms of consciousness.

http://krystleyez.com

Lunaya Shekinah

Healer, new media designer, educator, mystic, singer, healing artist, spiritual student-teacher and cultural creative, Lunaya Shekinah is a dynamic, awakening soul weaving an artful understanding of nature, spirit and technology together into the fabric of emerging planetary cultures. Lunaya founded Light Science, an arts, healing and education mystery hub, for the expansion of consciousness and acceleration of divine awakening for humanity and Earth.

http://lightscience.ca

Mark Henson

Mark was born in 1952 and raised in Northern California, and attended University there, receiving a degree in studio art in 1973. He has been a professional artist ever since, working with paint, metal, wood and computers. He has co-founded several gallery ventures, traveled extensively in Asia and elsewhere, and now operates Sacred Light Studio from his home out in the country north of San Francisco, and maintains a small artist's retreat in Costa Rica.

About his art and process he says:

"My art tends to be somewhat narrative- I like to tell a story or show a state of emotion or consciousness with my images. I often begin with some sort of idea or theme. This theme might be suggested by anything at all, or may just drift into the consciousness of an artistic inspiration during my day. I am often asked if I receive these images while dreaming asleep. Sometimes they do come in dreams, but usually they are found floating around in my mind, just waiting to be noticed..

While I tend to be a down-to -earth person, when activating my creative energies I begin by looking deeply within myself, and seek to express what I may find waiting there. My sincere wish is to tap into the Divine Source of Being, to Consciousness, to Spirit, or whatever you may call it- that place where existence comes from, and to bring into visual reality images manifesting the knowledge revealed while in this presence. Strangely enough, I have discovered that the more intensely personal my vision is, the more widely it resonates when presented to the world.

I believe that art can have the ability to catalyze positive social and cultural changes. In addition to stimulating our visual cortex, Art has the amazing magical power to evoke profound emotional intensity as well as to provoke intellectual thinking. My aspiration as an artist is to create compelling images of beauty and power that serve to promote our Conscious Evolution as human beings, and to show us how to live in a peaceful world. To this end I like to present images exploring themes of Awakening Consciousness, Divine Sexuality, Political Realities and Living in Harmony with Nature."

http://facebook.com/pages/Mark-Henson/47191016712?ref=hl

Cover & Introduction

The front and back cover was designed by Kelli Rua Klein, with digital collage art by Davin Skonberg and photography by City Repair

Opening Page Photo - Amanda Creighton, background transparent art by Krystleyez

p. 2-3 Amanda Creighton, background transparent art by Krystleyez

p. 4 background transparent art by Krystleyez

p. 6 Trees by Aaron Dorr

p. 7 She Sees All by Ka Amorastreya

p. 8,9 Pollination, Photo by Dana Wilson

p. 10,11 Dance of the Butterflies Photo by Dana Wilson

p. 10 Earth Day painting by Autumn Skye

How to Use This Book

p. 12-13 The 12 Icons were drawn and designed by Kelli Rua Klein in partnership with Davin Skonberg, La Laurrien and Jamaica Stevens

Chapter 1 Heart of Community

p. 14,15 Mystic Gaia, painting by Krystleyez

p. 16 Photo from Dustin Engelskind

p. 17 Photo from Dustin Engelskind

p. 17 Bee painting by Jessica Perlstein

p. 18-19 Photo from Debra Giusti

p. 20-24 Photos courtesy of Alberto Ruz

p. 20-25 Co-Creating the Future by Alberto Ruz

References:

Buenfil, Alberto Ruz. Rainbow Nation Without Borders: Towards an Ecotopian Millennium.

New Mexico: Bear & Company, 1991. Print.

p. 21 Snow Labyrinth Photo by Kelli Rua Klein

p. 26 Photo by Dana Wilson

p. 27 Photo used with permission from Mark Lakeman of City Repair

p. 28 Food preparation in Malaba, Photo by Dana Wilson

p. 30 Photo by Dustin Engelskind

p. 30-31 Neotribal Culture by Andrew Ecker

References:

Maffesoli, Michel. The Time of the Tribes: The Decline of Individualism in Mass Society. Sage Publications, 1990. Print.

p. 32 Fire Core by Davin Infinity & Dustin Engelskind

p. 33 Photo by Dustin Engelskind

p. 34 Photo by Aaron Dorr

p. 35 Photo by Dustin Engelskind

p. 37 Photo by Amanda Creighton

p. 38 Photo Courtesy of Adam Apollo

p. 38-43 - Guardians: An Allied Force for Global Transformation by Adam Apollo

References:

Brady, David A. and Harold G. White, Paul March, James T. Lawrence, and Frank J. Davies.

"Anomalous Thrust Production from an RF Test Device Measured on a Low-Thrust Torsion Pendulum." NASA Lyndon B. Johnson Space Center, Houston, Texas 77058. PDF file.

http://ntrs.nasa.gov/archive/nasa/casi.ntrs.nasa.gov/20140006052.pdf

Hambling, David. "NASA validates 'impossible' space drive." Wired UK, 31 July 2014. Web.

http://wired.co.uk/news/archive/2014-07/31/nasa-validates-impossible-space-drive

Bodanis, David. "Everything equals E=mc2." The Guardian, 30 April 2008. Web.

http://theguardian.com/science/2008/apr/30/peopleinscience.energy

Wikipedia March 2015 Revision. http://en.wikipedia.org/wiki/Spirituality

p. 39 Photo by Dustin Engelskind

p. 41 Idle no More, Photo by Dana Wilson

p. 42 Feathers from Sechelt Ceremony, Photo by Dana Wilson

p. 44 Artemis Project gathering, Photo by Kristen M. Rivers

p. 45 Artemis Project gathering, Photo by Kristen M. Rivers

p. 46 Photo by Aaron Dorr

Chapter 2 Health & Healing

p. 48 Wisdom of the Ages, painting by Autumn Skye Morrison

p. 50 Shelter for Opening, painting by Autumn Skye Morrison

p. 51 Photo provided by Oasis Express

p. 52 Sivananda Satsang with Satya, photo by Dana Wilson

p. 53 Dance of the Butterflies, photo by Dana Wilson

p. 54 Sivananda Satsang, photo by Dana Wilson

p. 56,57,58,59 Photos of Nature's Nests on event land projects provided by Nature Dreamweaver

p. 60,61,62,63 Photos provided by Paititi Institute

p. 64 Wilson Creek Forest Altar, Photo by Dana Wilson

p. 65 Car Free Day with Village Vancouver, Photo by Dana Wilson

p. 66 Photo by Damien Genardi- Permission from TCN

p. 68-69 Healing of Love by Ian Mackenzie

References:

Ryan, Christopher and Cacilda Jethá. Sex at Dawn: The Prehistoric Origins of Modern Sexuality. Harper, 2010. Print.

p. 68 Photo by Kelli Rua Klein

p. 69 Beloved, painting by Autumn Skye

p. 70 Photo by Dustin Engelskind

p. 72 Photo by Akira Chan

p. 74 Feminine Medicine art by Olivia Curry
http://dreamnectar.com

p. 76 Solar Heart by Davin Infinity

p. 76-79 Renewing the Sunfire Within the Heart of the Masculine by Davin Infinity

References:

Barnes, Andrew and Yvonne Lumsden. Heart of the Flower: Book of Yonis. 2010. Print.

Corsini, Graell. Raising a Goddess Temple. Graell Corsini, Raven Oak, 2013. Print.

Divine Nectar: An Explosive Journey Through the Sacred Feminine. Dir. Tullulah Sulis. Ed., Art Dir. Davin Infinity. 2006. Film.

Birth of the New World. Dir. Davin Infinity. 2009. Film.

p. 79 Mens Rites of Passage art by Davin Infinity

Chapter 3 Art & Culture

p. 80 La Semillera, painting by Mark Henson

p. 82 Photo by Elijah Parker

p. 83 Immaculate Conception art by Darren Minke

p. 84,87 Travel photos by Simon Yugler

p. 84-89 The Alchemy of Travel by Simon Yugler

References:

Prechtel, Martel. Interview by Derrick Jensen. "Saving The Indigenous Soul - An Interview With Martin Prechtel." The Sun Magazine. The Sun Magazine, April 2001. Web. http://thesunmagazine.org/issues/304/saving_the_indigenous_soul

Abram, David. The Spell of the Sensuous: Perception and Language in a More-Than-Human World. Vintage, 1997. Print.

p. 90 Photo by Dustin Engelskind

p. 92 Photo by Dustin Engelskind

p. 94,95,96 Photos courtesy of Wren La Feet

p. 94-97 Dancing the Village by Wren LaFeet

References:

1. Powers, Richard. "Dancing Makes You Smarter." Stanford Dance. http://socialdance.stanford.edu/syllabi/smarter.htm

2. The simplest way to start is for one person to place their right hand on their partner's left shoulder blade (this will be the "Leader"), and one person to rest their left arm on their partner's right arm (the "Follower"). Your free hands will meet palm to palm clasped with the "Leader's" fingers to the outside, and you'll simply start to sway from foot to foot in a "1-2, 1-2" rhythm to the music. Put a little bounce in your sway moving up from your knees on the 1 and 2, and you've got your basic blues pulse. Breathe. Feel the connection between your partner and you. Feel your feet pressing into the ground; pressing it away as you shift your weight. Feel the others around you. Listen to the music. Trust it and surrender your body to the movements that feel right. Leaders, be confident, clear, and patient. Followers, trust, listen,

and surrender to the movement. Release any story or agenda that might form in your mind and give yourself over to the present moment. There is nothing after this dance. Feeling is the secret. Let the spirit of the music guide you and inform your dance. It isn't what it looks like that's important; it's how it FEELS!

p. 98 Slithering Art, Photo by Dana Wilson

p. 99 Photography used with permission from Mark Lakeman and City Repair

p. 98-101 Anything but Temporary by Luke Holden

References:

Fuller, Buckminster. Operating Manual for Spaceship Earth. 1968. Print.

p. 100 Photo courtesy of Mark Lakeman

p. 102-105 Photos courtesy of Samantha Sweetwater

p. 106,107 Photos courtesy of Onedoorland by Elijah Parker

Chapter 4 Learning & Education

p.108 Harmony, painting by Jessica Perlstein

p.110 Seed, painting by Autumn Skye Morrison

p.111 Guardians of the Sacrament painting by Mark Henson

p. 112,113 Art by Laurrien Gilman

p. 112-113 Meta-Poetics: On Patterns and Language by La Laurrien

References:

Also known as Non Violent Communication or NVC created by Marshall Rosenberg.

Originated by Robert Tennyson Stevens.

Commonly referred to as NLP, Co-created by Richard Bandler and John Grinder.

https://en.wikipedia.org/wiki/Pattern_language

A pattern language for bringing life to meetings and other gatherings. http://groupworksdeck.org/

Jacke, David and Eric Toensmeier. Edible Forest Gardens (Volume 2). Chelsea Green Publishing, 2005. Print.

p. 116 Planting the Apple Tree at Rolling Earth Farm Retreat, Photo by Dana Wilson

p. 118 Photo courtesy of Robin Leipman

p. 120 Icons by Kym Chi, Heather Lippold & Shannon Reinhol

p. 122 Photos courtesy of GaiaCraft, design by Lunaya Shekinah

p. 123 Integral Quadrant Map (open source) developed by Ken Wilber

p. 124,125 Photos courtesy of CultureSeed

Chapter 5 Regional Resilience

p. 126 New Pioneers painting by Mark Henson

p. 128 Regeneration, painting by Autumn Skye Morrison

p. 130-131 Photos provided courtesy of Francesco Tripoli

p. 130 The Bullock's Permaculture Homestead on Orcas, Photo by Dana Wilson

p. 130-133 Visioning an Ecotopian Future by Carl Grether

References:

Edible City: Grow the Revolution. Dir. Andrew Hasse. Prod. Andrew Hasse and Carl Grether. 2012. Film.

Occupy the Farm. Dir. Todd Darling. Prod. Carl Grether and Steve Brown. 2014. Film.

Callenbach, Ernest. Ecotopia. 1975. Print.

p. 131 The Bullock's Permaculture Homestead on Orcas with Dave Boehnlein, Photo by Dana Wilson

p. 132 Rolling Earth Farm Retreat in Roberts Creek, Photo by Dana Wilson

p. 134 Heart of the Tree of Life by Davin Infinity

p. 137 Photo courtesy of Jamaica Stevens

p. 136 Photo courtesy of Jamaica Stevens

p. 137 Day out of Time with Pan and Love at the Roberts Creek Mandala, photo by Dana Wilson

p. 138 Photos courtesy of Gaiacraft

p. 139 Bio portraits by Ben Tour

p. 140 Photo by Amanda Creighton

p. 141 Photo by Dustin Engelskind

p. 140-143 Community Based Resilience Framework by Dave Pollard

References:

Hopkins, Rob. The Transition Handbook: From Oil Dependency to Local Resilience. Green Books, 2008. Print.

Orlov, Dmitry. Communities That Abide. CreateSpace, 2014. Print.

p. 143 Creating a Straw Bale Wall with the Mudgirls Natural Building Collective- Dana Wilson

p. 144 Photo by Kelli Klein

p. 146 The Call of the River by Autumn Skye Morrison

p. 144-147 ReInhabiting Watersheds by Heather Beckett and Chelsea Estep-Armstrong

References:

Berg, Peter. Reinhabiting a Separate Country: A Bioregional Anthology of Northern California. Planet Drum Foundation, 1978. Print.

p. 148,149 Photos provided by Outgrowing Hunger

Chapter 6 Inhabiting the Urban Village

p. 150-151 The Fifth Sacred Thing, Film Concept Art by Jessica Perlstein

p. 151-153 ReInhabiting the Urban Village Intro

References:

World's population increasingly urban with more than half living in urban areas. United Nations Department of Economic and Social Affairs. New York, 10 July 2014. Web. http://un.org/en/development/desa/news/population/world-urbanization-prospects.html

Statement at the Interactive Dialogue on Harmony with Nature. General Assembly of the United Nations. New York, 27 April 2015. Web. http://un.org/pga/270415_statement-interactive-dialogue-on-harmony-with-nature/

p. 152-153 Photo by Amanda Creighton

p. 154,155 Images by Dana Wilson

p. 156,157,158 Carl Grether Gardening Images

p. 156-159 Growing Edible Cities by Carl Grether

References:

Edible City: Grow the Revolution. Dir. Andrew Hasse. Prod. Andrew Hasse and Carl Grether. 2012. Film.

p. 160-165 Organic Gardens images provided by Matthew Finkelstein

p. 160-165 Gardening as a Tool for Transformation by Matthew Finklestein

References:

Louv, Richard. Last Child in the Woods: Saving Our Children From Nature-Deficit Disorder. Algonquin Books, 2005. Print.

Gershuny, Grace. Start With The Soil. Rodale Press, 1993. Print.

Elpel, Thomas J. Botany In A Day: The Patterns Method of Plant Identification. HOPS Press, 1998. Print.

Brenzel, Kathleen Norris (Editor). Sunset's Western Garden Book. Oxmoor House, 1986. Print.

Peaceful Valley. "Pruning 101." 2013. Video. http://groworganic.com/organic-gardening/videos/pruning-101

Hemenway, Toby. Gaia's Garden: A Guide to Home-Scale Permaculture. Chelsea Green, 2001. Print.

p. 166 Vancouver's Compost Demonstration Garden, Photos by Dana Wilson

p. 166(2) Photo by Kelli Klein

p. 167 Photos courtesy Sebastian Collet

p. 168 Photo used with permission by Mark Lakeman/City Repair

p. 170,171,172 Photos provided courtesy of Ryan Rising

p. 174,175 Photos used with permission from Mark Lakeman/City Repair

Chapter 7 Community Land Projects

p. 176 Lucidity Tree of Life, Marc Goerner with permission from Lucidity Festival, LLC.

p. 178-179 Photos by Amanda Creighton

p. 180,181 Photos Courtesy of Atlan

p. 182 Photo Courtesy of Dustin Engelskind p. 182 (2&3) Photos used with permission by Mark Lakeman/City Repair

p. 184,185,186,187 Photos and models courtesy of Geofrey Collins

p. 188 Photo by Aaron Dorr

p. 190 The Vale Village, Photo by Dana Wilson

p. 192 Love Grows at Atlan, Photo by Dana Wilson

Chapter 8 Holistic Event Production

p 194-195 Seeing Is Believing, by Jessica Perlstein

p. 196 (top) Photo by Aaron Dorr

p. 196 (bottom) Photo courtesy of Harmony Festival

p. 198 Photo courtesy of Boom Festival

p. 200- 203 Lucidity Festival photos by Dustin Engelskind

p. 205 Photo by Ingrid Langer

p. 200- 205 Lucidity Manifesto by Jonah Haas

References:

Eisenstein, Charles. Sacred Economics. Evolver Editions, 2011. Print.

Wilder, Barbara. Money is Love: Reconnecting with the Sacred Origins of Money. Wild Ox Press, 1998. Print.

Campbell, Joseph. The Hero with a Thousand Faces. 1949. Print.

p. 206 Chalking with Corrina Keeling, photo by Dana Wilson

p. 208 Photo by Daniel Zetterstrom

p. 210-211 Urban Festivals by Magenta Ceiba

References:

Atlee, Tom. Empowering Public Wisdom: A Practical Vision of Citizen-Led Politics. Evolver Editions, 2012. Print.

p. 210 Photo courtesy of Evolver

p. 212-217 Catching and Storing Festival Energy by Clayton Gaar

References:

[1] Mollison, Bill. Introduction to Permaculture. 1991. Tasmania, Australia: Tagari Publications.

[2] Mollison, Bill. Permaculture: A Designer's Manual. 1988. Tasmania, Australia: Tagari Publications. p. 2.

[3] Blair, Lawrence. Rhythms of Vision: The Changing Patterns of Belief. 1975. London: Croom Helm Ltd. ISBN 978-0-8052-3610-1.

[4] Holmgren, David. Permaculture: Principles and Pathways Beyond Sustainability. 2002. Holmgren Design.

[5] Lampikoski, Tommi. Green, Innovative, and Profitable: A Case Study of Managerial Capabilities at Interface Inc. Technology Innovation Management Review. Nov 2012. p. 4-12.

[6] Roland, Ethan. Eight Forms of Capital. Permaculture Magazine, No. 68. 2011. p. 58-61.

[7] Christian, Diana L. Creating a Life Together: Practical Tools to Grow Ecovillages and Intentional Communities. New Society Publishers. 2003.

p. 212,214 Photo courtesy of Clayton Gaar

p. 218-221 Photos of the Vision Council gathering by Ivan Sawyer Garcia

Chapter 9 Living Economy

p. 222,223 The Fifth Sacred Thing Concept Art by Jessica Perlstein

p. 224 Pope Francis Quote: http://w2.vatican.va/content/francesco/en/encyclicals/documents/papa-francesco_20150524_enciclica-laudato-si.html

p. 224-225 Anapurnadevi, art by George Atherton (Geo-Glyphiks)

p. 226 Photo by Kelli Klein

p. 228 Photo by Kelli Klein

p. 229 Holonomic Logo by Davin Infinity

p. 228-231 Holonomic Perspective by Steven Michael EhlingerII

References:

1 Trivett, Vincent. "25 US Mega Corporations: Where They Rank If They Were Countries." 27 June 2011. Web. www.businessinsider.com/25-corporations-bigger-tan-countries-2011-6.

2 Wilder, Barbara. Money Is Love: Reconnecting with the Sacred Origins of Money. Wild Ox Press, 1998. Print.

3 Bailey, Andrew Cameron and Connie Baxter Marlow. The Trust Frequency: Ten Assumptions for a New Paradigm.CameronBaxter Books, 2013. Print.

4 Lietaer, Bernard. Complimentary Yin/Yang Currencies: "A World in Balance?". Reflections: Journal of the Society for Organizational Learning, Summer 2003. Print.

5 Vaughan, Genevieve. The Foundation for a Compassionate Society. Web. www.gift-economy.com/gen_her.html.

6 Twist, Lynne. The Soul of Money: Reclaiming the Wealth of Our Inner Resources. W. W. Norton & Company, 2003. Print.

7 Sawaf, Ayman and Rowan Gabrielle. Heralding the Age of Consciousness, Sacred Commerce: A Blueprint for a New Humanity. EQ Enterprises, 2014. Print.

8 Eisenstein, Charles. The More Beautiful World Our Hearts Know Is Possible. North Atlantic Books, 2013. Print.

9 Workman, Eileen. Sacred Economics: The Currency of Life. Muse Harbor Publishing, 2011. Print.

10 Eisler, Riane. The Real Wealth of Nations: Creating a Caring Economics. Berrett-Koehler Publishers, 2007. Print.

p. 230 Photo by Dana Wilson

p. 232 Photo by Dana Wilson

p. 232-233 The Sharing Economy is Healing our World by Brandi Veil

References:

Tabarrok, Alex. Principles of Economics online series. Web. http://mruniversity.com/courses/principles-economics-microeconomics

Said, Carolyn. "S.F. ballot measure would severely limit short-term rentals: Group seeks cash, support for a November initiative." SF Gate. 29 April 2014. Web.

http://sfgate.com/news/article/S-F-ballot-measure-would-severely-limit-5436664.php

p. 234 Photo by Dana Wilson

p. 235 Solar Seed icon by Davin Infinity

p. 236 Photo courtesy of La Laurrien

p. 234- 237 Cultivating a Culture of Contribution by La Laurrien

References:

Tellinger, Michael. UBUNTU Contributionism. Web.

http://michaeltellinger.com/ubuntu-cont.php

p. 238 Photo by Dustin Engelskind

p. 239 Diagram by VillageLab

p. 240 Photo of diagram courtesy of TCN

p. 240 Network Map by RealEconomyLab.org

p. 241 Photo by Jamica Stevens

p. 243 Photo by Dana Wilson

p. 238- 243 Life Affirming Economy by Ferananda Ibara and Crystal Arnold

References:

Integral Life. "Integral Spotlight: Evolution's Purpose"- Steve McIntosh. Online video. YouTube. YouTube, Feb 26, 2013. Web. http://youtube.com/watch?v=GDxVKTtg6Zk

Pesce, Mark. Interview with Heather Schlegel in Future of Money Podcast. July 29, 2014.

Fuji Declaration. Web. http://fujideclaration.org/

Oates, J. Babylon, p 25. Thames & Hudson, 1979. Print.

p. 244,245 Photo by Dana Wilson

Chapter 10 Media & Storytelling

p. 246 Healing painting by Autumn Skye Morrison

p. 248 Photo by Wesley Pinkham

p. 248 (2) Photo by Dustin Engelskind

p. 249 photos by Republic of Light

p. 250 Digital art by Davin Infinity

p. 251 Transmedia Storytelling Infographic by Maya Zuckerman

p. 252 Maya Portrait by Davin Infinity

p. 250-253 Introduction to Transmedia by Maya Zuckerman

References:

Moore's Law. Wikipedia. Web. [1] 4 May 2015 < http://en.wikipedia.org/wiki/Moore%27s_law>

p. 253 Transmedia Recipe (info graphic) by Maya Zuckerman

p. 254 Digital art by Davin Infinity

p. 255 Consolidation info graphic provided by Maya Zuckerman

p. 256 Digital art by Davin Infinity

p. 254- 257 Global Evolution through Technology and Conscious Media by Maya Zuckerman

References:

[1] McLuhan, Marshall and Quentin Fiore. The Medium is the Massage: An Inventory of Effects. Bantam Books, 1967. Print.

Graves, Clare W. "The Emergent Cyclical Levels of Existence" Theory. 1974. http://clarewgraves.com/home.html

Sparks & Honey. "The Explosion of Conscious Media." GaiamTV, 2013. Web.

Brewer, Joe. "Why global warming is a lousy meme." Inside Passages, 14 May 2013. Web.

http://insidepassages.com/2013/05/14/joe-brewer-on-why-global-warming-is-a-lousy-meme/

Lutz, Ashley. "These 6 Corporations Control 90% Of The Media In America." Business Insider, 14 June, 2012. Web.

http://businessinsider.com/these-6-corporations-control-90-of-the-media-in-america-2012-6#ixzz3dadQzJ65

p. 258 Photos courtesy of Elevate Films

p. 259 Photo by Zipporah Lomax provided by Jeet Kai Leung

p. 260 Photos provided courtesy of Project Nuevo Mundo

p. 261 Photo by Dana Wilson

p. 263 Photo by Jamica Stevens

p. 264 Photo provided by Project NuMundo

p. 266 Photo by Kurt Klein

p. 266(2) Photo by Dustin Engelskind

p. 267 Photo by Aaron Dorr

p. 268 Photo by Aaron Dorr

p. 269 Photo by Dustin Engelskind

p. 266- 269 The Modern Fire Circle by Brad Nye

References:

McLuhan, Marshall. Understanding Media. Mit Press, 1964. Print.

p. 270,271 Photos courtesy of Evolver Network

Chapter 11 Appropriate Technology

p. 272,273 The Network by Davin Infinity

p. 274 Digital Art by George Atherton(GeoGlyphiks)

p. 276 Photo by Dana Wilson

p. 276, 277 Transformational Technologies by Julian Reyes, Natacha Pavlov and Maya Zuckerman

References:

Transformative Technology Lab @ Sofia University. "What is Transformative Technology?" Web. http://transtechlab.org/what-is-transformative-technology/

https://en.wikipedia.org/wiki/Neuromorphic_engineering

https://en.wikipedia.org/wiki/Artificial_intelligence

https://en.wikipedia.org/wiki/Augmented_reality

https://en.wikipedia.org/wiki/Energy_harvesting

[1] Schumacher, E.F. Small is Beautiful: Economics as if People Mattered. Blond & Briggs, 1973. Print.

[2] http://en.wikipedia.org/wiki/1970s_energy_crisis

[3] http://en.wikipedia.org/wiki/Open_source

[4] http://en.wikipedia.org/wiki/Open-source-appropriate_technology

http://en.wikipedia.org/wiki/Club_of_Rome

http://en.wikipedia.org/wiki/The_Limits_to_Growth

http://communitywiki.org/DoOcracy

p. 278,279 Photos by Dustin Engelskind

p. 282-284 Photos used by permission from Mark Lakeman/City Repair

p. 286 Solar Fractal by Davin Infinity

p. 286,287 Techn:ecology by A.Keala Young

References:

McDonough, William and Michael Braungart. "Design for the Triple Top Line: A New Definition of Quality." McDonough Innovation Design for the Circular Economy. 2003. Web. http://mcdonough.com/speaking-writing/design-for-the-triple-top-line/#.VYYq1_lVikp

Colao, J.J. "Welcome To The New Millennial Economy: Goodbye Ownership, Hello Access." Forbes, 11 Oct 2012. Web. http://forbes.com/sites/jjcolao/2012/10/11/welcome-to-the-new-millennial-economy-goodbye-ownership-hello-access/

Carey, Bjorn. "Stanford biologist warns of early stages of Earth's 6th mass extinction event." Stanford, 24 July 2014. Web. http://news.stanford.edu/news/2014/july/sixth-mass-extinction-072414.html

"It is a kind of leadership that unlocks the potential for greatness in the leader and those that follow the leader. Enlightened leadership is spiritual if we understand spirituality not as some kind of religious dogma or ideology but as the domain of awareness where we experience values like truth, goodness, beauty, love and compassion, and also intuition, creativity, insight and focused attention. These are the ingredients of success, really."

Chopra, Deepak. Interview by Jenna Goodreau. Forbes, 12 Jan 2011. Web.

http://forbes.com/sites/jennagoudreau/2011/01/12/deepak-chopra-on-enlightened-leadership-happiness-meaning-work-employee-engagement-president-barack-obama/

Fuller, Buckminster. "Operating Manual for Spaceship Earth." 1969. Design Science Lab. PDF file.

http://designsciencelab.com/resources/OperatingManual_BF.pdf

p. 288 Photo provided by Boom Festival

Chapter 12 Whole Systems

p. 290 Whole Systems art by Davin Infinity

p. 292 Painting by Ishka Lah, The Holy Trinity

p. 293 Painting by Eric Nez

p. 294 Photo by Kelli Klein

p. 296 AQAL Integral Map provided courtesy of formless-mountain.com

p. 297 Wheel of Co-Creation provided courtesy of Barbara Marx Hubbard

p. 298-300 Diagrams provided courtesy of Village Lab

p. 284-301 Mapping Whole Systems by Jeff Clearwater

References:

Gruber, Tom. "What is Ontology?" 1992. Web. http://www-ksl.stanford.edu/kst/what-is-an-ontology.html

Wilber, Ken. A Theory of Everything: An Integral Vision for Business, Politics, Science & Spirituality. Shambhala, 1996. Print.

p. 302 Digital Art by Davin Infinity

p. 303 Photo by Kelli Klein

p. 306 TCN logo by James Barnard & Davin Infinity

p. 306,307 Photos by Dustin Engelskind

p. 307 TCN Diagrams by Kelli Rua Klein

p. 308 Photo by Kelli Klein

p. 308 Image courtesy of Empowermentworks

p. 310 Photo by Dustin Engelskind

p. 312 Photo by Dana Wilson

p. 310-313 Moving Toward a Greater Evolution by Sheri Herndon and Patricia Ellsberg

References:

Hubbard, Barbara Marx. Emergence: The Shift from Ego to Essence. Hampton Roads Publishing Company, 2001. Print.

p. 314 Photo by Jamaica Stevens

p. 314-317 Images provided courtesy of Dr. Robert Gilman

p. 314-317 Inhabiting the Future by Robert Gilman

References:

"What Time Is It?" presentation videos. Context Institute, Feb 2014. Presentation. Web.

http://context.org/foundation-stones/what-time-is-it/

p. 319 art by Autumn Skye Morrison

p. 323 For Those Who Dare to Feel so Deeply by Ka Amorastreya

p. 324 art by Autumn Skye Morrison

p. 325 art by Krystleyez

p. 326 art by Ishka Lha

p. 328 art by Krystleyez

p. 347 art by Davin Infinity

Glossary

A theory of everything - a theory of everything (ToE) or final theory, ultimate theory, or master theory is a hypothetical single, all-encompassing, coherent theoretical framework of physics that fully explains and links together all physical aspects of the universe

Age of Aquarius - in popular culture, the Age of Aquarius refers to the advent of the New Age movement in the 1960s and 1970s

Agro-ecology - the study of ecological processes that operate in agricultural production systems

Alchemy - the process of taking something ordinary and turning it into something extraordinary, sometimes in a way that cannot be explained

Ancestories - a portmanteau fusing the meanings of stories and ancestors

Appropriate Technology - an ideological movement (and its manifestations) originally articulated by the economist Dr. E. F. Schumacher; technology that is designed with special consideration to the environmental, ethical, cultural, social, political, and economical aspects of the communities it is intended for

Ascension - the act of rising to an important position or a higher level

Augmented Reality - a technology that superimposes a computer-generated image on a user's view of the real world, thus providing a composite view

Back to Earth Movement - the back-to-the-land movement called for occupants of real property to grow food from the land on a small-scale basis for themselves or for others, and to perhaps live on the land while doing so

Bioconstruction - is the respect of nature while promoting health and wellness when building

Biodynamic design - is a method of organic farming originally developed by Rudolf Steiner that employs what proponents describe as "a holistic understanding of agricultural processes"

Biomimicry - the design and production of materials, structures, and systems that are modeled on biological entities and processes

Bioregionalism - advocacy of the belief that human activity should be harmonized with the distinct ecological and geographical regions

Bioremediation - the use of either naturally occurring or deliberately introduced microorganisms or other forms of life to consume and break down environmental pollutants, in order to clean up a polluted site

Bitcoin - a type of digital currency in which encryption techniques are used to regulate the generation of units of currency and verify the transfer of funds, operating independently of a central bank. See also cryptocurrency

Chakra system - in Sanātana/Hindu and tantric/yogic traditions and other belief systems, a chakra (Sanskrit: IAST: cakra) is an energy point or node in the subtle body. From the sanskrit for wheel

Community Supported Agriculture (CSA) - consists of a community of individuals who pledge support to a farm operation so that the farmland becomes, either legally or spiritually, the community's farm, with the growers and consumers providing mutual support and sharing the risks and benefits of food production

Compassionate communication - Nonviolent Communication (abbreviated NVC, also called Compassionate Communication or Collaborative Communication) is a communication process developed by Marshall Rosenberg

Compersion - the feeling of joy one experiences through witnessing another's joy

Conscious media - media that is characterized by encouraging of an awareness of one's environment and one's own existence, sensations, and thoughts

Conscious Tech Guild - A group of technologists committed to working together towards creating tech that is socially responsible

Corporatocratic - suggesting of rule and governance by corporations. From the greek words ocracy, meaning "power," "rule" and "government"

Coworking - a style of work that involves a shared working environment, often an office, and independent activity

Creationship - a portmanteau of create and relationship suggesting creation through relationship

Crowdfunding - the practice of funding a project or venture by raising many small amounts of money from a large number of people, typically via the Internet

Cryptocurrency - a digital currency in which encryption techniques are used to regulate the generation of units of currency and verify the transfer of funds, operating independently of a central bank

Cultural literacy - cultural literacy is a term coined by E. D. Hirsch , referring to the ability to understand and participate fluently in a given culture. Cultural literacy is an analogy to literacy proper (the ability to read and write letters)

Culture of Contribution - a cultural practice of service and generosity; the first theme in the Buddha's system of the ten paramis, one of the seven treasures, and the first of the three grounds for meritorious action

Decentralization - the process of redistributing or dispersing functions, powers, people or things away from a central location or authority

Dis-ease - this is a metapoetic application of punctuation to emphasize the root of the word ease. A reverse portmanteau where the separation of one word into two or more emphasizes multiple meanings. "Not at Ease"

Do-ocracy - a do-ocracy (also sometimes do-opoly, which is a more obvious pun on "duopoly") is an organizational structure in which individuals choose roles and tasks for themselves and execute them. Responsibilities attach to people who do the work, rather than elected or selected officials

Dystopia - an imagined place or state in which everything is unpleasant or bad, typically a totalitarian or environmentally degraded one

Eco-artivist - prefix eco meaning ecological and the portmanteau of artist and activist

Eco-feminism - a philosophical and political movement that combines ecological concerns with feminist ones, regarding both as resulting from the patriarchal domination of society

Eco-pedagogy - the ecopedagogy movement is a body of educational ideas and practices to develop a robust appreciation for the collective potentials of being human and to foster social justice throughout the world

Ecotopian - any ecologically ideal place or situation. From the name of an ecological utopia in a novel ("Ecotopia") by Ernest Callenbach

Ecovillages - a community whose inhabitants seek to live according to ecological principles, causing as little impact on the environment as possible

Elders - a respected leader or senior figure in a tribe, community or other group

EmWaves - electromagnetic waves are waves which can travel through the vacuum of outer space. Mechanical waves, unlike electromagnetic waves, require the presence of a material medium in order to transport their energy from one location to another

Ethnomystic - a metapoetic term for the mysticism mythos of particular Ethnic groups

Eustress - a useful form of stress and healthy framework to move a project to completion

Feedback loops - term commonly used in economics to refer to a situation where part of the output of a situation is used for new input. An example of a positive feedback loop would be one where success feeds success

Femvolution - a movement of feminine liberation for women worldwide; a portmanteau of feminine and revolution or evolution

Glossary

Freesponsibility - a portmanteau of freedom and responsibility suggesting willful or chosen responsibility. Also as freedom + response + ability, the ability to respond freely

Gaia - the name of the ancient Greek goddess of the Earth, and as a name revived by the hypothesis formed by James Lovelock and Lynn Margulis, who postulate that the whole biosphere is alive in that the Earth's life forms are collectively responsible for regulating the conditions that make life on the planet possible

Gaiaphile - Gaia (see above) with suffix meaning "a lover or admirer" of something specified; "one who loves Gaia/Nature"

Gender alchemy - the practice of balancing and integrating polaries of Yin and Yang or sometimes Feminine and Masculine characteristics

Gift exchange model / gifting economy - a mode of exchange where valuables are not sold, but rather given without an explicit agreement for immediate or future rewards

Global evolution - evolutionary forces of change and transformation experienced on a Global or worldwide scale

Groupmind - in social psychology, the collective mind of a group, class, or society; the stock of beliefs, customs, attitudes, etc., common to the members of such a group

Guardian - a defender, protector, steward or keeper

Guilds - an association of people for mutual aid or the pursuit of a common goal

Hackerspaces - community-operated physical places, where people share their interest in tinkering with technology, meet and work on their projects, and learn from each other

He'art - a metapoetic term of the 'Art from the Heart'

Healing sanctuary - a sacred place for renewal, healing, reflection, wellness

Heterarchy - a heterarchy is a system of organization where the elements of the organization are unranked (non-hierarchical) or where they possess the potential to be ranked a number of different ways

Higher intelligence - the acknowledgment of an intelligent design and order to the elegant Universe, the intelligence of Life that permeates all things

Ho'oponopono - ancient Hawaiian practice of reconciliation and forgiveness. Similar forgiveness practices were performed on islands throughout the South Pacific, including Samoa, Tahiti and New Zealand

Holocracy - social technology or system of organizational governance in which authority and decision-making are distributed throughout a holarchy of self-organizing teams rather than being vested in a management hierarchy

Holon - a holon is a process which is both a whole and a part

Holonomic perspective - mechanical systems in which all links are geometrical (holonomic)—that is, restricting the position (or displacement during motion) of points and bodies in the system but not affecting the velocities of these points and bodies

Holopticism /Holoptism - is opposed to panoptism. This conceptual distinction is used in the Collective Intelligence site of Jean-Francois Noubel, at http://thetransitioner.org/ic

Impact centers - an impact center is a land based project that offers individual transformation, regenerative living education and strives to leave a positive local impact. An impact center could be an ecovillage, organic farm, yoga retreat center, or even a hostel

Integral theory - a subset of integral studies developed by Ken Wilber; integral meaning comprehensive, inclusive, non-marginalizing, embracing

Intermediate Technology - technology suitable for use in developing countries, typically making use of locally available resources. Later referred to as Appropriate Technology

Interoperability - ability of a system to work with or use the parts or equipment of another system

Kibbutz - a communal settlement in Israel, typically a farm

Livingry - coined by Buckminster Fuller, juxtaposed to weaponry and "killingry"; meaning that which is in support of all human, plant, and Earth life

Meditation - to spend time in quiet or contemplative space for religious purposes or relaxation

Meditation Mobs - Groups of people who synchronize a meditation practice, usually in a public place and without pre-emptive publicity or promotion. Participants agree to join at a place and time and meditate to bring public awareness to the practice

Meme - a conception of essential units of cultural information modeled on the concept of genes, suggesting cultural dna

Meritocracy - government or the holding of power by people selected on the basis of their ability

Meshnets - network topology in which each node relays data for the network. All mesh nodes cooperate in the distribution of data in the network

Metanarrative - a term developed by Jean-François Lyotard to mean a theory that tries to give a totalizing, comprehensive account to various historical events, experiences, and social, cultural phenomena based upon the appeal to universal truth or universal values

Microfinancing - a source of financial services and support for entrepreneurs and small businesses

Mind maps - a diagram used to visually organize information. A mind map is often created around a single concept, drawn as an image in the center of a blank landscape page, to which associated representations of ideas such as images, words and parts of words are added

Moshavs - a type of Israeli town or settlement, in particular a type of cooperative agricultural community of individual farms

Mutualism - the doctrine that mutual dependence is necessary to social well-being

Neo-tribal - or modern tribalism is the ideology that human beings have evolved to live in tribal society, as opposed to mass society, and thus will naturally form social networks constituting new "tribes"

New village - a modern reflection of a Village or Village culture usually incorporating some element of Whole Systems Design

Non-transgenic seeds - seeds are seeds that have not been genetically engineered. Transgenic seeds are created through a breeding approach that uses recombinant DNA techniques to create plants with new characteristics. They are identified as a class of genetically modified organism (GMO)

Nuclear family - the concept of family that includes only the father, mother, and children

Om-ing - orgasmic meditation (OM) is a wellness practice (like yoga and pilates) that is designed for singles and couples to experience more connection, vitality, pleasure, and meaning in every aspect of their lives

Ontology - the branch of metaphysics dealing with the nature of being

Open-source appropriate technology (OSAT) - refers to technologies that are designed in the same fashion as free [1] and open-source software. These technologies must be "appropriate technology", definition above

Outernets - a wireless community network based in Poland, in which everyone owns their own node's hardware configured in a mesh network managed by OLSR

Parasitism - a relationship between two things in which one of them (the parasite) benefits from or lives off of the other; like fleas on a dog

Permaculture - a comprehensive design methodology with an ethical foundation that promotes care for the earth, care for people and the redistribution of surplus

Placemaking - a multi-faceted approach to the planning, design and management of public spaces. Placemaking capitalizes on a local community's assets, inspiration, and potential, with the intention of creating public spaces that promote people's health, happiness, and well being

Glossary

Portmanteau - a portmanteau word fuses both the sounds and the meanings of its components. Originally coined by Lewis Carrol in the book, "Through the Looking Glass"

Qigong - a Chinese system of physical exercises and breathing control related to tai chi

Rainbow people - multi-cultural, multi-ethnic people who tend to feel connected as a "tribe" or "family" unit without having a blood or ethnic connection

Regenerative whole systems map - regenerative design is a process-oriented systems theory based approach to design. The term "regenerative" describes processes that restore, renew or revitalize their own sources of energy and materials, creating sustainable systems that integrate the needs of society with the integrity of nature

ReInhabiting the Village - the practices, actions, and experience of 'belonging to a place and belonging to a people'. To live in accordance with principles of the "Village" which include an ecocentric, inclusive, intergenerational, regenerative, cooperative, interconnected group of people committed to living in balance as stewards of a place

Resources - a source or supply from which benefit is produced. Typically resources are materials, energy, services, staff, knowledge, or other assets that are transformed to produce benefit and in the process may be consumed or made unavailable

Responsibility - the state or fact of having a duty to deal with something or of having control over someone

Right relationship - aligning our individual behavior and social structures so that our way of life honors all of creation

Sacred commerce - the participation of the community in the exchange of information, goods and services that contribute to the revealing of the Divine (beauty, goodness and truth) in all

Self - a person's essential being that distinguishes them from others, especially considered as the object of introspection or reflexive action

Self-sustainability - the degree at which the system can sustain itself without external support. the fraction of time in which the system is self-sustaining

Sense of place - involves the human experience in a landscape, the local knowledge and folklore.Sense of place also grows from identifying oneself in relation to a particular piece of land on the surface of planet Earth

Sharing economy - also known as collaborative consumption, is a trending business concept that highlights the ability (and perhaps the preference) for individuals to rent or borrow goods rather than buy and own them

Social capital - refers to the collective value of all "social networks" [who people know] and the inclinations that arise from these networks to do things for each other ["norms of reciprocity"]

Solutionary - involves using our talents and passions to create solutions to environmental, social, and economic problems in our communities

Somatic - of or relating to the body, especially as distinct from the mind

Soul Evolution - growing in consciousness, steadily progressing through different levels or stages of consciousness

Spiral dynamics - theory of motivation that looks at the value systems that drive individuals' beliefs and actions

Sustainable living - lifestyle that attempts to reduce an individual's or society's use of the Earth's natural resources and personal resources. Practitioners of sustainable living often attempt to reduce their carbon footprint by altering methods of transportation, energy consumption, and diet

Symbiosis - interaction between two different organisms living in close physical association, typically to the advantage of both

Synchromysticism - a portmanteau of synchronisity (the concept, created by Carl Jung, that events are "meaningful coincidences" if they occur with no apparent causal relationship yet seem to be meaningfully related) and mysticism (Belief in direct experience of transcendent reality or God, especially by means of contemplation and asceticism instead of rational thought)

Synergenius - a portmanteau fusing the sounds and meanings of synergy and genius

Synergy - the interaction or cooperation of two or more organizations, substances, or other agents to produce a combined effect greater than the sum of their separate effects

Tai-chi - although originally a Martial Art it is mainly practiced today as an excellent form of exercise with many health benefits. The words Tai Chi Chuan mean Supreme Ultimate Boxing, used as an exercise for health it would loosely translate as Supreme Ultimate Exercise or Skill

Technomythology - the stories we use to remind ourselves of human accomplishment, merging the deification of scientific progress with enabling future generations the impetus to continually innovate

Transformational culture - a self identified culture whose practices, values, and beliefs revolve around the Transformation of mind, body, spirit of humanity

Transition Town - or more generally a transition initiative, is a grassroot community project that seeks to build resilience in response to peak oil, climate destruction, and economic instability by creating local groups that uphold the values of the transition network

Transmedia storytelling - is the technique of telling a single story or story experience across multiple platforms and formats including, but not limited to, games, books, events, cinema and television (also known as transmedia narrative or multiplatform storytelling, cross-media seriality etc)

Transparency - the quality or state of being transparent, a state of not concealing or hiding

Uniqual - a portmanteau fusing the sounds and meanings of unique and equal

Village - a self-contained district or community within a town, city or rural area, regarded as having features characteristic of village life

Visionary Arts - art that purports to transcend the physical world and portray a wider vision of awareness including spiritual or mystical themes, or is based in such experiences

Whole systems design / Whole systems theory / Whole-systems thinking - or just 'systems thinking' - is a method of analysis and decision-making that looks at the interrelationships of the constituent parts of a system rather than narrowly focusing on the parts themselves. By incorporating a range of perspectives, conditions, connections and capabilities into a dynamic analysis, practitioners design for interdependencies within any naturally occurring or human developed systems

Wiki - a website that allows collaborative editing of its content and structure by its users

Wisdom Keepers - one who learns from their life journey and affirms that this information is of value and important to share, especially with the next generation. Native American and other traditional people identify elders as wisdom keepers for their way of life

Yoga - a Hindu spiritual and ascetic discipline, a part of which, including breath control, simple meditation, and the adoption of specific bodily postures, is widely practiced for health and relaxation. The yoga widely known in the West is based on hatha yoga, which forms one aspect of the ancient Hindu system of religious and ascetic observance and meditation, the highest form of which is raja yoga and the ultimate aim of which is spiritual purification and self-understanding leading to samadhi or union with the divine

A special Thank You to YOU, the community who supported our successful Kickstarter crowdfunding campaign and helped make this project possible!
We honor your participation in this movement!

A Line Krs
Aaron Cyrus Dorr
Aaron Fehon
Adam Rubin
Ahni Radvanyi
Akira Chan
Aleks
Ali Shanti
ALIA
Aliah Selah & Lotus Heart Yoga
 Dancers and Sacred Jam Band,
 Spiral On
Alicia Tripp
Aline K
Alla Guelber
Alon Gecer
Amanda Sage
Amber Hartnell
Amir Niroumand
Amy & Emily Krohn
Andres Amador
Andrew
Andrew Ecker
Andrew Hammond
Andrew Kroma
Andy Fusso
Anne Malone
Anne Tierney Devlin
Annie Roderick Clare
Aowyn Jaylee Jones
Ariel White
Banks Family
Barbara Kozma
Beloved Festival
Beth Giansiracusa
Beth Leone
Bethany DeAngelis
Betsy Cacchione
Beverly LaFae
Bianca Heyming
Bianca Chung
Brandi Veil
Brandon Arling
Brandon Beachum
Bri Summers
Brian

Bridgette Brown
Brooke Rosen
Bruce Beeley
Bruce Caron
Bruno Treves
Bryan Sarpad
CA Foundation for Adv of
 Electronic Arts
Caitlin Sislin
Cameron Atkins
Carri Munn
Carrie Anne Huneycutt
Carrie Goodman
Cas Galiszewski
Casandra Tanenbaum
Cecilia Villaseñor Johnson
Chels Romero
Chris Blentzas
Chris Mussoline
Christi Kassity
Christina Sasser
Christopher Tuccitto
Chuck Woods
Claire Rumore
Cmwoods74
Cobalt & Crimson
Colleen Coyne
Collin Murphy
Cormac O'Brien
Cosmic Convergence
Craig Bailes
Cynthia Robinson
Cyrus Sutton
Dan Gambetta
Daniel Gorelick
Danielle (Dani Belle) Gennety
Darakshan Farber
Dashielle Vawter
David Cates
David Chen
David Lieberman
Dbgorelick
Deb Windham
Deborah Dove
Debra Giusti
Diggable Monkey Productions
Docta Nick
Doctor Otter

Doug Akin
Duke Dorje
Edica Pacha
Elizabeth Fitzgerald
Ellie Haagsman
Elliot Rasenick
Elric Centers
Ember DeQuincy
Emmet Grahn
Enzo Cappellini
Erika Logie
Erin Rosenthal
Escher VanKorlaar
Esteban - Steven Ehlinger
Eyefleye - Lisa Lonstron
Firebulb
Firefly Events
Flora Bowley
Flow Fests
Gabriella Coniglio
Gabrielle Sundra
Gary Foresman
Geo Glyphiks
Gigi Jennifer Moser
Gwendolyn Grace
Hannah Haas
Hannah Poirier
Hanre Chang
Hargobind Khalsa
Hart Sawyer
Hayes Starns
Heart Fractal
Holly Marie Suda
Hosam Khedr
Ian Douglas
Ian MacKenzie
Ignight Flow Conference
Ilan Mandel
Imagika Om
Inga Wahlander
Isabel Nelson
Ivy Elwyn
Jack downhill
Jack E Downhill Jr
Jake Musselman
James
James Barnard
James Davis

James Mercé Edwards
Jamie Janover
Jason Guille
Jeff Clearwater
Jeff Harris
Jen Pumo
Jen White
Jenn Azure
Jesse Fowler
Jesse McElwain
Jessica Plancich
Jill Littlewood
Joanne Holden
Joel Kuna
Joel Tiger de Ross
Joey Johnson
Johann Urb
John
John Augspurger
John C. Thomas
Jonah
Jonah Haas
Jordan New
Josh Davis
Julie Hightman
Karen Gabai
Karl Wood
Kate D.
Katrina Zavalney
Kaya Singer
Kelli Rua Klein
Kelly A. Smith
Kelly Crook
Kelly King
Kelly Neff
Keri Chang-McManus
Kerri Olson
Kirill Kireyev
Kris Bell
Kristopher Moller
Kurt Klein Jr.
Kyle Buckley
Laini Katheiser
Larry Weinberg
Lasers and Lights.com
Laurrien Gilman
Leo Shuster III
Lindsay Kent
Lisa Beck
Lisa Harding
Lisa Mahaffey
Lizzy & JahSun Martini
Lo Nathamundi

Lucidity Festival
Luke Holden
Lynn Gaar
Lynn Gravatt
Maeve Bandruid
Magalie Bonneau-Marcil
Mahaliyah
Manoj Matthews
Marco Pinter
Maria Allred
Maria Reidelbach, Corn
Cow Company
Marissa J Goettling
Mark Garrity
Marya Stark
Matheo James
Matt Siegel
Maya Ittah
Maya Zuckerman
Melanie St. James
Melissa Wright
Melodie Joy
Melody Brown
Michael DiGiorgio
Michael Manahan
   (Cascadia Festival/Seattle)
Michelle Dohrn
Michelle VanKlootwyk
Mikaela Schey
Mike Brent
Mikol Benjacob Fuller
Muffadal Saylawala
Namaste Foundation
Nancy Zamierowski
Natacha Pavlov
Natasha Lewis-Lively
Nathalie
Nathaniel James
Neil Goldstein
Nikki Drumm
Nils Hammerbeck
Nina von Feldman
DreamDance Global
Nlolis
NW Permaculture Institute
One Community
Our Sacred Acres
Patrick Hennessey
Patty Yuniverse
Paul Abad
Pete Silva
Photonic Bliss
Pineapple

Po Li
Polish Ambassador
Rebecca Stefani
Richard Learmont
Ripplestone
Robert Rivello
Robyn McClintock
Ron Glover
Ross Kirschenheiter
Ryan Walters
Ryan Ginn
Ryan Greendyk
Ryan Wren Barret
Saphir Lewis
Sara Victor
Sarakeylah Rose
Saralise Antara Nada Azrael
Saul Slotnick
Scott Davison
Seamunko
Sean Culman
Sean Frenette
Sean Levahn
Sean Oliver
Sean Terence Basalyga
Seyma Tepe Basbuyuk
Shamanatrix Missy Galore
Shannon R Brake
Sharon Joy
Shelley Nottingham Woodworth
Sheri Herndon
Shunyata O'Duibhir
Steven Ator
Steven Walker
Suns of the Earth
Syd Fredrickson
Tammy FireFly
Terra Celeste
Thayne Taylor
The Fritz
Theo Dore
Todji Kurtzman
Tomas Cormons
Tucker Garrison
Veerpal
Veronica & Baruch
Vikki Gilmore
Village in a Box
Wayne Meador
Wesley Wolfbear Pinkham
Xilla

Honoring our Partners and Allies!

TRIBAL CONVERGENCE NETWORK

ELEVATE

EMPOWERMENT WORKS

BRIDGE walkers

navigating change with people, process and collaboration

THE EVOLVER NETWORK

Building Community for the New Planetary Culture

CultureSeed

NuMundo

enlightened structure

SolPurpose

LIVING MANDALA

SolFood FESTIVAL

Sebastopol Village Building Convergence

emergence e:e earth

BIONEERS

R∃VOLution from the Heart of Nature

ONE COMMUNITY
www.OneCommunityGlobal.org

AWAKEN
Visionary Leadership Summit

HARMONY CONNECTIONS

Activated Villages
Co-creating intentional community properties

paititi INSTITUTE

FESTIVAL FIRE

ALISTHUB

344

Holonomic

A SPECIAL Thank you to our Sponsors!

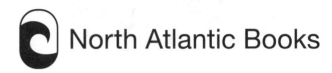
North Atlantic Books

To become a Sponsor for ReInhabiting the Village Project
contact us at reinhabitingthevillage@gmail.com

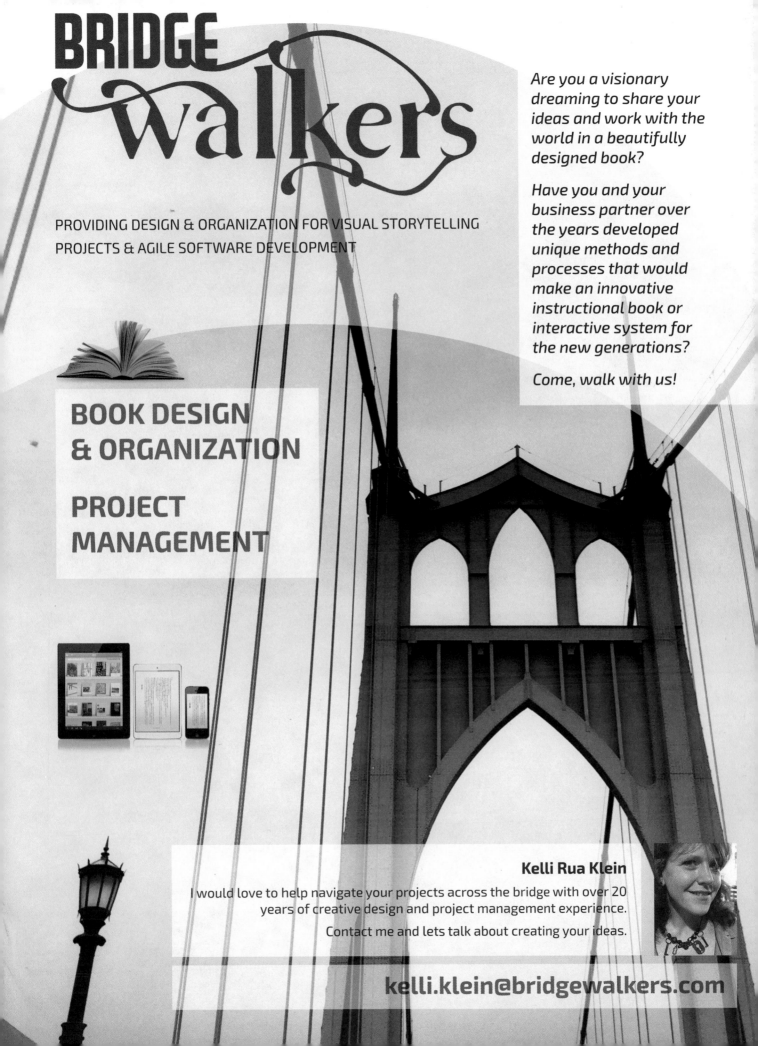